Nitrous Oxide and Climate Change

Nitrous Oxide and Climate Change

Edited by
Keith Smith

from Routledge

First published by Earthscan in the UK and USA in 2010

For a full list of publications please contact:
Earthscan
2 Park Square, Milton Park, Abingdon, Oxfordshire OX14 4RN
711 Third Avenue, New York, NY 10017, USA

First issued in paperback 2016

Earthscan is an imprint of the Taylor & Francis Group, an informa business

Notices
Practitioners and researchers must always rely on their own experience and knowledge in evaluating and using any information, methods, compounds, or experiments described herein. In using such information or methods they should be mindful of their own safety and the safety of others, including parties for whom they have a professional responsibility.

Product or corporate names may be trademarks or registered trademarks,and are used only for identification and explanation without intent to infringe.

ISBN 13: 978-1-138-97722-8 (pbk)
ISBN 13: 978-1-84407-757-1 (hbk)

Typeset by FiSH Books, Enfield
Cover design by Susanne Harris

For citation purposes it is suggested that author names be formatted according to their appearance on the Contents page (opposite)

A catalogue record for this book is available from the British Library

Library of Congress Cataloging-in-Publication Data

Nitrous oxide and climate change / edited by Keith Smith.
  p. cm.
Includes bibliographical references and index.
ISBN 978-1-84407-757-1 (hardback)
1. Atmospheric nitrous oxide–Environmental aspects. 2. Nitrous oxide–Environmental aspects. 3. Greenhouse gas mitigation. 4. Agricultural pollution. 5. Climatic changes.
I. Smith, Keith A., 1940-
TD885.5.N5N58 2010
363.738'74–dc22
                    2009051706

Contents

1

Introduction

Keith Smith

Nitrous oxide, N_2O, is present in earth's atmosphere at a trace level – its current 'mixing ratio' (i.e. the concentration in dry air) is of the order of 320 parts per billion (ppb). This mixing ratio has been increasing linearly over the last few decades (as can be seen in Plate 4.3) as a consequence of the introduction of N_2O into the atmosphere at a rate greater than its rate of removal by natural processes.

N_2O is environmentally important in two quite distinct respects. First, its capacity to absorb infrared radiation is about 300 times greater than that of carbon dioxide, CO_2, and therefore, although its mixing ratio is a thousand times less than that of CO_2, it contributes significantly to the greenhouse effect and thus to climate change. Second, when N_2O reaches the stratosphere it contributes, along with some halogen-containing gases, to the loss of ozone that acts as a barrier to the penetration of ultraviolet radiation to Earth's surface, with consequences for human health.

A generation ago, the impact of N_2O emissions on the ozone layer was the main environmental concern associated with this gas, but since then the increasing recognition that global warming is a major threat to life on Earth as we know it has led to a wide-ranging investigation of the factors that contribute to the warming, in particular the anthropogenic emissions of the long-lived greenhouse gases CO_2, methane and N_2O to the atmosphere. The desirability of explaining how and why N_2O has become important in this context, the past, present and likely future trends in emissions, and how these emissions to the atmosphere might be mitigated, have been the motivation for producing this book.

N_2O is a natural product of mainly microbial origin, as a result of the bio-geochemical processes occurring within the nitrogen (N) cycle. Emissions and natural destruction (mainly in the stratosphere) were broadly in balance until the advent of the industrial age, resulting in a fairly constant concentration in the atmosphere. However, emissions have been increasing, as a consequence of adding reactive forms of nitrogen into the biosphere beyond those natural additions from, principally, biological nitrogen fixation (by leguminous plants,

plants with other forms of symbiotic association with microorganisms, and free-living N-fixing bacteria), and electrical discharges – lightning flashes – in the atmosphere. This introduction comes about chiefly through adding synthetic nitrogenous fertilizers and animal manures to agricultural land; creating new agricultural land from natural forests and grasslands, and thus liberating nitrogen from relatively inert forms in the soil; and releasing reactive nitrogen compounds into the atmosphere, which are subsequently deposited onto land and water. These compounds are predominantly NO_x from industrial sources, power stations and vehicles, and ammonia from animal manures.

Chapter 2 by Elizabeth Baggs and Laurent Phillipot, and Chapter 3 by Hermann Bange and his co-authors, Alina Freing, Annette Kock and Carolin Löscher, describe and discuss the biochemical pathways within the nitrogen cycle that lead to the emission of N_2O from terrestrial soils and marine environments, respectively, and thus provide the process understanding that underpins the remaining chapters.

The largest N_2O source is now agriculture, driven mainly by the use, globally, of >80 million tonnes of N annually as synthetic nitrogen fertilizers, as well as biological nitrogen fixation by leguminous crops. Natural ecosystems also receive N compounds formed from the release into the atmosphere of NO_x from fossil fuel combustion and biomass burning, and ammonia from livestock manure. Together, these inputs of reactive nitrogen compounds to the biosphere have virtually doubled the mainly natural inputs existing at the beginning of the industrial age, and this increase has been matched by a corresponding increase in N_2O emissions. The relationship at the global scale between the magnitudes of reactive N inputs and the consequent N_2O outputs, including the implications of agricultural expansion to provide crop-based biofuels, is reassessed in Chapter 4 by the present author in conjunction with Paul Crutzen, Arvin Mosier and Wilfried Winiwarter. It has been a pleasure to work with them on this chapter, much of the inspiration for which came from our earlier collaboration led by Paul, on the implications for global warming of the production of first-generation biofuels from agricultural crops (Crutzen et al, 2008).

The dominance of agriculture and land use as a source of N_2O provides the justification for including three chapters focusing on this sector. In Chapter 5, Lex Bouwman, Elke Stehfest and Chris van Kassel cover the topic of emissions from arable land, ways of measuring emission factors, modelling and mitigation possibilities, while in Chapter 6, Cecile de Klein, Richard Eckard and Tony van der Weerden deal with analogous issues relating to N_2O emissions from livestock-based agriculture. Chapter 7, by Franz Conen and Albrecht Neftel, reviews the complex subject of how changes in land use and management affect the scale of N_2O emissions in different parts of the world.

A substantial proportion of the nitrogen applied to agricultural land in the form of synthetic fertilizers, animal manures and crop residues, and some of the N released from old soil organic matter by cultivation, is leached from land in drainage water into groundwater and into streams, rivers, estuaries and finally seas and oceans. Part of Chapter 8, by Reinhard Well and Klaus Butterbach-Bahl, deals with the problems of estimating the so-called indirect N_2O emissions resulting from denitrification of this leached N. The remainder deals with emissions from natural and semi-natural land resulting from aerial deposition of reactive N compounds in the atmosphere; this deposition follows emission of NO_x from industry and combustion sources, and ammonia from livestock farming, leading to short- medium- and long-range transport of these gases and their atmospheric reaction products, before deposition on the surface.

Combustion processes are responsible for direct emissions of N_2O, not merely for emissions of gases such as NO_x that can provide substrates for microbial N_2O production. Two processes in the chemical industry are direct sources of N_2O release to the atmosphere. The first of these is the production of nitric acid, and the second is the production of adipic acid, used chiefly in nylon manufacture. These non-biological sources, and the success of abatement measures employed to minimize them, are described by Peter Wiesen in Chapter 9.

Action is being taken to curb the industrial point-source emissions of N_2O, but measures to limit or reduce agricultural emissions are inherently more difficult to devise. Thus as we enter an era in which measures are being explored to reduce fossil fuel use and/or capture or sequester the CO_2 emissions from the fuel, it is likely that the relative importance of N_2O in the 'Kyoto basket' of greenhouse gases will increase, because comparable mitigation measures for N_2O are inherently more difficult, and because current and expected future expansion of the land area devoted to crops is likely to lead to an increase in N fertilizer use, and thus N_2O emission, worldwide. These issues are examined briefly in Chapter 10.

I have already mentioned my co-authors in Chapter 4. May I take this opportunity also to thank all the other authors who have contributed to this book for their efforts. I am only too aware that all of them are very busy people, for whom the request to write a book chapter has come on top of a pile of other commitments; yet they have taken part in this project willingly, have acceded to editorial requests without complaint, and have helped to deliver what I hope readers will consider to be a valuable contribution to knowledge about one of the key gases that is affecting the environment.

Reference

Crutzen, P. J., Mosier, A. R., Smith, K. A. and Winiwarter, W. (2008) 'N_2O release from agro-biofuel production negates global warming reduction by replacing fossil fuels', *Atmospheric Chemistry and Physics*, vol 8, pp389–395

2

Microbial Terrestrial Pathways to Nitrous Oxide

Elizabeth Baggs and Laurent Philippot

Introduction

Terrestrial systems are major sources of atmospheric N_2O, which accounts for about 6 per cent of the current greenhouse effect (IPCC, 2007). Production of N_2O in soil is predominantly biological, with bacteria possessing N_2O-genic enzymes, whilst the involvement of archaea or fungi is currently uncertain. Here we explore the biogeochemical pathways in which these microbes can produce and reduce N_2O, consider the approaches available for determining the predominant N_2O-producing process under certain conditions, highlight any current uncertainties in microbial sources of N_2O to direct future research, and examine how understanding the N_2O source can aid us in managing terrestrial systems to lower emissions of this greenhouse gas.

Why do we need to know the microbial source of nitrous oxide?

The ultimate aim in determining the microbial source of N_2O is to better constrain the global N_2O budget and to inform mitigation strategies. This is essential for the formulation of appropriate and more targeted mitigation strategies, which at the time of writing are urgently required if industrialized countries are to reach the Kyoto and post-Kyoto targets for reduction in greenhouse gas emissions, for example in Europe a reduction of up to 80 per cent by 2050 compared to emissions in 1990. Soil is a complex heterogeneous matrix that renders attributing N_2O production to different processes a challenge, as different processes may occur simultaneously in different microsites of the same soil (Robertson and Tiedje, 1987). Simply relying on a net emission of N_2O from a system does not tell us the responses of the key component groups of the microbial community. However, it is essential to understand these responses, because when imposing strategies for mitigation of N_2O it is important that there is no detrimental effect on the long-term functioning of the ecosystem facilitated through changes in key microbial functional groups. To ensure this we should link source partitioning – i.e. determination of the microbial sources of N_2O – to the understanding of activity and ecology of the underpinning microbial community. This means that efforts to measure N_2O from different environments and under different management regimes should more regularly consider all N_2O processes, with the starting point being the identification of conditions under which particular processes predominate. Unless the controls on enzyme regulation associated with these processes, and associated product ratios (primarily N_2O and N_2) are determined, then it will not be possible to develop more targeted mitigation strategies. The different processes involved in N_2O production respond differently to environmental parameters or imposed management, and the enzymatic systems of each process are regulated differently. This means that the down-regulation of N_2O production in one process as a result of management practice or change in environmental conditions may well lead to the up-regulation of N_2O production in another. Thus appropriate management for one process, may not be appropriate for another, and may well need to be flexible depending on the system, the prevailing environmental conditions, and the management options available.

What is producing nitrous oxide in soil?

Soil microbial N_2O production occurs via nitrification (ammonia oxidation) and nitrate dissimilation (denitrification and nitrate ammonification) pathways (Figure 2.1). These processes rarely occur in isolation, with possible competition for substrates under near-limiting conditions, and the possibility of transfer of N_2O or intermediary products from one process to another depending on prevailing environmental conditions, microbial community structure and location within the soil matrix. This means that N_2O produced during several processes may form one pool before being reduced to N_2 during denitrification. This representation of biological pathways of N_2O production reflects fairly recent advances in understanding; until the 1980s N_2O emissions in terrestrial systems were solely attributed to denitrification. Whilst denitrification may often be the predominant source of N_2O, particularly following rainfall events or application of NO_3^--based fertilizers to moist soil, these are not always the prevailing conditions, and there is a growing body of evidence in support of ammonia oxidation contributing to N_2O fluxes.

The processes presented in Figure 2.1 are the main focus of this chapter, reflecting our current understanding of N_2O pathways in terrestrial systems. However, it is not beyond the realms of possibility that there are other as yet 'undiscovered' processes or interactions between cycles that facilitate N_2O production, and in their discovery we are currently constrained by our biogeochemical and microbiological terminology and technical ability. We already know that the representation of N_2O sources in Figure 2.1 is oversimplified, with 'new' interactions between processes, and an increasing range of microbial groups being able to denitrify, the occurrence of aerobic

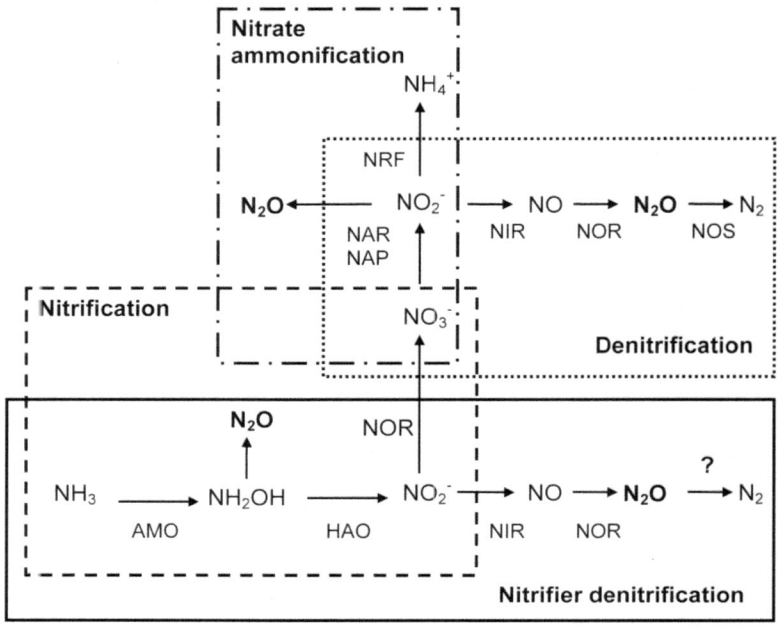

Figure 2.1 Microbial sources of N_2O in soil

Source: Adapted from Baggs (2008)

denitrification (Lloyd et al, 1987; Bell and Ferguson, 1991) and co-denitrification (Laughlin and Stevens, 2002), interactions with the carbon (C) cycle, such as methanotroph-dependent denitrification (Islas-Lima et al, 2004), and the potential for microbial groups to switch function under certain conditions, such as N_2O production by methanotrophic bacteria in the presence of NH_3 (Ren et al, 2000).

Nitrate-reducing processes

Denitrification

Denitrification is the reduction of NO_3^- or NO_2^- to N_2 under anaerobic conditions with N_2O and NO emitted as intermediary gaseous products (Robertson and Tiedje, 1987; Bremner, 1997). The process is catalysed by the enzymes nitrate reductase, nitrite reductase, nitric oxide reductase and nitrous oxide reductase, and the transport of electrons to NO_3^- or other N oxides is coupled to the synthesis of ATP (adenosine triphosphate) (Hochstein and Tomlinson, 1988). The nature and regulation of the reductase enzymes involved in denitrification have been well characterized (Zumft, 1997).

Denitrifiers are predominantly heterotrophic bacteria with the denitrification trait being widespread in more than 50 genera (Smith and Zimmerman, 1981; Knowles, 1982). Some archaea and fungi, such as *Fusarium* (Shoun et al, 1992), were also shown to be capable of denitrifying.

Since the ability to denitrify is sporadically distributed both within and between different genera, the genes encoding the catalytic sub-unit of the different denitrification reductases are commonly used as molecular markers in studies investigating the diversity of denitrifiers. Thus membrane-bound and periplasmic nitrate reductase encoding genes (*napA*, *narG*) (Gregory et al, 2000; Philippot et al, 2002), cytochrome cd_1 and Cu-containing nitrite reductase encoding genes (*nirS*, *nirK*) (Hallin and Lindgren, 1999) and N_2O reductase encoding genes (*nosZ*) (Stres et al, 2004) have successfully been used in terrestrial environments for fingerprinting analyses to identify the environmental factors driving the denitrifier community composition.

The activity of denitrifying bacteria is controlled by C availability, O_2 concentration (mainly through soil water content), N availability, pH and temperature. The contribution of denitrification to N_2O emissions from soils is traditionally thought to be greatest under sub-oxic conditions, and in soils at a water-filled pore space (WFPS) >70 per cent where soil NO_3^- and available C are non-limiting (Davidson, 1991; Bateman and Baggs, 2005). High O_2 concentrations are known to suppress the activity and synthesis of the denitrification reductases. The N_2O reductase is thought to be the most sensitive to O_2 (Otte et al, 1996), reflected in a higher N_2O/N_2 ratio with increasing O_2 availability (Weier et al, 1993). When re-exposed to O_2 after an anaerobic phase, all denitrification enzymes but the N_2O reductase remain active (Morley et al, 2008), and conversely when aerobic soils become anaerobic, such as following heavy rainfall, the NO_3^- and NO_2^- reductases are typically activated sooner than the N_2O reductase, so that the denitrifier N_2O/N_2 ratio is higher up to one or two days after rainfall (Knowles, 1982). Several bacteria isolated from soils and sediment are capable of denitrifying in the presence of O_2 (Lloyd et al, 1987; Patureau et al, 2000). Although N_2 and N_2O production is common during aerobic denitrification by cultured isolates (Patureau et al, 2000; Takaya et al, 2003) and aerobic denitrifiers are thought to be present in high numbers in soils (Patureau et al, 2000), their contribution to N_2O emissions from soils has yet to be proven, and in the meantime, denitrification is still modelled as an anaerobic process.

Denitrification rates are influenced by the availability of N electron acceptors, and this is why they are stimulated by addition of N, usually in the form of inorganic fertilizers (Eichner, 1990). N availability also affects the N_2O/N_2 ratio, which can be lowered with NO_3^- concentrations >10µg g^{-1} soil, where NO_3^- is preferred over N_2O as an electron acceptor (Blackmer and Bremner, 1978; Baggs et al, 2003). The enzyme kinetics and affinity for substrates by denitrifiers is highly variable, possibly reflecting the wide phylogenetic diversity of this group, with K_m values for N_2O reduction of between 0.1 and 100µM (Conrad, 1996; Holtan-Hartwig et al, 2000).

As most denitrifiers are heterotrophic, the availability of organic C, which can be used as an electron donor, is also a major factor determining denitrification rates (Knowles, 1982; Bremner, 1997). However, organic C availability is the least well-understood factor controlling denitrifier N_2O production and reduction, despite the fact that it has traditionally been thought a reliable index for predicting a soil's denitrification capacity (Burford and Bremner, 1975). The effects of C are not only mediated directly, but also indirectly through the creation of anaerobiosis generated as a result of microbial respiration (Azam et al, 2002). With regards to direct effects, there is likely to be strong coupling between plant exudate-C and denitrification, through the provision of the reductant for denitrification, and changing the size, structure and activity of the denitrifier community. Indeed, Smith and Tiedje (1979) found a positive effect of roots on denitrification when soil NO_3^- concentrations were high, which was thought to be due to release of organic C from the roots, with denitrifying activity decreasing rapidly in the first few millimetres away from the roots. Denitrification has been shown to increase following addition of easily oxidizable organic matter to soils, often very rapidly, and significant correlations have been reported between total organic C (Baggs and Blum, 2004), 'available' C (Stanford et al, 1975), water-soluble and 'mineralizable' C (Burford and Bremner, 1975; Paul and Beauchamp, 1989) and soluble C-to-N ratio of applied residues (Millar and Baggs, 2005). However, little is known about the effect of individual C compounds on the regulation of the enzymes producing or reducing N_2O, and we are still unsure whether the type of C compounds is an important driver of the denitrifier community. In a recent study, Henry et al (2008) applied artificial root exudates composed of different combinations of sugars, organic acids and amino acids to soil microcosms, and showed that a higher proportion of sugars resulted in a lower N_2O/N_2 ratio. Such results warrant further attention before we can propose that manipulating plant-derived C flow has the potential to increase reduction of N_2O to N_2 (Philippot et al, 2009b).

Among the other factors influencing denitrification, special emphasis has been placed on soil pH. Denitrification rates are usually thought to decrease with pH, but significant denitrification rates can be still observed below pH 4.9 (Ellis et al, 1998). This can reflect differences in the composition of the denitrifier community (Enwall et al, 2005) and adaptation of denitrifiers to low soil pH (Parkin et al, 1985). Interestingly, soil pH does not only affect denitrification rates but also the N_2O/N_2 ratio. Thus, several studies reported that the N_2O reductase enzyme is more sensitive to low pH than the other denitrification reductases, resulting in higher ratios of N_2O to N_2 as pH declines (Firestone et al, 1980; Nägele and Conrad, 1990; Thomsen et al, 1994; Šimek and Cooper, 2002).

Fungal denitrification

The potential for fungi to produce N_2O has been demonstrated in woodland

(Laverman et al, 2000) and in grassland soils (Laughlin and Stevens, 2002). N_2O release from acid coniferous forest soils is often low and attributed to nitrification (Martikainen et al, 1993; Sitaula and Bakken, 1993), but there is evidence that fungal denitrification may actually be important in these systems (Kester et al, 1996; Laverman et al, 2000). Several fungal strains have been shown to possess cytochromes $P450_{nir}$ or $P450_{nor}$ that enable them to denitrify and release N_2O when anaerobically incubated with NO_2^- in culture (Shoun et al, 1992; Takaya and Shoun, 2000). Fungal *nir* appears to be expressed in the mitochondrion and is analogous to the bacterial copper-containing enzyme (Kobayashi and Shoun, 1995), whereas the fungal *nor* ($P450_{nor}$) gene differs from that found in bacteria. In fungal denitrification the enzymes accept two electrons from NADH or NADPH without involving a membrane-bound electron transfer chain as in bacterial denitrification. In most cases the main product of fungal denitrification is N_2O, as many fungi lack a N_2O reductase (Shoun et al, 1992), although it is possible that the $P450_{nor}$ gene has two functions and may also facilitate N_2 production (Takaya and Shoun, 2000).

Nitrifier denitrification

In nitrifier denitrification, the oxidation of NH_3 to NO_2^- by ammonia-oxidizing bacteria (AOB) is followed by the reduction of NO_2^- to N_2O and possibly also to N_2 (Wrage et al, 2001), although there is still no direct evidence that ammonia oxidizers can produce N_2 during this process. The entire process is undertaken by ammonia-oxidizing bacteria (Kuai and Verstraete, 1998) and therefore contrasts with 'coupled nitrification-denitrification' where coexisting nitrifiers and denitrifiers can together transform NH_3 to NO_3^- and NO_3^- to N_2, respectively. The enzymes involved in nitrifier denitrification are essentially those for ammonia oxidation and denitrification, namely ammonia monooxygenase, hydroxylamine oxidoreductase, nitrite reductase, nitric oxide reductase and, possibly, nitrous oxide reductase (Jiang and Bakken, 1999). This process has been demonstrated in a number of studies on mixed and pure cultures of *Nitrosomonas*, with NH_3 converted under oxygen limitation to N_2O and N_2, with NO_2^- and NO as intermediates (Abeliovich and Vonshak, 1992; Bock et al, 1995; Kuai and Verstraete, 1998). Shaw et al (2006) provided direct evidence for reduction of exogenously applied ^{15}N-NO_2^- to ^{15}N-N_2O in seven strains of ammonia-oxidizing bacteria representative of clusters 0, 2 and 3 in the cultured *Nitrosospira* lineage, with up to 13.5 per cent of measured N_2O derived from the applied ^{15}N-NO_2^- (Table 2.1). This suggests that the ability to denitrify may be a widespread trait amongst betaproteobacterial ammonia-oxidizing bacteria, but its significance as a N_2O source in soil has still to be proven as it requires advances in techniques for its quantification (see below). Ammonia-oxidizing archaea have been shown to possess a nitrite reductase enzyme, and although they have been demonstrated to oxidize ammonia (Prosser and Nicol, 2008), any ability to denitrify is as yet unproven, but may well be akin to nitrifier denitrification.

Table 2.1 Ammonia oxidizer $^{14+15}$N-N$_2$O and ^{15}N-N$_2$O production rates and molar yields expressed as a percentage of the NO$_2^-$ and $^{14+15}$N-N$_2$O production rate, respectively

AOB strain	Total N$_2$O production rate (amol* $^{14+15}$N-N$_2$O h^{-1} cell^{-1})[†]	Yield of $^{14+15}$N-N$_2$O on a nitrite basis[‡] (%)	^{15}N-N$_2$O production rate (amol* ^{15}N-N$_2$O h^{-1} cell^{-1})[‡]	Yield of ^{15}N$_2$O on a total $^{14+15}$N-N$_2$O basis[‡] (%)
N. europaea ATCC 19718	58.0[a]*	0.45[a]	3.29[a]	7.8[a]
N. europaea ATCC 25978	15.5[b]	0.19[b]	0.34[b]	5.6[b]
N. briensis strain 128	4.2[b]	0.03[c]	0.02[c]	3.5[c]
N. multiformis ATCC 25196	7.6[b]	0.08[d]	0.20[d]	3.1[c]
N. tenuis strain NV12	2.0[b]	0.06[d]	0.06[e]	4.5[d]
Nitrosospira sp strain 40KI	4.6[b]	-**	0.88[f]	13.5[e]
Nitrosospira sp strain En13	5.7[b]	0.70[c]	0.66[g]	12.4[f]
Nitrosospira sp strain NpAV	3.9[b]	0.29[f]	0.33[b]	8.9[g]

Note: Mean values in columns superscripted by different letters are significantly different (P<0.05). *amol = attomol = 10^{-18} mol.
† Values are the means of three independent experiments. ‡ Values are the means (n = 4) for cells harvested from a single independent flask. **Rate of nitrite production by Nitrosospira sp 40KI was not significantly different from 0 (P>0.05); therefore percentage yield of $^{14+15}$N-N$_2$O on a nitrite basis was not calculated.
Source: Shaw et al (2006)

Little is known about the environmental regulation of nitrifier denitrification and effects on N$_2$O production during this process. Several studies provide evidence for greater N$_2$O production through nitrifier denitrification as oxygen concentration decreases (Goreau et al, 1980; Lipschultz et al, 1981; Hynes and Knowles, 1984; Kester et al, 1996; Jiang and Bakken, 1999; Dundee and Hopkins, 2001), most likely because NO$_2^-$ is used as an alternative to O$_2$ as an electron acceptor for microorganisms temporarily subjected to anaerobic conditions (Ritchie and Nicholas, 1972). Decreasing pH may also influence nitrifier denitrification (Jiang and Bakken, 1999; Wrage et al, 2001). From our ongoing experiments we estimate that nitrifier denitrifier-N$_2$O accounts for a daily loss of fertilizer-N of up to 0.2 per cent and have found that although N$_2$O emissions were raised following amendment of soil with amino acids and artificial root exudates, the proportional contribution of nitrifier denitrification did not vary with the different C amendments.

Co-denitrification

Here one N atom from NO or N$_2$O combines with one atom from another source (a co-substrate) forming a hybrid product (Tanimoto et al, 1992; Su et al, 2004) N$_2$ can also be produced solely from the co-substrate (Tanimoto et al, 1992) although it is not known how common this is. Co-denitrification has been shown to occur in both fungi (Tanimoto et al, 1992) and bacteria (Garber and Hollocher, 1982) but the reasons why, or the conditions under which, an organism would utilize a co-substrate are unknown. Co-denitrification and conventional denitrification have been shown to occur simultaneously in some denitrifying bacteria such as *Pseudomonas stutzeri* and denitrifying fungi (Shoun et al, 1992; Tanimoto et al, 1992). Su et al (2004) demonstrated the $P450_{nor}$ in *Fusarium oxysporum* to be multifunctional and able to catalyse denitrification or co-denitrification. There is some evidence for co-substrate preference between different fungal strains, which may indicate different mechanisms. For example, *Fusarium oxysporum* only produced N_2O during co-denitrification when azide or salicylhydroxamic acid was available (Tanimoto et al, 1992; Su et al, 2004) but other fungi, for example *Fusarium solani* and *Cylindrocarpon tonkinense* (now *Fusarium lichenicola*), have been shown to emit N_2 as a co-denitrification product when amino acids are available (Shoun et al, 1992).

The contribution of co-denitrification to N_2O emission from soils is unknown, as there has been no targeted source determination for this process in situ, in part due to difficulties in distinguishing co-denitrification from denitrification. The only existing evidence from soil is from Laughlin and Stevens (2002) who estimated 92 per cent of measured N_2 production in a grassland soil (sandy loam, pH 6.3) amended with glucose (up to 15mg per g soil) to result from co-denitrification, with the remaining 8 per cent attributed to conventional denitrification.

Nitrate ammonification or dissimilatory nitrate reduction to ammonium (DNRA)

During nitrate-dependent ammonification NO_3^- is reduced to NO_2^- and NH_4^+ (Figure 2.1) thereby providing a 'short circuit' in the nitrogen cycle bypassing denitrification and N_2 fixation (Mohan et al, 2004). The process of nitrate ammonification is in many cases coupled to a respiratory electron transport system through which energy conservation (and thereby ATP synthesis) can be achieved. However, a range of electron donors can be used by ammonifiers, including fermentation products such as lactate, formate and hydrogen, as well as NADH produced from intermediary metabolism, and the nature of electron donor influences the level of energy conservation. This process is undertaken by both Gram-negative and Gram-positive bacteria, including obligate anaerobes (e.g. *Clostridium* spp), facultative anaerobes (e.g. *Enterobacter* spp) and aerobes (e.g. *Bacillus* spp) (Fazzolari et al, 1990), which are widely distributed in many different environments. Some of these bacteria are capable of reducing NO_2^- to N_2O during nitrate ammonification, as well as to NH_4^+ (Kelso et al, 1997). Two biochemically distinct nitrate reductases, the membrane-bound nitrate reductase (Nar) and the periplasmic nitrate reductase (Nap), are found in ammonifying bacteria. These two enzymes can also catalyse the first step of the denitrification pathway. Nitrite is then reduced to ammonium by NrfA, which can be N_2O-genic (Jackson et al, 1991). However, the biochemistry that determines the amount of N_2O formed is still unknown. The contribution of nitrate ammonification to N_2O emissions from soils is also pretty much unknown, mainly because of technical limitations in its determination, which are discussed below.

It is possible that both nitrate ammonification and denitrification can occur simultaneously in anaerobic micro-sites in soils (Fazzolari et al, 1990) and produce N_2O at the same time (Smith and Zimmerman, 1981; Stevens et al, 1998) and may compete for NO_3^-, depending on environmental conditions as well as the soil microbial population. Compared to denitrification, very little is known about the regulation of nitrate ammonification. Intensively reduced and C-rich environments are traditionally thought to favour ammonification, as it is a more efficient electron acceptor where C availability is non-limiting (Buresh and Patrick, 1978; Tiedje et al, 1982). Addition of glucose to anoxic soil has been shown to result in ammonification of 43 per cent of nitrate applied (Caskey and Tiedje, 1979), and there is evidence that nitrate ammonification may be significant in agricultural soils, even those to which no C has been added (Chen et al, 1995a, 1995b).

Nitrification

Ammonia-oxidizing bacteria (AOB) convert ammonia to nitrite in a two-step process with hydroxylamine as an intermediate (Figure 2.2). The membrane-bound ammonia mono-oxygenase enzyme (AMO) catalyses the oxidation of ammonia to hydroxylamine, and hydroxylamine is oxidized to nitrite by the periplasm-associated enzyme, hydroxylamine oxidoreductase (Hooper et al, 1997). The production of N_2O occurs at this stage. Two of the four electrons derived from the oxidation of hydroxylamine to nitrite are required during the oxidation of ammonia to hydroxylamine, the other two are available for energy production and the reduction of oxygen to water (Colliver and Stephenson, 2000). All known bacterial autotrophic ammonia oxidizers in soils belong to the genus *Nitrosomonas* and *Nitrosospira*, which form a monophyletic cluster within the subclass of *Proteobacteria*. Recent findings have revolutionized the diversity of ammonia oxidizers by showing that archaea are also capable of oxidizing ammonia to nitrite and that they could dominate over AOB in soils (Prosser and Nicol, 2008). However their contribution to nitrification is still under debate and their role in N_2O production is unknown.

The rate of ammonia oxidation is influenced by ammonia availability. This is closely linked to the protonation and lower availability of NH_3 at low pH. Surprisingly, ammonia oxidizers seem to be able to survive under conditions of NH_3 starvation although the ability to compete for NH_3 and the ability to respond to NH_3 after a period of starvation have been shown to vary between different AOB (Frijlink et al, 1992; Gerards et al, 1998; Bollmann et al, 2002). Furthermore, the theoretical sensitivity to low pH also varies between different AOB, with some possessing mechanisms to overcome this problem so that nitrification is not always restricted by low pH environments (De Boer and Kowalchuk, 2001). Much of the information on the physiology of AOB comes from studies on *Nitrosomonas* spp due to the relative ease of culturing this

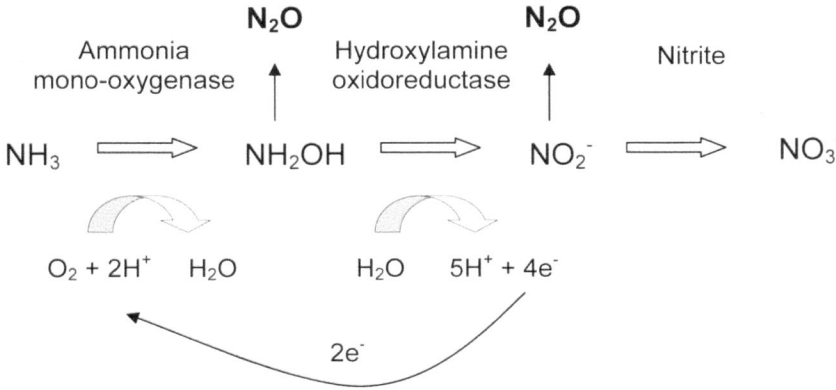

Figure 2.2 The pathway of nitrification, showing stages at which N_2O can be produced

Source: Adapted from Colliver and Stephenson (2000) and Wrage et al (2001)

microorganism in the laboratory. However, molecular investigations into AOB communities have shown that the more predominant AOB in the soil, especially soils treated with ammonia-based fertilizers, are members of the genus *Nitrosospira* (Kowalchuk and Stephen, 2001; Avrahami et al, 2002).

Nitrifier-N_2O production has been measured in cultures of nitrifiers under reduced O_2 potential (Goreau et al, 1980), and there is recent evidence to suggest that ammonia oxidation can significantly contribute to net N_2O emission from soil (e.g. Abbasi and Adams, 2000; Bateman and Baggs, 2005; Avrahami and Bohannan, 2009; Wan et al, 2009). Bateman and Baggs (2005) showed nitrification to be the predominant N_2O-producing process in a silt loam soil held at 35–60 per cent WFPS, accounting for up to 81 per cent of N_2O emitted at 60 per cent WFPS, indicating the significance of this process for global warming, despite its role having often been underplayed compared to that of denitrification.

Nitrification can also be carried out by a wide range of heterotrophic microbes (bacteria and fungi) using organic substrates such as urea as well as NH_3 (Papen et al, 1989). Several different pathways of heterotrophic nitrification have been proposed, including an inorganic pathway similar to that of the autotrophic nitrifiers (Killham, 1986), and this capacity has been demonstrated in pure-culture studies of a range of heterotrophic microbes. For example, many heterotrophs isolated from soil (including *Absidia cylindro-spora*, *Pseudomonas putida* and *Paracoccus denitrificans*) have the ability to nitrify NH_3 in culture (Stroo et al, 1986; Moir et al, 1996a; Daum et al, 1998). Brierley and Wood (2001) isolated heterotrophic bacteria and fungi from an acid forest soil capable of nitrifying both inorganic (ammonium acetate) and organic (b-alanine, peptone) N in pure culture and in inoculated soil solution. The potential for heterotrophic nitrification is further supported by the characterization of enzymes (ammonia and hydroxylamine reductases) capable of catalysing oxidation reactions typical of the autotrophic pathway in heterotrophs (Moir et al, 1996a, 1996b). The exact mechanisms of both organic and mineral heterotrophic nitrification pathways remain to be clarified, and the evidence seems to suggest that a combination of an appropriate N source and suitable soil environmental conditions combine to dictate whether the process occurs (Killham, 1986).

The regulation and magnitude of N_2O production during these pathways are less well constrained, although the production of N_2O by heterotrophic nitrifiers has been demonstrated in culture (Papen et al, 1989), and it has been shown under aerobic culture conditions that some heterotrophic nitrifiers, such as *Alcaligenes faecalis*, can produce much more N_2O per cell than the autotrophic nitrifier *Nitrosomonas europaea* (Papen et al, 1989; Anderson et al, 1993). The oxygen concentration at which most N_2O was produced was higher for the heterotroph *Alcaligenes faecalis* than the autotroph *Nitrosomonas europaea* (Anderson et al, 1993). To our knowledge the only direct investigation of N_2O production in soil by heterotrophic nitrification was undertaken by Bateman and Baggs (2005), using C_2H_2 to inhibit ammonia oxidation. However, due to high variability, there was no conclusive evidence for heterotrophic contribution in their arable soil.

Abiotic nitrous oxide production

Chemodenitrification is generally considered to encompass all non-biological processes that produce NO, N_2O or N_2. Its occurrence in terrestrial systems is rarely considered, and no attempts have been made to include the process in studies aiming to discriminate between all sources of N_2O. Chemodenitrification is thought to occur when NO_2^- accumulates and reacts with organic compounds (thought to be phenolics) to produce NO and N_2O (Bremner, 1997), and may be a significant source of N_2O in acid soils <pH 5.5 where the NO_2^-:HNO_2 equilibrium shifts in favour of nitrous acid (HNO_2), which either self-decomposes to form NO, HNO_3 and H_2O, or reacts with constituents of soil organic matter to form N_2, N_2O and CH_3ONO (Chalk and Smith, 1983; Van Cleemput and Samater, 1996; Venterea and Rolston, 2000). Under acidic conditions chemodenitrification may be a more significant source of NO than N_2O (Nelson, 1982; Yamulki et al, 1997). The occurrence of chemodenitrification is generally verified in sterilized soils, assuming complete inactivation of the microbial population, and thereby the impossibility of microbial production of N_2O (Yamulki et al, 1997; Clough et al, 2001; Morkved et al, 2007). For example, Morkved et al (2007) measured a 0.4 per cent conversion of NO_2^- to N_2O in sterilized peat soil at pH 4.1–4.2, which they attributed to chemodenitrification.

Nitrite may also in some environments be reduced to NO and/or N_2O, where Fe(II) is used as an electron donor. Cooper et al (2003) observed an eightfold increase in N_2O emissions in the presence of Fe(II) in a wastewater treatment, and whilst Parkes et al (2007) stated that such reactions may have contributed to nitrogen removal in a municipal waste leachate treatment facility, similar relationships have yet to be observed in soils.

Nitrous oxide emission from plants – a source or a conduit of nitrous oxide?

There are several reports of plant-mediated N_2O emission, mainly from rice or temperate arable systems, but there is current controversy as to the mechanism involved in this emission. Whilst emissions from the plants themselves have been shown to account for up to 87 per cent of emissions measured from plant–soil systems (Pihlatie et al, 2005), the mechanisms involved are unknown. Aerenchyma are thought to be implicated in the transfer of N_2O from soil (Rusch and Rennenberg, 1998), with significant transfer of N_2O reported through rice plants (Yan et al, 2000). Chang et al (1998) and Pihlatie et al (2005) also proposed that, as N_2O is a soluble gas, it can be transferred from roots in the transpiration stream of upland plants, to leaves and then emitted to the atmosphere. Pihlatie et al (2005) demonstrated using a ^{15}N-enrichment approach that all of the ^{15}N-N_2O emitted from *Fagus sylvatica* leaves was derived from soil-applied $^{15}NH_4^{15}NO_3$. These authors pertinently point out that if N_2O emission from plants is a common phenomenon, then inventories based solely on emissions from the soil surface may underestimate ecosystem fluxes, and if transpiration is the main mechanism, then it may be possible to predict canopy fluxes from soil N_2O production rates and mineral N availability. However, it is unknown how the spatial arrangement of N_2O production in the rhizosphere relates to the potential for plant N_2O uptake, the zone of root uptake of N_2O, or the ability of plants to take up this N_2O under dry conditions.

It is also possible that N_2O can be formed in leaves during nitrite photo-assimilation (Smart and Bloom, 2001; Hakata et al, 2003) or by plant endophytic bacteria. If endophytic N_2O production occurs it is likely that this is denitrification, as a number of bacteria that form associations with plants are denitrifiers, particularly in root nodules, and may in part account for reported N_2O uptake from the rhizosphere. Thus Daniel et al (1980) and Fernandez et al (2008) reported denitrification ability in a large number of rhizobia isolates. However, this source of N_2O is unlikely to significantly contribute to net N_2O emission, compared with transport of N_2O through the plant arising from stimulation of denitrification in the rhizosphere following inorganic N addition.

Techniques for source partitioning

Acetylene inhibition

Inhibitors such as nitrapyrin and chlorate, and biocides such as cyclohexamide (De Boer and Kowalchuk, 2001) have been applied to soil to distinguish the relative contributions of denitrification and heterotrophic and autotrophic nitrification to N_2O emissions. Acetylene (C_2H_2) gas at 10kPa was first recognized as an inhibitor of the N_2O reductase involved in denitrification (Yoshinari and Knowles, 1976), then the inhibition of nitrification at lower concentrations (10Pa or 0.01 per cent v/v) was noted (Hynes and Knowles, 1978). The suppression by C_2H_2 of N_2O reduction to N_2 by denitrifying bacteria is due to non-competitive inhibition of the enzyme N_2O reductase, and requires high partial pressures (10kPa) to have an effect (Webster and Hopkins, 1996); however, recovery is rapid after C_2H_2 removal.

These actions of C_2H_2 have led to the established procedure of applying 10Pa C_2H_2 to soil to inhibit nitrification, and 10kPa C_2H_2 to inhibit both nitrification and N_2O reduction in denitrification (Klemedtsson et al, 1988). Many studies have used this C_2H_2 inhibition methodology to demonstrate the simultaneous production of N_2O by denitrification and nitrification (e.g. Maag and Vinther, 1996; Webster and Hopkins, 1996), and to show how they are differently affected by soil water content and soil texture. For example, the proportion of N_2O produced by nitrification has been found to increase with soil water content (Maag and Vinther, 1996), and nitrification has been identified as the predominant source of N_2O from dry soils and denitrification as the predominant source above 100 per cent soil water-holding capacity (Webster and Hopkins, 1996). Both the initiation of inhibition and recovery (Kester et al, 1996) are dependent on the diffusion capacity of C_2H_2 within the soil, which is related to the soil texture and water content. However, there are well-recognized limitations of the acetylene inhibition technique, which have led to the development and adoption of alternative stable isotope approaches. The limitations with C_2H_2 include an underestimation of denitrification by preventing the supply of nitrifier-NO_3^-, C_2H_2 itself can be decomposed and used as a substrate for denitrification if C is limiting, it inhibits nitrate ammonification, the extra pair of electrons that would have been used to reduce N_2O to N_2 can increase the reduction of NO_3^- in denitrification, and its diffusion into fine-textured or saturated soil can be slow (Groffman et al, 2006).

Stable isotope approaches

Stable isotope techniques have a crucial role to play in the attribution of N_2O emissions to different microbial processes. This may be done using estimation techniques (variations in natural abundance, site preference) or quantification techniques (using isotopic enrichment), which employ the ^{15}N and ^{18}O signatures of N_2O determined by isotope ratio mass spectrometry. Here we give an outline of these approaches, but guide the reader to Baggs (2008) for theoretical details behind each of these approaches.

Natural abundance approaches rely on the biological fractionation against ^{15}N and ^{18}O. Fractionation during nitrification is generally higher than for denitrification, meaning that N_2O produced during nitrification is more depleted (more negative δ) in ^{15}N and ^{18}O relative to substrates than that produced during denitrification (Wahlen and Yoshinari, 1985; Yoshida, 1988). Due to the different oxygen sources for NH_3 oxidation to hydroxylamine (soil air, $δ^{18}O$ 23.5 per mil (i.e. parts per thousand)) and oxidation of NH_2OH to NO_2^- (soil water, $δ^{18}O$ 10 per mil) the $δ^{18}O$ in N_2O from nitrifier denitrification should be intermediate in value (Sutka et al, 2006). The fractionation during nitrate ammonification has yet to be determined. Fractionation during ammonia oxidation has been shown to differ amongst the nitrifying bacteria. The differences in $δ^{15}N$ in part arise because reduction of N_2O to N_2 enriches the remaining N_2O in ^{15}N (Webster and Hopkins, 1996; Barford et al, 1999). This approach has been applied to the determination of the microbial source of N_2O in a range of ecosystems and controlled environment experiments, and is most advantageous in natural or unfertilized systems (e.g. Webster and Hopkins, 1996; Wrage et al, 2004; Perez et al, 2006).

Enrichment approaches have been developed aimed at quantifying the individual sources of N_2O in situ. To date, these have mostly focused on distinguishing between nitrification and denitrification following application of ^{15}N-labelled fertilizer. Application of $^{15}N-NH_4^+$ and/or $^{15}N-NO_3^-$ to soil and attribution of the $^{15}N-N_2O$ fluxes to nitrification or denitrification depending on the ^{15}N source applied negates the need for C_2H_2 inhibition (Baggs et al, 2003; Bateman and Baggs, 2005; Mathieu et al, 2006). For example, Baggs et al (2003) used this ^{15}N-enrichment approach to verify that increased N_2O emission under elevated atmospheric CO_2 at the Swiss FACE experiment was mainly due to increased denitrification, with greater below-ground C allocation stimulating both denitrifier-N_2O and -N_2 production (Figure 2.3). Unfortunately, this approach is unable to distinguish denitrification from nitrate ammonification or nitrifier denitrification. A combined ^{15}N-, ^{18}O-enrichment approach has been proposed by Wrage et al (2005), involving application of ^{18}O-labelled water to determine N_2O production during nitrifier denitrification. However, quantification of nitrate ammonification N_2O by the enrichment approach remains elusive.

Consideration of the isotopomer site preference of ^{15}N in N_2O has been applied to determine the microbial source of N_2O in terrestrial systems (Yamulki et al, 2001; Bol et al, 2003; Well et al, 2006). N_2O is a linear molecule, N-N-O, and the $^{14}N/^{15}N$ ratios of the central and outer N atoms can naturally vary. Site preference (SP) is termed as the difference in $δ^{15}N$ between the central and outer N atoms in N_2O, with different microbial processes and functional groups thought to exhibit distinct ^{15}N-SPs (Bol et al, 2003; Sutka et al, 2003, 2004). Sutka et al (2006) demonstrated in cultures the potential for SP to distinguish between ammonia oxidation and nitrifier denitrification.

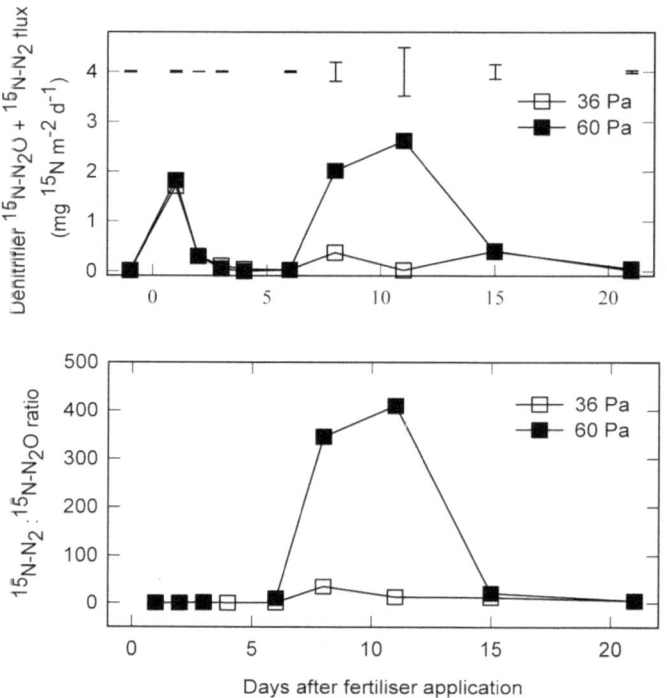

Figure 2.3 Total denitrifier ^{15}N-$(N_2O + N_2)$ production and the ratio of ^{15}N-N_2 to ^{15}N-N_2O following application of 11.2g N m^{-2} (1 atom per cent excess ^{15}N) to *Lolium perenne* swards under ambient (36Pa; empty symbols) and elevated (60Pa; solid symbols) atmospheric CO_2

Source: Baggs et al (2003)

However, similarities between conventional denitrification and nitrifier denitrification mean that this approach alone will be unable to distinguish between these processes, and denitrification may need to first be quantified with a ^{15}N-enrichment approach. This SP approach would seem to be insufficient to distinguish between all N_2O sources on its own, but it provides a powerful means to partition between all processes when used in combination with other approaches.

Linking processes to the underpinning microbiology

Understanding how changes in the size or in the diversity of microbial communities producing or reducing N_2O in response to environmental conditions or agricultural practices are related to N_2O fluxes is key to understanding controls on process rates and the microbial source of N_2O. However, until recently, diversity and activity of microbial communities involved in N-cycling were most often investigated in separate studies, which resulted in an artificial dichotomy between research on microbial processes and microbial community ecology. This gap is now slowly being bridged by combining molecular-based analysis of microbial diversity with measurements of process rates.

Only a few studies have investigated the relationships between activity and diversity of nitrifiers with any focus on nitrification rates and nitrifier-N_2O production being most often neglected (Kowalchuk et al, 2000; De Bie et al, 2001). In one of the most comprehensive studies, which used molecular and cultivation-based approaches of ammonia oxidizer diversity together with measurements of changes in nitrate and ammonium concentrations, Webster et al (2005) demonstrated a direct link between bacterial community structure and nitrification activity. More recently, Avrahami and Bohanann (2009) observed a significant relationship between diversity of AOB and N_2O emission rates and suggested that N_2O emissions may vary locally depending on local AOB community composition. However, other studies relating nitrifier-N_2O production to the nitrifier community have concluded that any change in activity, for example following addition of N, or with differing water content, was probably not the result of a major change in the ammonia-oxidizing community but of a physiological response (Mendum et al, 1999; Avrahami et al, 2002; Bateman, 2005). Bateman (2005) found changes in the AOB community structure over time after application of NH_4NO_3 to soils of varying WFPS, but these changes occurred after the main period of flux activity (Figure 2.4).

Relationships between denitrifier community composition and N_2O production have also been suggested by Cavigelli and Robertson (2000), who found that the enzymes involved in N_2O production or reduction had different sensitivities to oxygen or pH in denitrifying communities from a conventionally tilled agricultural field and a never-tilled successional field. Further studies have both confirmed and contradicted such links between denitrifier community composition and N_2O production. The addition of different types of artificial root exudates to soil microcosms resulted in differences in the N_2O/N_2 ratio, which were not related to differences in denitrifier community size or composition (Henry et al, 2008). Similarly, comparison of agricultural soil, riparian soil and creek sediment showed that denitrifier community composition and N_2O production were uncoupled across these agroecosystems (Rich and Myrold, 2004). In contrast, within these same systems, Philippot et al (2009a) found a significant correlation between the distribution of the percentage of bacteria capable of reducing N_2O within the total bacterial community and potential N_2O emissions, both being also strongly correlated to soil pH (Plate 2.1).

Whilst major progress has been made in characterizing the soil microbial community, understanding the relationships between diversity and activity of

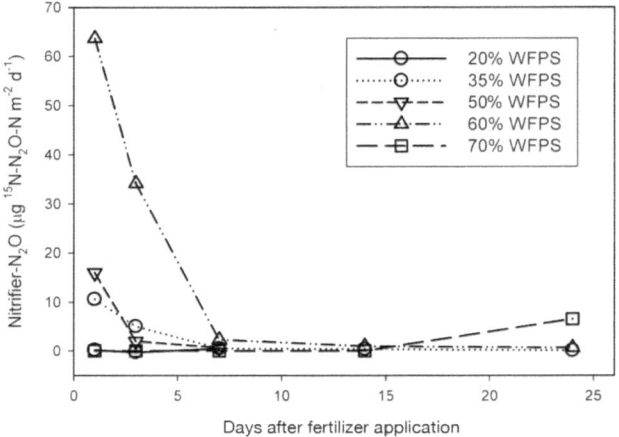

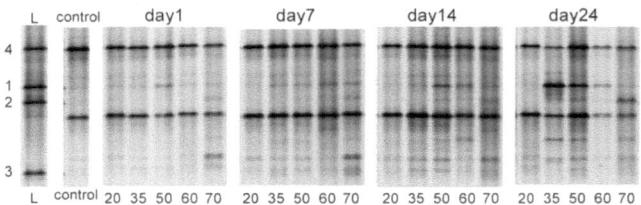

Figure 2.4 Nitrifier-^{15}N-N$_2$O production after application of fertilizer (20g Nm^{-2}) to a silt loam soil at 20–70 per cent WFPS, and the AOB community profiled by denaturing gradient gel electrophoresis (DGGE) of 16S rRNA gene PCR products amplified from the soil at different WFPS before and following fertilizer application

Note: L represents a control sample containing a recognized AOB from each cluster designation 1–4.
Source: Adapted from Bateman (2005)

a functional community still remains a major challenge in microbial ecology (Philippot and Hallin, 2005). In the contradictory studies described above, the number of ecosystems/conditions studied is rather limited. However, relationships between microbial diversity and processes are probably very complex and interwoven with many parameters, and therefore a comprehensive understanding can only be achieved by the analysis of microbial diversity and activity under a much broader range of ecosystems or of environmental gradients using high-throughput approaches.

Alternatively, the development of approaches targeting the active fraction of functional communities such as stable isotope probing (SIP) provides an elegant means of distinguishing organisms contributing to the observed processes. ^{13}C-SIP has successfully been applied to soil to identify methane-oxidizing bacteria (Bull et al, 2000; Morris et al, 2002; Radajewski et al, 2002) and AOB (Whitby et al, 2001). Unfortunately the application of ^{13}C-SIP to identify active denitrifiers is limited by the fact that denitrifiers can also assimilate the labelled C during respiratory processes other than denitrification. The identification of active denitrifiers using ^{15}N-SIP is likely to remain elusive since N is dissimilated. However, there is the potential for links to be made between plant-derived C flow and denitrifiers by identifying heterotrophic microorganisms capable of denitrification and utilizing the plant-C, through ^{13}C-CO$_2$ pulsing of vegetation (Johnson et al, 2002; Rangel-Castro et al, 2005). Whilst issues about turnover and recovery of applied ^{13}C, and cross-feeding of this C (Manefield et al, 2007) pose uncertainties for this approach, there are nevertheless some exciting opportunities for further characterization of active microorganisms through the use of stable isotopes to determine microbial sources.

Nitrous oxide uptake by soil

Based on average emission factors of N$_2$O and N$_2$ following addition of N, Vieten et al (2009) estimate the microbial N$_2$O sink by reduction of N$_2$O to N$_2$ to be 0.8 to 0.9 times the soil N$_2$O source. However, under certain conditions the capacity of soil to act as a sink for N$_2$O is thought to be greater than its capacity to emit N$_2$O, resulting in negative fluxes of N$_2$O (Blackmer and Bremner, 1976; Freney et al, 1978; Ryden, 1981). Interpretation of such findings, and a consensus of environmental and soil conditions conducive to such strong reduction, are confounded by our limited knowledge of the regulation of the denitrifier N$_2$O reductase (Richardson et al, 2009). Net negative fluxes indicative of a N$_2$O sink have been reported in several systems and controlled environment experiments (reviewed by Chapuis-Lardy et al, 2007). Since these net negative fluxes can be significant in magnitude (Flechard et al, 2005) they should never be ignored when calculating mean fluxes, or seasonal or annual emissions. However, the underlying reasons for these negative fluxes are rarely explored, with them often being attributed to experimental, sampling or analysis artefacts.

Uptake of N$_2$O in soil is often attributed to reduction of overlying atmospheric N$_2$O to N$_2$ during denitrification, the only proven terrestrial biological N$_2$O sink, and most likely occurs where N$_2$O production is low (Chapuis-Lardy et al, 2007). However, there is no direct evidence of atmospheric N$_2$O being drawn down into the soil pore space. The range of conditions under which negative fluxes have been reported, including those not typically associated with anaerobic denitrification (Khalil et al, 2002), suggest that abiotic mechanisms, or as yet unidentifed biological sinks, may also be operating. N$_2$O consumption appears to be negatively correlated with O$_2$ availability and pH, and is greater under low N availability (Chapuis-Lardy et al, 2007). The latter poses problems when relying on application of ^{15}N-labelled substrates or ^{15}N-N$_2$O to quantify N$_2$O reduction, as application of N may lower a soil's sink potential. Clough et al (1999) showed that two-thirds of pore space ^{15}N-N_2O was reduced to ^{15}N-N_2 before it had diffused upwards to the soil surface, and conversely it is possible for N_2O produced within the surface layers of soil to move downward by passive diffusion, convective movement with rainfall, or transport in solution in soil leachates (Clough et al, 2005). The longer N_2O remains in the soil, either due to production in deep soil layers or due to factors slowing diffusion, the more N_2O is consumed. The potential and rate for this consumption will depend on the proportion of N_2O-reducing micro-sites as predominantly controlled by O_2 concentration, O_2 demand, aggregate size, water content and soil pH (Šimek and Cooper, 2002; Fujita and Dooley, 2007; Vieten et al, 2009). Michaelis-Menten kinetics indicates the potential for N_2O consumption to account for 1.6 per cent of total respiration in soil at pH 7, and 0.9 per cent at pH 2.9 (Vieten et al, 2009). Greater understanding of the regulation of microbial N_2O reduction and abiotic mechanisms of N_2O sinks is urgently required before this N_2O uptake can be included in the budgets of the Intergovernmental Panel on Climate Change (IPCC) (Mosier et al, 1998), or for us to propose management strategies to mitigate N_2O emission by increasing its consumption.

Microbial sources of nitrous oxide in models

To ensure that any proposed management practices for mitigation of N_2O (see below) have no detrimental effects on future climate, it is essential that we can also predict emissions from each process under different climate scenarios, from a range of ecosystems and under different management options. Most predictive models only consider nitrification and anaerobic denitrification as sources of N_2O, and at the time of writing, nitrogen budgets and models ignore the contribution of nitrate ammonification. Models differ in the way they simulate nitrification and denitrification. For example, DNDC (Li et al, 1992) and ECOSYS (Grant, 1995) simulate microbial growth and associated N_2O production (Li, 2007), whereas DAYCENT (Del Grosso et al, 2000; Parton et al, 2001) and CERES (Gabrielle et al, 2006) use empirical formulae to simulate the N_2O production. No attempt has been made to include nitrifier denitrification or nitrate ammonification as N_2O sources, in part because of a lack of data on the controls of these processes for model parameterization or validation.

Scale is a primary consideration when source partitioning and quantifying N_2O emissions, because the prediction of emissions and formulation of policy decisions are made on regional, national or even global scales – fundamentally different scales from that at which most experiments are conducted to understand underlying processes (Standing et al, 2007). We discuss below how, with better understanding of the controls and sources of N_2O production in terrestrial systems, it should be possible to develop appropriate mitigation strategies. However, for such strategies to be appropriate and effective, outstanding issues of scale may need to be resolved. This is a major challenge, as technical constraints and reliance on stable isotope approaches have meant that efforts to quantify the contributions of different microbial processes to measured N_2O emissions have been focused on the microscale to plot scale, but most predictive models rely on the aggregated response offered at field or landscape scale. Linear relationships are often applied when upscaling from the microscale, thereby losing the level of detail offered at the microscale, and assuming that the hierarchy of controlling parameters on N_2O emission is the same across all scales. It is possible that uncertainty in these models can be lowered by consideration of the regulation of microbial sources of N_2O and the conditions under which different processes predominate, and inherent in that is the need for integration from the microscale. Modelling approaches should also be able to guide us in both temporally and spatially representative sampling for source partitioning.

Opportunities for mitigation

One of the main aims of determining the microbial sources of N_2O is to provide a scientific basis for more targeted mitigation strategies. Data on the environmental regulation of N_2O production during the different microbial processes are scarce, particularly for nitrifier denitrification and nitrate ammonification. More effort needs now to be placed on considering all potential sources of N_2O when measuring and interpreting N_2O fluxes from different environments and under different management regimes. Richardson et al (2009) argue that understanding the regulation of the denitrifier N_2O reductase is central to the development of management options to lower net N_2O emissions by enhancing its reduction to N_2, rather than trying to eliminate denitrification. They consider the effect on the regulation of the N_2O reductase of management options such as application of nitrogenous or copper fertilizer to regulate Cu availability for this Cu-based enzyme, soil organic matter (SOM) management and liming of crops or grasslands with specific characterized carbon and nitrogen traits.

Reducing the application of inorganic N fertilizers has traditionally been considered as the most successful option for lowering net emission of N_2O. However, source partitioning by Baggs et al (unpublished data) suggests that nitrifier denitrification may be increased under low-N conditions, and therefore lower N application may not necessarily be the most appropriate strategy in all systems. There are 'new' options being proposed for mitigation of greenhouse gases that may also hold the potential for enhancing C sequestration. One of these is application of biochar, which is a good example of where the ability to determine the N_2O source becomes essential. Although the mechanism by which biochar can lower net emission of N_2O is still open to debate (Renner, 2007; Gaunt and Lehmann, 2008), its chemical composition and structure suggest that it may be due to uptake of NH_4^+ into the char structure, and a change in pH, rather than by enhancing the reduction of N_2O to N_2. If this is the case then the largest effect will be in lowering nitrifier-N_2O, with only indirect effects on nitrate-reducing processes. This can be verified by coupling approaches for determining microbial sources of N_2O with rigorous investigations of the effect of biochar on the regulation of different enzymatic steps of each process and on microbial community structure. Such studies are required before biochar can be recommended as a reliable management option for lowering greenhouse gas emissions. This is just one example of where the effects on microbial sources of N_2O should also be considered in the light of possible interactions with the C cycle, and most pertinently with methane oxidation and methanotroph activity (Philippot et al, 2009b).

Our ability to determine the microbial source of N_2O in terrestrial systems, and even to quantify the contribution from each source, is improving with the advent and development of new techniques. This offers us exciting opportunities for targeted management options to optimize mitigation potential, but we still have some way to go before this can become a reality. Better understanding is required of the regulation of all processes, and particularly the uncertainties associated with the conditions conducive to N_2O production during nitrate ammonification and nitrifier denitrification. To be able to determine the microbial source of N_2O with any degree of accuracy in natural, unfertilized or fragile ecosystems we need to refine methodologies, or combine established and new methodologies, moving us away from reliance on application of ^{15}N-labelled substrates, which may artificially favour one process over others. Any mitigation approach should be grounded in predictions of future emissions from different management scenarios under a continuing changing climate. To achieve this, models require further development to encompass all microbial sources of N_2O, and should also consider abiotic sources and plant-mediated emissions. This will be facilitated by advances in techniques to unite source partitioning with upscaling of emissions from the micro-plot. The benefit in lowering model uncertainty that parameterizing the underpinning microbiology would provide is uncertain, but to ensure sustainability for the future it is essential to ascertain that any management option imposed to lower emissions has no adverse effect on the diversity and functioning of the microbial community.

References

Abbasi, M. K. and Adams, W. A. (2000) 'Gaseous N emission during simultaneous nitrification-denitrification associated with mineral N fertilisation to a grassland soil under field conditions', *Soil Biology and Biochemistry*, vol 32, pp1251–1259

Abeliovich, A. and Vonshak, A. (1992) 'Anaerobic metabolism of *Nitrosomonas europaea*', *Archives of Microbiology*, vol 158, pp267–270

Anderson, I. C., Poth, M., Homstead, J. and Burdige, D. (1993) 'A comparison of NO and N_2O production by the autotrophic nitrifier *Nitrosomonas europaea* and the heterotrophic nitrifier *Alcaligenes faecalis*', *Applied and Environmental Microbiology*, vol 59, pp3525–3533

Avrahami, S. and Bohannan, B. J. M. (2009) 'N_2O emission rates in a California meadow soil are influenced by fertilizer level, soil moisture and the community structure of ammonia-oxidizing bacteria', *Global Change Biology*, vol 15, pp643–655

Avrahami, S., Conrad, R. and Braker, G. (2002) 'Effect of soil ammonium concentration on N_2O release and on the community structure of ammonia oxidizers and denitrifiers', *Applied and Environmental Microbiology*, vol 68, pp5685–5692

Azam, F., Muller, C., Weiske, A., Benckiser, G. and Ottow, J. C. G. (2002) 'Nitrification and denitrification as sources of atmospheric nitrous oxide – role of oxidizable carbon and applied nitrogen', *Biology and Fertility of Soils*, vol 35, pp54–61

Baggs, E. M. (2008) 'A review of stable isotope techniques for N_2O source partitioning in soils: Recent progress, remaining challenges and future considerations', *Rapid Communications in Mass Spectrometry*, vol 22, pp1664–1672

Baggs, E. M. and Blum, H. (2004) 'CH_4 oxidation and CH_4 and N_2O emissions from *Lolium perenne* swards under elevated atmospheric CO_2', *Soil Biology and Biochemistry*, vol 36, pp713–723

Baggs, E. M., Richter, M., Cadisch, G. and Hartwig, U. A. (2003) 'Denitrification in grass swards is increased under elevated atmospheric CO_2', *Soil Biology and Biochemistry*, vol 35, pp729–732

Barford, C. C., Montoya, J. P., Altabet, M. A. and Mitchell, R. (1999) 'Steady-state nitrogen isotope effects of N_2 and N_2O production in *Paracoccus denitrificans*', *Applied and Environmental Microbiology*, vol 65, pp989–994

Bateman, E. J. (2005) 'The contribution of nitrification to nitrous oxide emissions from soils', PhD thesis, University of London, London

Bateman, E. J. and Baggs, E. M. (2005) 'Contributions of nitrification and denitrification to N_2O emissions from soils at different water-filled pore space', *Biology and Fertility of Soils*, vol 41, pp379–388

Bell, L. C. and Ferguson, S. J. (1991) 'Nitric oxide and nitrous oxide reductases are active under aerobic conditions in cells of *Thiosphaera pantotropha*', *Biochemical Journal*, vol 273, pp423–427

Blackmer, A. M. and Bremner, J. M. (1976) 'Potential of soil as a sink for atmospheric nitrous oxide', *Geophysical Research Letters*, vol 3, pp739–742

Blackmer, A. M. and Bremner, J. M. (1978) 'Inhibitory effect of nitrate on reduction of N_2O to N_2 by soil microorganisms', *Soil Biology and Biochemistry*, vol 10, pp187–191

Bock, E., Schmidt, I., Stuven, R. and Zart, D. (1995) 'Nitrogen loss caused by denitrifying *Nitrosomonas* cells using ammonium or hydrogen as electron donors and nitrite as electron acceptor', *Archives of Microbiology*, vol 163, pp16–20

Bol, R., Toyoda, S., Yamulki, S., Hawkins, J. M. B., Cardenas, L. M. and Yoshida, N. (2003) 'Dual isotope and isotopomer ratios of N_2O emitted from a temperate grassland soil after fertiliser application', *Rapid Communications in Mass Spectrometry*, vol 17, pp2550–2556

Bollmann, A., Bar-Gilissen, M. J. and Laanbroek, H. J. (2002) 'Growth at low ammonium concentrations and starvation response as potential factors involved in niche differentiation among ammonia-oxidizing bacteria', *Applied and Environmental Microbiology*, vol 68, pp4751–4757

Bremner, J. M. (1997) 'Sources of nitrous oxide in soils', *Nutrient Cycling in Agroecosystems*, vol 49, pp7–16

Brierley, E. D. R. and Wood, M. (2001) 'Heterotrophic nitrification in an acid forest soil: Isolation and characterisation of a nitrifying bacterium', *Soil Biology and Biochemistry*, vol 33, pp1403–1409

Bull, I. D., Parekh, N. R., Hall, G. H., Ineson, P. and Evershed, R. P. (2000) 'Detection and classification of atmospheric methane oxidizing bacteria in soil', *Nature*, vol 405, pp175–178

Buresh, R. J. and Patrick, W. H. (1978) 'Nitrate reduction to ammonium in anaerobic soil', *Soil Science Society of America Journal*, vol 42, pp913–918

Burford, J. R. and Bremner, J. M. (1975) 'Relationships between denitrification capacities of soils and total, water-soluble and readily decomposable soil organic matter', *Soil Biology and Biochemistry*, vol 7, pp389–394

Caskey, W. H. and Tiedje, J. M. (1979) 'Nitrate stimulated mineralisation of ammonium in anaerobic soils', *Oecologia*, vol 41, pp339–341

Cavigelli, M. A. and Robertson, G. P. (2000) 'The functional significance of denitrifier community composition in a terrestrial ecosystem', *Ecology*, vol 81, pp1402–1414

Chalk, P. M. and Smith, C. J. (1983) 'Chemodenitrification', in J. R. Freney and J. R. Simpson (eds) *Gaseous Loss of Nitrogen from Plant-Soil Systems*, Martinus Nijhoff and W. Junk, The Hague, pp65–90

Chang, C., Janzen, H. H., Cho, C. M. and Nakonechny, E. M. (1998) 'Nitrous oxide emission via plant transpiration', *Soil Science Society of America Journal*, vol 62, pp35–38

Chapuis-Lardy, L., Wrage, N., Metay, A., Chotte, J-C. and Bernoux, M. (2007) 'Soil a sink for N$_2$O? A review', *Global Change Biology*, vol 13, pp1–17

Chen, D. L., Chalk, P. M. and Freney, J. R. (1995a) 'Distribution of reduced products of N-15-labelled nitrate in anaerobic soils', *Soil Biology and Biochemistry*, vol 27, pp1539–1545

Chen, Y., Tessier, S., Mackenzie, A. F. and Laverdiere, M. R. (1995b) 'Nitrous oxide emission from an agricultural soil subjected to different freeze-thaw cycles', *Agriculture, Ecosystems and Environment*, vol 55, pp123–128

Clough, T. J., Jarvis, S. C., Dixon, E. R., Stevens, R. J., Laughlin, R. J. and Hatch, D. J. (1999) 'Carbon induced subsoil denitrification of ^{15}N-labelled nitrate in 1m deep soil columns', *Soil Biology and Biochemistry*, vol 31, pp31–41

Clough, T., Stevens, R. J., Laughlin, R. J., Sherlock, R. R. and Cameron, K. C. (2001) 'Transformations of inorganic-N in soil leachate under differing storage conditions', *Soil Biology and Biochemistry*, vol 33, pp1473–1480

Clough, T. J., Sherlock, R. R. and Rolston, D. E. (2005) 'A review of the movement and fate of N$_2$O in the subsoil', *Nutrient Cycling in Agroecosystems*, vol 72, pp3–11

Colliver, B. B. and Stephenson, T. (2000) 'Production of nitrogen oxide and dinitrogen oxide by autotrophic nitrifiers', *Biotechnology Advances*, vol 18, pp219–232

Conrad, R. (1996) 'Soil microorganisms as controllers of atmospheric trace gases (H$_2$, CO, CH$_4$, OCS, N$_2$O, and NO)', *Microbiological Reviews*, vol 60, pp609–640

Cooper, D. C., Picardal, F. W., Schimmelmann, A. and Coby, A. J. (2003) 'Chemical and biological interactions during nitrate and geothite reduction by *Shewanella putrefaciens* 200', *Applied and Environmental Microbiology*, vol 69, pp3517–3525

Daniel, R. M., Smith, I. M., Philipp, A. D., Ratcliffe, H. D., Drozd, J. W. and Bull, A. T. (1980) 'Anaerobic growth and denitrification by *Rhizobium japonicum* and other rhizobia', *Journal of General Microbiology*, vol 120, pp517–521

Daum, M., Zimmer, W., Papen, H., Kloos, K., Nawrath, K. and Bothe, H. (1998) 'Physiological and molecular biological characterization of ammonia oxidation of the heterotrophic nitrifier *Pseudomonas putida*', *Current Microbiology*, vol 37, pp281–288

Davidson, E. A. (1991) 'Fluxes of nitrous oxide and nitric oxide from terrestrial ecosystems', in J. E. Rogers and W. B. Whitman (eds) *Microbial Production and Consumption of Greenhouse Gases: Methane, Nitrogen Oxides, and Halomethanes*, American Society for Microbiology, Washington, DC, pp219–236

De Bie, M. J. M., Speksnijder, A. G. C. L., Kowalchuk, G. A., Schuurman, T., Zwart, G., Stephen, J. R., Diekmann, O. E. and Laanbroek, H. J. (2001) 'Shifts in the domiant populations of ammonia-oxidizing β-subclass proteobacteria along the eutrophic Schelde estuary', *Aquatic Microbial Ecology*, vol 23, pp225–236

De Boer, W. and Kowalchuk, G. A. (2001) 'Nitrification in acid soils: Micro-organisms and mechanisms', *Soil Biology and Biochemistry*, vol 33, pp853–866

Del Grosso, S. J., Parton, W. J., Mosier, A. R., Ojima, D. S., Kulmala, A. E. and Phongpan, S. (2000) 'General model for N_2O and N_2 gas emissions from soils due to denitrification', *Global Biogeochemical Cycles*, vol 14, pp1045–1060

Dundee, L. and Hopkins, D. W. (2001) 'Different sensitivities to oxygen of nitrous oxide production by *Nitrosomonas europaea* and *Nitrosolobus multiformis*', *Soil Biology and Biochemistry*, vol 33, pp1563–1565

Eichner, M. J. (1990) 'Nitrous oxide emissions from fertilized soils – summary of available data', *Journal of Environmental Quality*, vol 19, pp272–280

Ellis, S., Howe, M. T., Goulding, K. W. T., Mugglestone, M. A. and Dendooven, L. (1998) 'Carbon and nitrogen dynamics in a grassland soil with varying pH: Effect of pH on the denitrification potential and dynamics of the reduction enzymes', *Soil Biology and Biochemistry*, vol 30, pp359–367

Enwall, K., Philippot, L. and Hallin, S. (2005) 'Activity and composition of the denitrifying bacterial community respond differently to long-term fertilization', *Applied and Environmental Microbiology*, vol 71, pp8335–8343

Fazzolari, E., Mariotti, A. and Germon, J. C. (1990) 'Dissimilatory ammonia production vs denitrification in vitro and in inoculated agricultural soil samples', *Canadian Journal of Microbiology*, vol 36, pp786–793

Fernandez, L. A., Perotti, E. B., Sagardoy, M. A. and Gomez, M. A. (2008) 'Denitrification activity of *Bradyrhizobium* sp isolated from Argentine soybean cultivated soils', *World Journal of Microbiology and Biotechnology*, vol 24, pp2577–2585

Firestone, M. K., Firestone, R. B. and Tiedje, J. M. (1980) 'Nitrous oxide from soil denitrification – factors controlling its biological production', *Science*, vol 208, pp749–751

Flechard, C. R., Neftel, A., Jocher, M., Ammann, C. and Fuhrer, J. (2005) 'Bi-directional soil/ atmosphere N_2O exchange over two mown grassland systems with contrasting management practices', *Global Change Biology*, vol 11, pp2114–2127

Freney, J. R., Denmead, O. T. and Simpson, J. R. (1978) 'Soil as a source or sink for atmospheric nitrous oxide', *Nature*, vol 273, pp530–532

Frijlink, M. J., Abee, T., Laanbroek, H. J., De Boer, W. and Konings, W. N. (1992) 'The bioenergetics of ammonia and hydroxylamine oxidation in *Nitrosomonas europaea* at acid and alkaline pH', *Archives of Microbiology*, vol 157, pp194–199

Fujita, K. and Dooley, D. M. (2007) 'Insights into the mechanism of N_2O reduction by reductively activated N_2O reductase from kinetics and spectroscopic studies of pH effects', *Inorganic Chemistry*, vol 46, pp613–615

Gabrielle. B., Laville, P., Duval, O., Nicoullaud, B., Germon, J. C. and Henault, C. (2006) 'Process-based modeling of nitrous oxide emissions from wheat-cropped soils at the sub-regional scale', *Global Biogeochemical Cycles*, vol 20, GB4018

Garber, E. A. and Hollocher, T. C. (1982) 'Positional isotopic equivalence of nitrogen in nitrous oxide produced by the denitrifying bacterium *Pseudomonas stutzeri*: indirect evidence for a nitroxyl pathway', *Journal of Biological Chemistry*, vol 257, pp4705–4708

Gaunt, J. L. and Lehmann, J. (2008) 'Energy balance and emissions associated with biochar sequestration and pyrolysis bioenergy production', *Environmental Science and Technology*, vol 42, pp4152–4158

Gerards, S., Duyts, H. and Laanbroek, H. J. (1998) 'Ammonium-induced inhibition of ammonium-starved *Nitrosomonas europaea* cells in soil and sand slurries', *FEMS Microbiology Ecology*, vol 26, pp269–280

Goreau, T. J., Kaplan, W. A., Wofsy, S. C., McElroy, M. B., Valois, F. W. and Watson, S. W. (1980) 'Production of NO_2^- and N_2O by nitrifying bacteria at reduced concentrations of oxygen', *Applied and Environmental Microbiology*, vol 40, pp526–532

Grant, R. F. (1995) 'Mathematical modelling of nitrous oxide evolution during nitrification', *Soil Biology and Biochemistry*, vol 27, pp1117–1125

Gregory, L. G., Karakas-Sen, A., Richardson, D. J. and Spiro, S. (2000) 'Detection of genes for membrane-bound nitrate reductase in nitrate-respiring bacteria and in community DNA', *FEMS Microbiology Letters*, vol 183, pp275–279

Groffman, P. M., Altabet, M. A., Bohlke, J. K., Butterbach-Bahl, K., David, M. B., Firestone, M. K., Giblin, A. E., Kana, T. M., Nielsen, L. P. and Voytek, M. A. (2006) 'Methods for measuring denitrification: Diverse approaches to a difficult problem', *Ecological Applications*, vol 16, pp2091–2122

Hakata, M., Takahashi, M., Zumft, W., Sakamoto, A. and Morikawa, H. (2003) 'Conversion of the nitrate nitrogen and nitrogen dioxide to nitrous oxide in plants', *Acta Biotechnologica*, vol 23, pp249–257

Hallin, S. and Lindgren, P. E. (1999) 'PCR detection of genes encoding nitrile reductase in denitrifying bacteria', *Applied and Environmental Microbiology*, vol 65, pp1652–1657

Henry, S., Texier, S., Hallet, S., Bru, D., Dambreville, C., Cheneby, D., Bizouard, F., Germon, J. C. and Philippot, L. (2008) 'Disentangling the rhizosphere effect on nitrate reducers and denitrifiers: Insight into the role of root exudates', *Environmental Microbiology*, vol 10, pp3082–3092

Hochstein, L. and Tomlinson, G. A. (1988) 'The enzymes associated with denitrification', *Annual Review of Microbiology*, vol 42, pp231–261

Holtan-Hartwig, L., Dorsch, P. and Bakken, L. R. (2000) 'Comparison of denitrifying communities in organic soils: Kinetics of NO_3^- and N_2O reduction', *Soil Biology and Biochemistry*, vol 32, pp833–843

Hooper, A. B., Vannelli, T., Bergmann, D. J. and Arciero, D. M. (1997) 'Enzymology of the oxidation of ammonia to nitrite by bacteria', *Antonie Van Leeuwenhoek International Journal of General and Molecular Microbiology*, vol 71, pp59–67

Hynes, R. K. and Knowles, R. (1978) 'Inhibition by acetylene of ammonia oxidation in *Nitrosomonas europaea*', *FEMS Microbiology Letters*, vol 4, pp319–321

Hynes, R. K. and Knowles, R. (1984) 'Production of nitrous oxide by *Nitrosomonas europaea* – effects of acetylene, pH and oxygen', *Canadian Journal of Microbiology*, vol 30, pp1397–1404

IPCC (Intergovernmental Panel on Climate Change) (2007) *Climate Change 2007: Synthesis Report. Contribution of Working Groups I, II and III to the Fourth Assessment Report of the Intergovernmental Panel on Climate Change*, Core Writing Team, R. K. Pachauri and A. Reisinger (eds), IPCC, Geneva, Switzerland

Islas-Lima, S., Thalasso, F. and Gomez-Hernandez, J. (2004) 'Evidence of anoxic methane oxidation coupled to denitrification', *Water Research*, vol 38, pp13–16

Jackson, M. A., Tiedje, J. M. and Averill, B. A. (1991) 'Evidence for an NO-rebound mechanism for production of N_2O from nitrite by the copper containing nitrite reductase from *Achromobacter cycloclastes*', *FEBS Letters*, vol 291, pp41–44

Jiang, Q. Q. and Bakken, L. R. (1999) 'Nitrous oxide production and methane oxidation by different ammonia-oxidizing bacteria', *Applied and Environmental Microbiology*, vol 65, pp2679–2684

Johnson, D., Leake, J. R., Ostle, N., Ineson, P. and Read, D. J. (2002) 'In situ (CO_2)-C-13 pulse-labelling of upland grassland demonstrates a rapid pathway of carbon flux from arbuscular mycorrhizal mycelia to the soil', *New Phytologist*, vol 153, pp327–334

Kelso, B. H. L., Smith, R. V., Laughlin, R. J. and Lennox, S. D. (1997) 'Dissimilatory nitrate reduction in anaerobic sediments leading to river nitrite accumulation', *Applied and Environmental Microbiology*, vol 63, pp4679–4685

Kester, R. A., de Boer, W. and Laanbroek, H. J. (1996) 'Short exposure to acetylene to distinguish between nitrifier and denitrifier nitrous oxide production in soil and sediment samples', *FEMS Microbiology Ecology*, vol 20, pp111–120

Khalil, M. I., Rosenani, A. B., Van Cleemput, O., Fauziah, C. I. and Shamshuddin, J. (2002) 'Nitrous oxide emissions from an ultisol of the humid tropics under maize-groundnut rotation', *Journal of Environmental Quality*, vol 31, pp1071–1078

Killham, K. (1986) 'Heterotrophic nitrification', in J. I. Prosser (ed) *Nitrification*, Special Publications of the Society For General Microbiology, Volume 20, IRL Press Ltd, Oxford, pp117–126

Klemedtsson, L., Svensson, B. H. and Rosswall, T. (1988) 'A method of selective inhibition to distinguish between nitrification and denitrification as sources of nitrous oxide in soil', *Biology and Fertility of Soils*, vol 6, pp112–119

Knowles, R. (1982) 'Denitrification', *Microbiological Reviews*, vol 46, pp43–70

Kobayashi, M. and Shoun, H. (1995) 'The copper containing dissimilatory nitrite reductase involved in the denitrifying system of the fungus *Fusarium oxysporum*', *Journal of Biological Chemistry*, vol 270, pp4146–4151

Kowalchuk, G. A. and Stephen, J. R. (2001) 'Ammonia-oxidizing bacteria: A model for molecular microbial ecology', *Annual Review of Microbiology*, vol 55, pp485–529

Kowalchuk, G. A., Stienstra, A. W., Heilig, G. H., Stephen, J. R. and Woldendorp, J. W. (2000) 'Molecular analysis of ammonia-oxidising bacteria in soil of successional grasslands of the Drentsche A (The Netherlands)', *FEMS Microbiology Ecology*, vol 31, pp207–215

Kuai, L. F. and Verstraete, W. (1998) 'Ammonium removal by the oxygen-limited autotrophic nitrification-denitrification system', *Applied and Environmental Microbiology*, vol 64, pp4500–4506

Laughlin, R. J. and Stevens, R. J. (2002) 'Evidence for fungal dominance of denitrification and codenitrification in a grassland soil', *Soil Science Society of America Journal*, vol 66, pp1540–1548

Laverman, A. M., Zoomer, H. R., Engelbrecht, D., Berg, M. P., van Straalen, N. M., van Verseveld, H. W. and Verhoef, H. A. (2000) 'Soil layer-specific variability in net nitrification and denitrification in an acid coniferous forest', *Biology and Fertility of Soils*, vol 32, pp427–434

Li, C. (2007) 'Quantifying greenhouse gas emissions from soils: Scientific basis and modelling approach', *Soil Science and Plant Nutrition*, vol 53, pp344–352

Li, C. S., Frolking, S. and Frolking, T. A. (1992) 'A model of nitrous oxide evolution from soil driven by rainfall events: 1. Model structure and sensitivity', *Journal of Geophysical Research*, vol 97, pp9759–9776

Lipschultz, F., Zafiriou, O. C., Wofsy, S. C., McElroy, M. B., Valois, F. W. and Watson, S. W. (1981) 'Production of NO and N_2O by soil nitrifying bacteria', *Nature*, vol 294, pp541–643

Lloyd, D., Boddy, L. and Davies, K. J. P. (1987) 'Persistence of bacterial denitrification capacity under aerobic conditions – the rule rather than the exception', *FEMS Microbiology Ecology*, vol 45, pp185–190

Maag, M. and Vinther, F. P. (1996) 'Nitrous oxide emission by nitrification and denitrification in different soil types and at different soil moisture contents and temperatures', *Applied Soil Ecology*, vol 4, pp5–14

Manefield, M., Griffiths, R., McNamara, N. P., Sleep, D., Ostle, N. and Whiteley, A. (2007) 'Insights into the fate of a C-13 labelled phenol pulse for stable isotope probing (SIP) experiments', *Journal of Microbiological Methods*, vol 69, pp340–344

Martikainen, P. J., Lehtonen, M., Lang, K., De Boer, W. and Ferm, A. (1993) 'Nitrification and nitrous oxide production potentials in aerobic soil samples from the soil profile of a Finnish coniferous site receiving high ammonium deposition', *FEMS Microbiology Ecology*, vol 13, pp113–121

Mathieu, O., Hénault, C., Lévêque, J., Baujard, E., Milloux, M. J. and Andreux, F. (2006) 'Quantifying the contribution of nitrification and denitrification to the nitrous oxide flux using N-15 tracers', *Environmental Pollution*, vol 144, pp933–940

Mendum, T. A., Sockett, R. E. and Hirsch, P. R. (1999) 'Use of molecular and isotopic techniques to monitor the response of autotrophic ammonia-oxidizing populations of the beta subdivision of the class Proteobacteria in arable soils to nitrogen fertilizer', *Applied and Environmental Microbiology*, vol 65, pp4155–4162

Millar, N. and Baggs, E. M. (2005) 'Relationships between N_2O emissions and water-soluble C and N contents of agroforestry residues after their addition to soil', *Soil Biology and Biochemistry*, vol 37, pp605–608

Mohan, S. B., Schmid, M., Jetten, M. and Cole, J. (2004) 'Detection and widespread distribution of the nrfA gene encoding nitrite reduction to ammonia, a short circuit in the biological nitrogen cycle that competes with denitrification', *FEMS Microbiology Ecology*, vol 49, pp433–443

Moir, J. W. B., Wehrfritz, J. M., Spiro, S. and Richardson, D. J. (1996a) 'The biochemical characterization of a novel non-haem-iron hydroxylamine oxidase from *Paracoccus denitrificans* GB17', *Biochemical Journal*, vol 319, pp823–827

Moir, J. W. B., Crossman, L. C., Spiro, S. and Richardson, D. J. (1996b) 'The purification of ammonia monooxygenase from *Paracoccus denitrificans*', *FEBS Letters*, vol 387, pp71–74

Morley, N., Baggs, E. M., Dorsch, P. and Bakken, L. (2008) 'Production of NO, N_2O and N_2 by extracted soil bacteria, regulation by NO_2^- and O_2 concentrations', *FEMS Microbiology Ecology*, vol 65, pp102–112

Morkved, P. T., Dorsch, P. and Bakken, L. R. (2007) 'The N_2O product ratio of nitrification and its dependence on long-term changes in soil pH', *Soil Biology and Biochemistry*, vol 39, pp2048–2057

Morris, S. A., Radajewski, S., Willison, T. W. and Murrell, J. C. (2002) 'Identification of the functionally active methanotroph population in a peat soil microcosm by stable-isotope probing, *Applied and Environmental Microbiology*, vol 68, pp1446–1453

Mosier, A., Kroeze, C., Nevison, C., Oenema, O., Seitzinger, S. and van Cleemput, O. (1998) 'Closing the global N_2O budget: Nitrous oxide emissions through the agricultural nitrogen cycle', *Nutrient Cycling in Agroecosystems*, vol 52, pp225–248

Nägele, W. and Conrad, R. (1990) 'Influence of pH on the release of NO and N_2O from fertilised and unfertilised soil', *Biology and Fertility of Soils*, vol 10, pp139–144

Nelson, D. W. (1982) 'Gaseous losses of nitrogen other than through denitrification', in F. J. Stevenson (ed) *Nitrogen in Agricultural Soils*, American Society of Agronomy, Madison, WI, pp327–364

Otte, S., Grobben, N. G., Robertson, L. A., Jetten, M. S. M. and Kuenen, J. G. (1996) 'Nitrous oxide production by *Alcaligenes faecalis* under transient and dynamic aerobic and anaerobic conditions', *Applied and Environmental Microbiology*, vol 62, pp2421–2426

Papen, H., Vonberg, R., Hinkel, I., Thoene, B. and Rennenberg, H. (1989) 'Heterotrophic nitrification by *Alcaligenes faecalis* – NO_2^-, NO_3^-, N_2O and NO production in exponentially growing cultures', *Applied and Environmental Microbiology*, vol 55, pp2068–2072

Parkes, S. D., Jolley, D. F. and Wilson, S. R. (2007) 'Inorganic nitrogen transformations in the treatment of landfill leachate with a high ammonium load: A case study', *Environmental Monitoring and Assessment*, vol 124, pp51–61

Parkin, T. B., Sexstone, A. J. and Tiedje, J. M. (1985) 'Adaptation of denitrifying populations to low soil pH', *Applied and Environmental Microbiology*, vol 49, pp1053–1056

Parton, W. J., Holland, E. A., Del Grosso, S. J., Hartman, M. D., Martin, R. E., Mosier, A. R., Ojima, D. S. and Schimel, D. S. (2001) 'Generalized model for NO_x and N_2O emissions from soils', *Journal of Geophysical Research*, vol 106, pp17403–17419

Patureau, D., Zumstein, E., Delgenes, J. P. and Moletta, R. (2000) 'Aerobic denitrifiers isolated from diverse natural and managed ecosystems', *Microbial Ecology*, vol 39, pp145–152

Paul, J. W. and Beauchamp, E. G. (1989) 'Effect of carbon constituents in manure on denitrification in soil', *Canadian Journal of Soil Science*, vol 69, pp49–61

Perez, T., Garcia-Montiel, D., Trumbore, S., Tyler, S., De Camargo, P., Moreira, M., Piccolo, M. and Cerri, C. (2006) 'Nitrous oxide nitrification and denitrification N-15 enrichment factors from Amazon forest soils', *Ecological Applications*, vol 16, pp2153–2167

Philippot, L. and Hallin, S. (2005) 'Finding the missing link between diversity and activity using denitrifying bacteria as a model functional community', *Current Opinion in Microbiology*, vol 8, pp234–239

Philippot, L., Piutti, S., Martin-Laurent, F., Hallet, S. and Germon, J. C. (2002) 'Molecular analysis of the nitrate-reducing community from unplanted and maize-planted soils', *Applied and Environmental Microbiology*, vol 68, pp6121–6128

Philippot, L., Cuhel, J., Saby, N. P. A., Cheneby, A., Chronáková, A., Bru, D., Arrouays, D., Martin-Laurent, F. and Šimek, M. (2009a) 'Mapping field-scale spatial distribution patterns of size and activity of the denitrifier community', *Environmental Microbiology*, vol 11, pp1518–1526

Philippot, L., Hallin, S., Borjessen, G. and Baggs, E. M. (2009b) 'Biochemical cycling in the rhizosphere having an impact on global change', *Plant and Soil*, doi:10.1007/s11104-008-9796-9

Pihlatie, M., Ambus, P., Rinne, J., Pilegaard, K. and Vesala, T. (2005) 'Plant-mediated nitrous oxide emissions from beech (*Fagus sylvatica*) leaves', *New Phytologist*, vol 168, pp93–98

Prosser, J. I. and Nicol, G. W. (2008) 'Relative contributions of archaea and bacteria to aerobic ammonia oxidation in the environment', *Environmental Microbiology*, vol 10, pp2931–2941

Radajewski, S., Webster, G., Reay, D. S., Morris, S. A., Ineson, P., Nedwell, D. B., Prosser, J. I. and Murrell, J. C. (2002) 'Identification of active methylotroph populations in an acidic forest soil by stable isotope probing', *Microbiology-SGM*, vol 148 pp2331–2342

Rangel-Castro, J. I., Killham, K., Ostle, N., Nicol, G. W., Anderson, I. C., Scrimgeour, C. M., Ineson, P., Meharg, A. and Prosser, J. I. (2005) 'Stable isotope probing analysis of the influence of liming on root exudate utilization by soil microorganisms', *Environmental Microbiology*, vol 7, pp828–838

Ren, T., Roy, R. and Knowles, R. (2000) 'Production and consumption of nitric oxide by three methanotrophic bacteria', *Applied and Environmental Microbiology*, vol 66, pp3391–3897

Renner, R (2007) 'Rethinking biochar', *Environmental Science and Technology*, vol 41, pp5932–5933

Rich, J. J. and Myrold, D. D. (2004) 'Community composition and activities of denitrifying bacteria from adjacent agricultural soil, riparian soil, and creek sediment in Oregon, USA', *Soil Biology and Biochemistry*, vol 36, pp1431–1441

Richardson, D., Felgate, H., Watmough, N., Thomson, A. and Baggs, E. (2009) 'Mitigating release of the potent greenhouse gas N_2O from the nitrogen cycle – could enzymic regulation hold the key?', *Trends in Biotechnology*, vol 27, pp388–397

Ritchie, G. A. and Nicholas, D. J. (1972) 'Identification of sources of nitrous oxide produced by oxidative and reductive processes in *Nitrosomonas europaea*', *Biochemical Journal*, vol 126, pp1181–1191

Robertson, G. P. and Tiedje, J. M. (1987) 'Nitrous oxide sources in aerobic soils – nitrification, denitrification and other biological processes', *Soil Biology and Biochemistry*, vol 19, pp187–193

Rusch, H. and Rennenberg, H. (1998) 'Black Alder (*Alnus glutinosa* (L.) Gaertn.) trees mediate methane and nitrous oxide emission from the soil to the atmosphere', *Plant and Soil*, vol 201, pp1–7

Ryden, J. C. (1981) 'N_2O exchange between a grassland soil and the atmosphere', *Nature*, vol 292, pp235–237

Shaw, L. J., Nicol, G. W., Smith, Z., Fear, J., Prosser, J. I. and Baggs, E. M. (2006) '*Nitrosospira* spp. can produce nitrous oxide *via* a nitrifier denitrification pathway', *Environmental Microbiology*, vol 8, pp214–222

Shoun, H., Kim, D. H., Uchiyama, H. and Sugiyama, J. (1992) 'Denitrification by fungi', *FEMS Microbiology Letters*, vol 73, pp277–281

Šimek, M. and Cooper, J. E. (2002) 'The influence of soil pH on denitrification: Progress towards the understanding of this interaction over the last 50 years', *European Journal of Soil Science*, vol 53, pp345–354

Sitaula, B. K. and Bakken, L. R. (1993) 'Nitrous oxide release from spruce forest soil – relationships with nitrification, methane uptake, temperature, moisture and fertilisation', *Soil Biology and Biochemistry*, vol 25, pp1415–1421

Smart, D. R. and Bloom, A. J. (2001) 'Wheat leaves emit nitrous oxide during nitrate assimilation', *Proceedings of the National Academy of Sciences of the USA*, vol 98, pp7875–7878

Smith, M. S. and Tiedje, J. M. (1979) 'Effect of roots on soil denitrification', *Soil Science Society of America Journal*, vol 43, pp951–955

Smith, M. S. and Zimmerman, K. (1981) 'Nitrous oxide production by non-denitrifying soil nitrate reducers', *Soil Science Society of America Journal*, vol 45, pp865–871

Standing, D., Baggs, E. M., Smith, P., Wattenbach, M. and Killham, K. (2007) 'Meeting the challenge of scaling up processes in the plant–soil–microbe system', *Biology and Fertility of Soils*, vol 44, pp245–257

Stanford, G., Vanderpol, R. A. and Dzienia, S. (1975) 'Denitrification rates in relation to total and extractable soil carbon', *Soil Science Society of America Journal*, vol 39, pp284–289

Stevens, R. J., Laughlin, R. J. and Malone, J. P. (1998) 'Measuring the mole fraction and source of nitrous oxide in the field', *Soil Biology and Biochemistry*, vol 30, pp541–543

Stres, B., Mahne, I., Avgustin, G. and Tiedje, J. M. (2004) 'Nitrous oxide reductase (nosZ) gene fragments differ between native and cultivated Michigan soils', *Applied and Environmental Microbiology*, vol 70, pp301–309

Stroo, H. F., Klein, T. M. and Alexander, M. (1986) 'Heterotrophic nitrification in an acid forest soil and by an acid-tolerant fungus', *Applied and Environmental Microbiology*, vol 52, pp1107–1111

Su, F., Takaya, N. and Shoun, H. (2004) 'Nitrous oxide-forming codenitrification catalyzed by cytochrome P450nor', *Bioscience Biotechnology and Biochemistry*, vol 68, pp473–475

Sutka, R. L., Ostrom, N. E., Ostrom, P. H., Gandhi, H. and Breznak, J. A. (2003) 'Nitrogen isotopomer site preference of N_2O produced by *Nitrosomonas europaea* and *Methylococcus capsulatus* Bath', *Rapid Communications in Mass Spectrometry*, vol 17, pp738–745

Sutka, R. L., Ostrom, N. E., Ostrom, P. H., Gandhi, H. and Breznak, J. A. (2004) 'Nitrogen isotopomer site preference of N_2O produced by *Nitrosomonas europaea* and *Methylococcus capsulatus* Bath', *Rapid Communications in Mass Spectrometry*, vol 18, pp1411–1412

Sutka, R. L., Ostrom, N. E., Ostrom, P. H., Breznak, J. A., Gandhi, H., Pitt, A. J. and Li, F. (2006) 'Distinguishing nitrous oxide production from nitrification and denitrification on the basis of isotopomer abundances', *Applied and Environmental Microbiology*, vol 72, pp638–644

Takaya, N. and Shoun, H. (2000) 'Nitric oxide reduction, the last step in denitrification by Fusarium oxysporum, is obligatorily mediated by cytochrome P450nor', *Molecular and General Genetics*, vol 263, pp342–348

Takaya, N., Catalan-Sakairi, M. A. B., Sakaguchi, Y., Kato, I., Zhou, Z. M. and Shoun, H. (2003) 'Aerobic denitrifying bacteria that produce low levels of nitrous oxide', *Applied and Environmental Microbiology*, vol 69, pp3152–3157

Tanimoto T., Hatano, K. and Kim, D. H. (1992) 'Co-denitrification by the denitrifying system of the fungus *Fusarium oxysporum*', *FEMS Microbiology Letters*, vol 93, pp177–180

Thomsen, J. K., Geest, T. and Cox, R. P. (1994) 'Mass spectrometric studies of the effect of pH on the accumulation of intermediates in denitrification by *Paracoccus denitrificans*', *Applied and Environmental Microbiology*, vol 60, pp536–541

Tiedje, J. M., Sexstone, A. J., Myrold, D. D. and Robinson, J. A. (1982) 'Denitrification – ecological niches, competition and survival', *Antonie van Leeuwenhoek Journal of Microbiology*, vol 48, pp569–583

Van Cleemput, O. and Samater, A. H. (1996) 'Nitrite in soils: Accumulation and role in the formation of gaseous N compounds', *Fertilizer Research*, vol 45, pp81–89

Venterea, R. T. and Rolston, D. E. (2000) 'Mechanisms and kinetics of nitric and nitrous oxide production during nitrification in agricultural soil', *Global Change Biology* vol 6, pp303–316

Vieten, B., Conen, F., Neftel, A. and Alewell, C. (2009) 'Respiration of nitrous oxide in suboxic soil', *European Journal of Soil Science*, vol 60, pp332–337

Wahlen, M. and Yoshinari, T. (1985) 'Oxygen isotope ratios in N_2O from different environments', *Nature*, vol 313, pp780–782

Wan, Y. J., Ju, X. T., Ingwersen, J., Schwarz, U., Stange, C. F., Zhang, F. S. and Streck, T. (2009) 'Gross nitrogen transformations and related nitrous oxide emissions in an intensively used calcareous soil', *Soil Science Society of America Journal*, vol 73, pp102–112

Webster, E. A. and Hopkins, D. W. (1996) 'Nitrogen and oxygen isotope ratios of nitrous oxide emitted from soil and produced by nitrifying and denitrifying bacteria', *Biology and Fertility of Soils*, vol 22, pp326–330

Webster, E. A., Halpin, C., Chudek, J. A., Tilston, E. L. and Hopkins, D. W. (2005) 'Decomposition in soil of soluble, insoluble and lignin-rich fractions of plant material from tobacco with genetic modifications to lignin biosynthesis', *Soil Biology and Biochemistry*, vol 37, pp751–760

Weier, K. L., Doran, J. W., Power, J. F. and Walters, D. T. (1993) 'Denitrification and the dinitrogen nitrous oxide ratio as affected by soil water, available carbon and nitrate', *Soil Science Society of America Journal*, vol 57, pp67–72

Well, R., Kurganova, I., Lopes de Gerenyu, V. and Flessa, H. (2006) 'Isotopomer signatures of soil-emitted N_2O under different moisture conditions – A microcosm study with arable loess soil', *Soil Biology and Biochemistry*, vol 38, pp2923–2933

Whitby, C. B., Hall, G., Pickup, R., Saunders, J. R., Ineson, P., Parekh, N. R. and McCarthy, A. (2001) 'C-13 incorporation into DNA as a means of identifying the active components of ammonia-oxidizer populations', *Letters in Applied Microbiology*, vol 32, pp398–401

Wrage, N., Velthof, G. L., van Beusichem, M. L. and Oenema, O. (2001) 'Role of nitrifier denitrification in the production of nitrous oxide', *Soil Biology and Biochemistry*, vol 33, pp1723–1732

Wrage, N., Lauf, J., del Prado, A., Pinto, M., Pietrzak, S., Yamulki, S., Oenema, O. and Gebauer, G. (2004) 'Distinguishing sources of N_2O in European grasslands by stable isotope analysis', *Rapid Communications in Mass Spectrometry*, vol 18, pp1201–1207

Wrage, N., van Groeningen, J. W., Oenema, O. and Baggs, E. M. (2005) 'Distinguishing between soil sources of N_2O using a new [15]N- and ^{18}O-enrichment method', *Rapid Communications in Mass Spectrometry*, vol 19, pp3298–3306

Yamulki, S., Harrison, R. M., Goulding, K. W. T. and Webster, C. P. (1997) 'N_2O, NO and NO_2 fluxes from a grassland: Effect of soil pH', *Soil Biology and Biochemistry*, vol 29, pp1199–1208

Yamulki, S., Toyoda, S., Yoshida, N., Veldkamp, E., Grant, B. and Bol, R. (2001) 'Diurnal fluxes and the isotopomer ratios of N_2O in a temperate grassland following urine amendment', *Rapid Communications in Mass Spectrometry*, vol 15, pp1263–1269

Yan, X., Du Shi, S. L. and Xing, G. (2000) 'Pathways of N_2O emission from rice paddy soil', *Soil Biology and Biochemistry*, vol 32, pp437–440

Yoshida, N. (1988) 'N-15 depleted N_2O as a product of nitrification', *Nature*, vol 335, pp528–529

Yoshinari, T. and Knowles, R. (1976) 'Acetylene inhibition of nitrous oxide reduction by denitrifying bacteria', *Biochemical and Biophysical Research Communications*, vol 69, pp705–710

Zumft, W. G. (1997) 'Cell biology and molecular basis of denitrification', *Microbiology and Molecular Biology Reviews*, vol 61, pp533–616

3

Marine Pathways to Nitrous Oxide

Hermann Bange, Alina Freing, Annette Kock and Carolin Löscher

Introduction

There is no doubt that oceanic N_2O emissions play a major role in the atmospheric N_2O budget. The quantification of the oceanic N_2O emissions and the identification of the marine pathways of N_2O formation and consumption have received increasing attention during the last few decades. The very first study of oceanic N_2O (in the South Pacific Ocean) was published by Craig and Gordon (1963), followed by studies in the North Atlantic Ocean by Junge and Hahn during the late 1960s and early 1970s (Junge and Hahn, 1971; Hahn, 1974). Junge and Hahn were the first to quantify the oceanic source of atmospheric N_2O. In 1976 Yoshinari published his now 'classical' study of N_2O profiles in the Sargasso and Caribbean Seas, which turned out to be groundbreaking because it was the first study to report the inverse correlation between N_2O and O_2 concentrations in the water column (Yoshinari, 1976). He also introduced the term ΔN_2O (for a definition see below) as a measure of the 'apparent N_2O production' and found a linear correlation between ΔN_2O and AOU (apparent oxygen utilization) (Yoshinari, 1976). Based on this, he suggested that 'N_2O production in the sea is related in some way to the oxidation sequence of organic matter', which was an early hint of nitrification (i.e. microbial oxidation of NH_4^- to NO_3^-) as a major N_2O formation process in the ocean. Cohen and Gordon (1978), Cohen (1978) and Elkins et al (1978) were the first to report a significant N_2O consumption in the oxygen minimum zone in the subsurface waters of the eastern tropical Pacific Ocean and the anoxic waters of the Saanich Inlet basin (off Vancouver Island). They attributed the N_2O loss to microbial reduction of N_2O to N_2 (i.e. denitrification). In order to verify the marine pathways to N_2O, isotope studies have been introduced in recent years: first, measurements of the $\delta^{15}N$ value of dissolved N_2O were presented by Yoshida et al (1984) and nine years later Kim and Craig (1993) published the first measurement of the dual isotope signature ($\delta^{15}N$ and $\delta^{18}O$) of oceanic N_2O. This was followed by the publication of the isotopomeric signature of N_2O (which makes it possible to distinguish the position of ^{15}N within the asymmetric N_2O molecule: NNO) by Popp et al (2002) and Toyoda et al (2002).

In this chapter we present a short overview of the current knowledge about the role of the ocean as a source of N_2O and a short description about oceanic N_2O distribution. It is followed by a discussion of the major marine pathways to N_2O. In the concluding section we discuss possible consequences of climate change for both the marine pathways to N_2O and the oceanic emissions of N_2O. More information about N_2O in the ocean can be found in a recently published overview article by Bange (2008).

The role of the ocean for the global budget of atmospheric nitrous oxide

The oceanic N_2O emissions play a major role in the atmospheric N_2O budget (see for example Bange, 2006a). In the Fourth Assessment Report of the IPCC, mean annual N_2O emissions (ranges are given in parenthesis) of 3.8 (1.8–5.8) $\times 10^{12}g$ (Tg) N and 1.7 (0.5–2.9) Tg N were attributed to the open ocean and coastal areas (including rivers), respectively (IPCC, 2007). According to the IPCC report, open ocean and coastal areas (including rivers) represent about 21 per cent and 10 per cent of the total natural and anthropogenic N_2O sources of 17.7Tg N yr^{-1}, respectively (IPCC, 2007). There are various reasons for the considerable ranges of uncertainty in the global N_2O emission estimates (Bange, 2008):

- different methodological approaches (empirical models versus extrapolation of measurements),
- the application of different air–sea exchange models and
- the fact that the applied classification of coastal areas is not uniform.

Nitrous oxide in the ocean

Concentrations of dissolved N_2O are usually expressed as nmol $litre^{-1}$ or nmol kg^{-1}. The degree of N_2O saturation (given in per cent) is defined as the ratio of the measured N_2O concentration to the theoretical N_2O equilibrium concentration. The equilibrium concentration in turn depends on the water temperature, salinity, ambient air pressure and the atmospheric N_2O dry mole fraction at the time when the water mass was last in contact with the atmosphere (Weiss and Price, 1980). An N_2O surface saturation of 100 per cent indicates that the water phase is in equilibrium with the overlying atmosphere. N_2O saturation values <100 per cent indicate undersaturation (i.e. uptake of N_2O into the water phase when measured in the ocean surface layer) whereas saturation values >100 per cent stand for supersaturation (i.e. N_2O

release from the water phase to the atmosphere when measured in the ocean surface layer). The N_2O excess (or N_2O anomaly) is defined as the difference between the measured N_2O and the theoretical N_2O equilibrium value. It can be expressed either as a difference in concentration units, $\Delta[N_2O]$, or as a difference in partial pressures, ΔpN_2O.

Surface ocean

Nevison et al (1995) calculated a global mean N_2O surface saturation of 103.5 per cent, which indicates that the ocean, on a global scale, is supersaturated with N_2O and acts as a net source of N_2O to the atmosphere. N_2O saturations in the ocean surface layer are not uniform and can show considerable seasonal variability (Nevison et al, 1995). However, the current data coverage does not make it possible to decipher the seasonality in most parts of the ocean. Global maps of ΔpN_2O in the upper 10m of the world's oceans have been computed by Nevison et al (1995) and Suntharalingam and Sarmiento (2000). Common features of both maps (see Plate 3.1) are: (1) enhanced N_2O anomalies in the equatorial upwelling regions of the eastern Pacific and Atlantic Oceans, enhanced N_2O anomalies along coastal upwelling regions such as along the west coasts of North and Central America, off Peru, off Northwest Africa and in the northwestern Indian Ocean (Arabian Sea); and (2) N_2O anomalies close to zero (i.e. near equilibrium) in the North and South Atlantic Ocean, the South Indian Ocean and the central gyres of the North and South Pacific Oceans.

Differences in the two maps result mainly from different computation methods. Additionally, both maps are biased by insufficient data coverage in some parts of the ocean (for example in the Indian and western Pacific Oceans).

Coastal areas

In general, enhanced N_2O emissions in coastal areas are found in upwelling systems and nitrogen-rich estuaries (Seitzinger et al, 2000; Nevison et al, 2004). However, as in the case of the open ocean emissions, flux estimates from coastal areas are heavily biased by a seasonal variability, which is, in the majority of the studies, only inadequately resolved.

The narrow bands of coastal upwelling systems such as those found in the northwestern Indian Ocean (Arabian Sea) and in the southeastern Pacific Ocean (off central Chile) have been identified as 'hot spots' for extremely high N_2O anomalies with N_2O saturations of up to 8250 per cent and 2426 per cent, respectively (Naqvi et al, 2005; Cornejo et al, 2007). In nitrogen-rich estuarine systems, high N_2O anomalies are usually only found in the estuaries themselves, whereas the adjacent shelf waters, which are not influenced by the river plumes, are close to equilibrium with the atmosphere. Bange (2006b), for example, computed mean N_2O saturations of 113 per cent and 467 per cent for European shelf and estuarine systems, respectively.

Nitrous oxide distribution in the water column

The shapes of N_2O profiles generally fall into three categories:

- Cat. I profiles from oceanic regions with dissolved oxygen concentrations $[O_2] > 10\mu mol$ litre^{-1} throughout the water column (for example in the Atlantic Ocean, the South Indian Ocean and the central North Pacific and central South Pacific Oceans);
- Cat. II profiles from regions with sub-oxic environments ($0 < [O_2] < 2-10\mu mol$ litre^{-1}, Codispoti et al, 2005) such as found in intermediate water depths from about 200m to about 800m in the Arabian Sea and the eastern North/South Pacific Ocean;
- Cat. III profiles from regions with anoxic deep water masses with $[O_2] = 0\mu mol$ litre^{-1} and hydrogen sulphide present. Anoxic water masses are found only in a few regions of the world's oceans. Perennial anoxic environments occur in the Black Sea and the Cariaco Basin off Venezuela. Temporarily occurring anoxic conditions have been reported from the deep basins of the central Baltic Sea.

Typical N_2O profiles illustrating Cat. I-III profiles are shown in Figure 3.1 (additional examples and references can be found in Bange, 2008). It is obvious that the shapes of the N_2O profiles undergo a significant change when $[O_2]$ falls below the threshold for sub-oxic conditions. For instance, the one-peaked profiles (Cat. I) observed in the southern Arabian Sea turn into two-peaked profiles in the central Arabian Sea where sub-oxic conditions are found in the intermediate layers (Bange et al, 2001). Cat. III show no pronounced N_2O peak at the boundary of the oxic and anoxic water masses (see for example Hashimoto et al, 1983; Walter et al, 2006b; Westley et al, 2006).

The characteristics of the profiles described above are valid for 'static' oceanic systems under steady-state conditions with turnover times much longer than one year. Some coastal areas, however, show a dynamic behaviour, with a rapid seasonal overturning from oxic via sub-oxic to anoxic conditions and vice versa (for example the shelf off West India, the western Baltic Sea, the shelf off Chile, an upwelling area off southwest Africa and the Gulf of Mexico). In these kinds of transient systems, significant amounts of N_2O can accumulate temporarily during the short transition time when the system is about to change its oxygen regime. Interestingly, the timing of the N_2O accumulation occurs at different transition stages and seems to be characteristic for different coastal systems: in the southwestern Baltic Sea, N_2O only accumulates when the system is shifting from anoxic to oxic conditions (Figure 3.2) (Schweiger, 2006), whereas N_2O accumulates when the systems are shifting from oxic to sub-oxic (off central Chile) or to anoxic (off West India) conditions (Naqvi et al, 2006; Cornejo et al, 2007). During the transition stages, the accumulation of N_2O does not occur in the anoxic zone itself but at the oxic/anoxic boundaries. In anoxic zones, N_2O is usually found at very low or even undetectable concentrations.

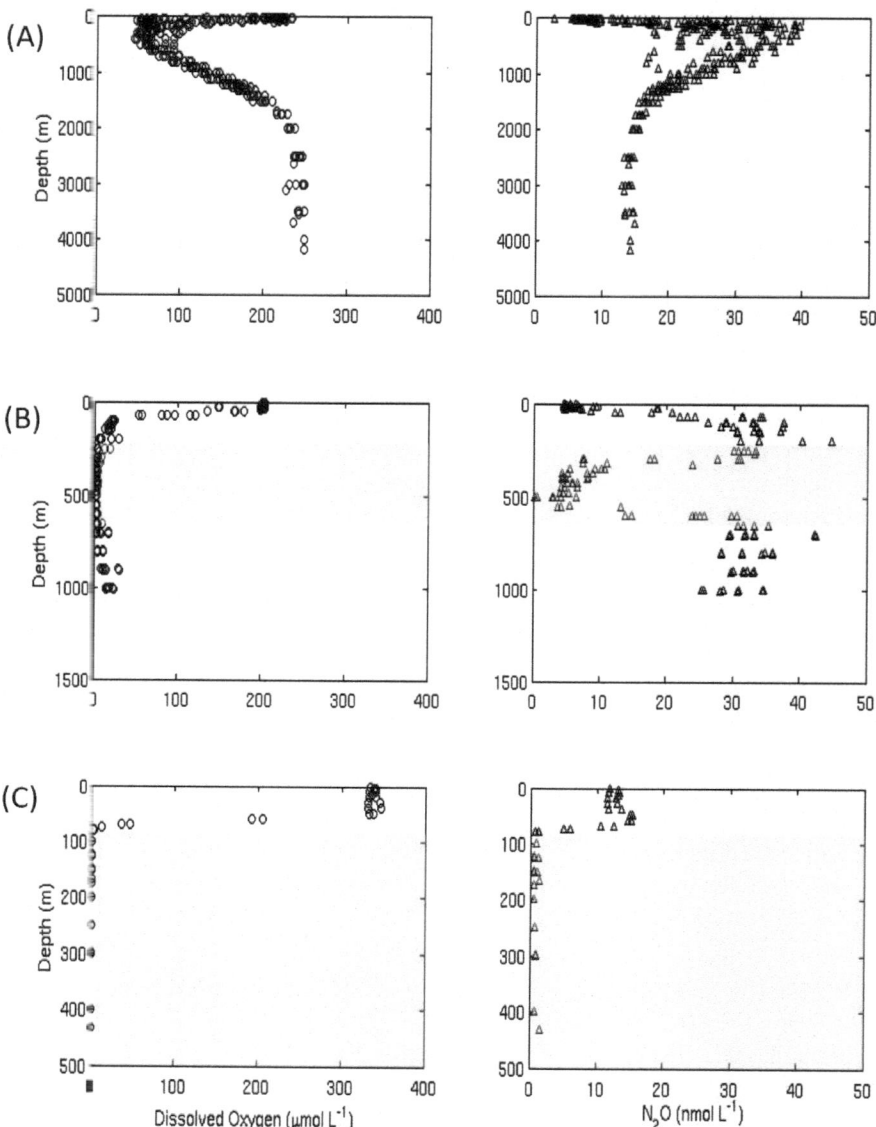

Figure 3.1 Typical N$_2$O profiles (right column) and dissolved O$_2$ (left column): (panel A) Cat. I profiles from the tropical North Atlantic Ocean; (panel B) Cat. II profiles from the Guinea Dome in eastern Tropical North Pacific Ocean; (panel C) Cat. II profile from the Landsort Deep in the western Gotland Basin (central Baltic Sea)

Note: Sub-oxic (panel B) and anoxic (panel C) layers are indicated by the shading.
Source: Panel A from Walter et al (2006a); panel B from Bange and Walter (2007); panel C from Walter et al (2006b)

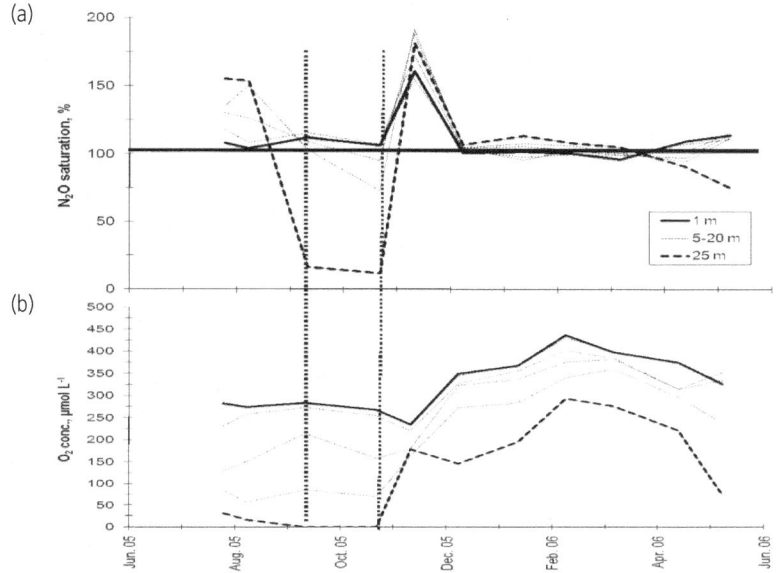

Figure 3.2 (a) N_2O saturations and (b) O_2 concentrations at the time series station Boknis Eck (southwestern Baltic Sea, 54°31′N, 10°02′E, max. depth 28m) measured on a monthly basis from July 2005 to May 2006

Note: The anoxic event is marked by the vertical dashed lines. The horizontal bold line in (A) depicts the theoretical N_2O saturation at equilibrium with the atmosphere.
Source: Based on unpublished data from Schweiger (2006)

Major pathways

Today's prevailing view is that there are only two dominating microbial processes, i.e. bacterial nitrification and bacterial denitrification, during which oceanic N_2O is formed either as a by-product or as an intermediate (Figures 3.3 and 3.4). The global budget of oceanic N_2O sources and sinks given in Bange and Andreae (1999) indicates that about 35 per cent of the oceanic N_2O is produced during denitrification, with the rest resulting from nitrification.

Bacterial denitrification

Denitrification results in a loss of bioavailable (fixed) nitrogen in the form of gaseous products such as N_2O and N_2 (for details on denitrification see the overview article by Devol, 2008):

$$NO_3^- \rightarrow NO_2^- \rightarrow NO \rightarrow N_2O \rightarrow N_2.$$

As can be seen from the denitrification reaction sequence, N_2O is an intermediate, with its concentration at any time determined by the balance between

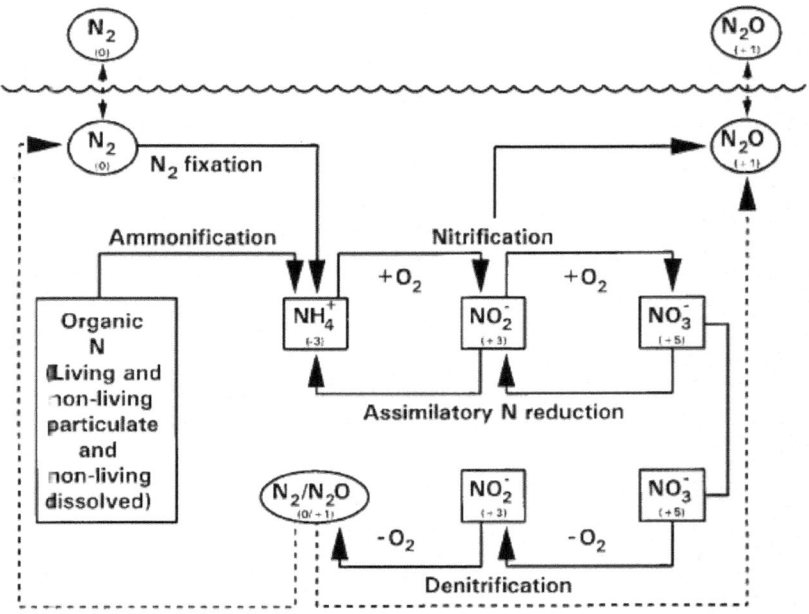

Figure 3.3 Simplified sketch of the oceanic nitrogen cycle

Source: Karl et al (2002)

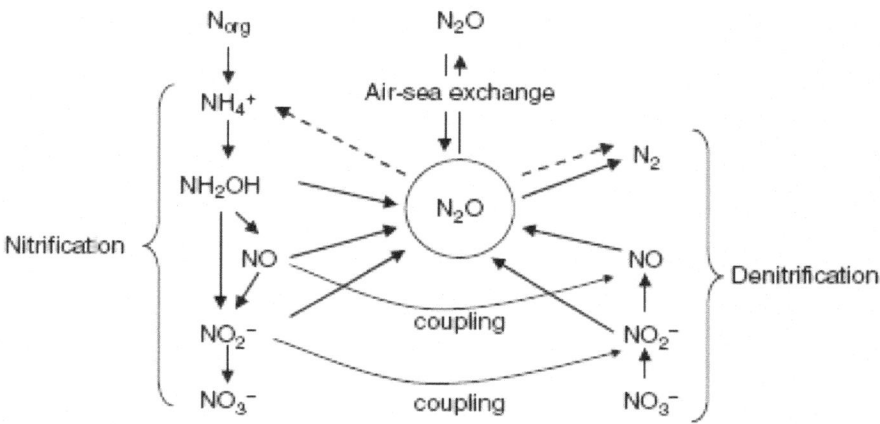

Figure 3.4 Overview of processes that influence the N_2O distribution in the ocean

Note: The dashed arrows indicate N_2O reduction during N_2 fixation. Note that NO is not an obligate intermediate of the nitrification sequence.
Source: Bange (2008)

production and consumption to N_2. The net accumulation of dissolved N_2O depends on the dissolved O_2 concentrations (see below). Under extreme O_2 depletion (such as found in the intermediate depths of the eastern tropical Pacific Ocean and the Arabian Sea, see above) there is a net N_2O consumption during denitrification, resulting in low N_2O concentrations. Denitrification is a well-known feature of many different bacteria species in terrestrial and oceanic environments. Denitrifiers are facultative anaerobic bacteria, which can reduce NO_3^- when oxygen becomes limiting. Thus the occurrence of denitrification is favoured under sub-oxic (0 $<O_2$ <2–$10\mu mol$ litre^{-1}, Codispoti et al, 2005) conditions. Denitrification does not occur under anoxic conditions (O_2 = $0\mu mol$ litre^{-1}, hydrogen sulphide present).

Bacterial nitrification

Nitrification is the oxidation of ammonium, NH_4^+, to NO_3^- via hydroxyl-amine, NH_2OH, and nitrite, NO_2^-. For details about nitrification see the overview article by Ward (2008). Autotrophic nitrification represents the final step of the remineralization of nitrogen containing organic matter and is performed in two steps by ammonium-oxidizing bacteria (AOB) (for example *Nitrosomonas, Nitrosospira and Nitrosococcus*) and nitrite-oxidizing bacteria (NOB) (for example *Nitrobacter* and *Nitrospira*), respectively:

AOB: $NH_4^+ \rightarrow NH_2OH (\rightarrow NO) \rightarrow NO_2^-$

NOB: $NO_2^- \rightarrow NO_3^-$

NO is not known to be an obligatory intermediate during ammonium oxidation. It can be produced by AOB but the mechanism is not well understood. During autotrophic nitrification N_2O can be formed by AOB either via the pathways $NH_2OH \rightarrow N_2O$ and $NO \rightarrow N_2O$ or via the pathway $NO_2^- \rightarrow NO \rightarrow N_2O$ (the latter is part of the so-called nitrifier-denitrification process). Nitrification is an aerobic process; however, under low-oxygen (sub-oxic) conditions, N_2O yields are enhanced. Alternatively, N_2O can be formed during heterotrophic nitrification (i.e. nitrification linked to aerobic denitrification) via the reaction $NO_2^- \rightarrow NO \rightarrow N_2O$ as well, but the enzymes involved in the heterotrophic reaction sequence are different from those involved in the autotrophic pathway. Under oxic conditions, N_2O yields from heterotrophic nitrification are higher than those from autotrophic nitrification. However, the relevance of heterotrophic nitrification in the marine environment is not known yet.

Both nitrification and denitrification as sources and sinks of oceanic N_2O have been described in the water column, in the sediments and in association with suspended particles (for example Schropp and Schwarz, 1983; Seitzinger, 1990; Michotey and Bonin, 1997; Nevison et al, 2003; Codispoti et al, 2005). N_2O yields from nitrification range from 0.004 per cent to 0.4 per cent, whereas the N_2O yield from the denitrifying sub-oxic zone in the Arabian Sea was estimated to be about 2 per cent (see overview in Bange, 2008). Culture studies with strains of nitrifiers revealed that the N_2O yield from nitrification is significantly enhanced (up to 10 per cent) under sub-oxic conditions (Goreau et al, 1980). N_2O yields from sedimentary denitrification range from 0.1 per cent to 0.5 per cent, with values up to 6 per cent in nutrient-rich regions (see overview in Seitzinger and Kroeze, 1998).

Nitrous oxide–oxygen gas relationship

The relationship between oceanic N_2O production/consumption and dissolved O_2 concentrations is shown schematically in Figure 3.5. While the influence of O_2 concentrations on the N_2O production via nitrification is still lacking a mechanistic explanation, the influence of O_2 on denitrification and thus N_2O production results from two factors: (1) the redox potential of NO_3^- respiration favours denitrification under reduced O_2 concentrations (see for example Falkowski et al, 2008) and (2) the enzyme involved in N_2O consumption, N_2O reductase, is sensitive to O_2 concentrations (Firestone and Tiedje, 1979). For example, Naqvi et al (2000) attributed the accumulation of N_2O off West India to the onset of denitrification at low O_2 concentrations, with the assumption that the activity of the N_2O reductase could not be established because of frequent aeration of the shallow shelf waters (so-called stop-and-go denitrification).

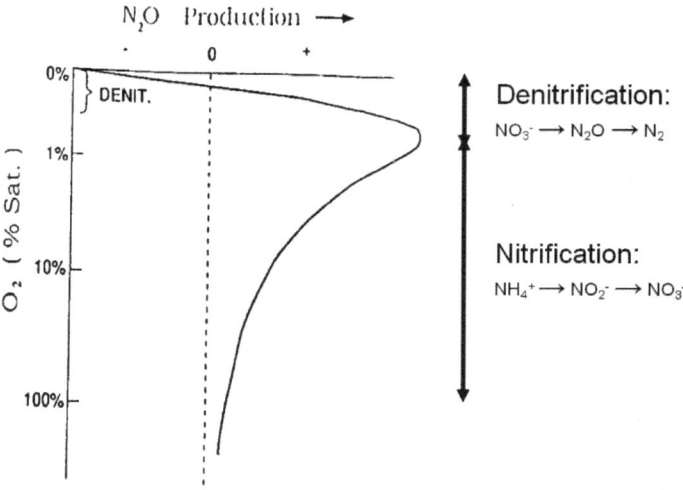

Figure 3.5 N_2O production versus O_2 saturation in the ocean

Note: The approximate regimes of nitrification and denitrification are indicated. Note that there is no clear threshold between nitrification and the onset of denitrification; nitrification and denitrification can occur at the same O_2 saturation levels.
Source: Modified from the original figure in Codispoti et al (1992)

The apparent oxygen utilization (AOU) is a measure of the amount of O_2 consumed during organic matter remineralization (oxidation) in the ocean. Because nitrification is part of the organic matter oxidation sequence, plots of ΔN_2O versus AOU have been used to identify the prevailing formation and consumption processes of N_2O in the water column. The overwhelming majority of the Cat. I profiles (see above) show positive linear $\Delta N_2O/AOU$ relationships, suggesting that nitrification is the main N_2O formation process in most parts of the oceans (Bange and Andreae, 1999). This is supported by the fact that in most oxic water columns N_2O is positively correlated with dissolved nitrate (NO_3^-), the final product of nitrification (see for example Walter et al, 2006a). However, there are caveats against a straightforward interpretation of the linear $\Delta N_2O/AOU$ relationship as an indicator for N_2O formation via nitrification because a linear $\Delta N_2O/AOU$ relationship may not necessarily result from nitrification alone: most recently, based on N_2O isotopomer data (see below), Yamagishi et al (2005) argued that net N_2O formation in the oxygen minimum zone (OMZ) of the western North Pacific Ocean mainly results from denitrification with only a small contribution from nitrification. They showed that this N_2O, when diffusing into deep waters, produces a reasonably linear $\Delta N_2O/AOU$ relationship. Moreover, by applying a two-end-member mixing model, Nevison et al (2003) showed that isopycnal mixing of water masses with different preformed N_2O and O_2 concentrations can result in a linear $\Delta N_2O/AOU$ relationship, which can mask the 'true' biological N_2O production. They state:

> we find that the biological N_2O yield per mole O_2 consumed cannot be calculated with great confidence from cross-plot correlation slopes. The essential problem is that the N_2O yield is spatially variable. As a result, strong mixing gradients exist in the data that can overwhelm more subtle N_2O production terms.

A linear $\Delta N_2O/AOU$ relationship does not exist in sub-oxic and anoxic water masses (i.e., Cat. II and Cat. III profiles, see above) indicating a complex interplay between N_2O formation and consumption during denitrification and/or a coupling of nitrification and denitrification at the upper boundary of the sub-oxic zones (see for example Bange et al, 2005; Walter et al, 2006b; Westley et al, 2006; Farías et al, 2007; Yamagishi et al, 2007).

Isotope studies

The isotope ratio $^{15}N/^{14}N$ of N_2O is expressed as $\delta^{15}N_{atm}$ relative to atmospheric N_2:

$$\delta^{15}N_{atm} \text{ (sample) } [\text{‰}] = ((^{15}N/^{14}N)_{sample} / (^{15}N/^{14}N)_{std} - 1) \times 1000. \quad (3.1)$$

In the same way, the isotope ratio $^{18}O/^{16}O$ of N_2O is usually expressed as $\delta^{18}O_{VSMOW}$ relative to Vienna standard mean ocean water (VSMOW). However, in some cases $\delta^{18}O_{atm}$ relative to O_2 in the atmosphere is reported. $\delta^{18}O_{VSMOW}$ can be converted to $\delta^{18}O_{atm}$ with the equation:

$$^{18}O_{atm} = -23.0 + {}^{18}O_{VSMOW} / 1.0235 \qquad (3.2)$$

(Kim and Craig, 1993). Mean $\delta^{15}N_{atm}$ and $\delta^{18}O_{VSMOW}$ of N_2O in the troposphere are 6.72 ± 0.12 per mil and 44.62 ± 0.21 per mil, respectively (Kaiser et al, 2003).

The isotopic composition of oceanic N_2O is determined by its atmospheric imprint, the isotopic signals of biological sources and sinks, and mixing processes within the ocean. This, in turn, implies that there are characteristic signals of enrichment or depletion (so-called fractionation), which can be attributed to different biological processes as well as physical processes. The isotopic signature of biologically derived N_2O depends on the isotopic composition of the substrates such as NO_3^- (denitrification) and NH_4^+ (nitrification) and the isotopic depletion/enrichment during these processes. An overview of the isotopic depletion/enrichment of N_2O from culture experiments is shown in Figure 3.6. It is obvious that the range of the resulting nitrogen depletion in N_2O during denitrification and nitrification is similar. The isotopic signal of oxygen in N_2O produced during nitrification is introduced by the $\delta^{18}O$ value of both dissolved O_2 and H_2O (Ostrom et al, 2000). The isotopic signal resulting from air–sea exchange is small compared to the biological processes; therefore, biological N_2O formation should yield a clear isotopic signature in oceanic N_2O. However, the identification of nitrification or denitrification as N_2O-producing processes strongly depends on the knowledge of the isotopic signatures of the substrates, which can vary temporarily and spatially. A detailed overview of studies of the isotopic signature of oceanic N_2O is given in Bange (2008). The main results of actual N_2O isotopic studies are summarized in the following sections.

Repeated measurements of N_2O depth profiles at the Hawaii ocean time series station ALOHA in the subtropical North Pacific Ocean revealed that $\delta^{15}N$ and $\delta^{18}O$ of N_2O were in equilibrium with tropospheric N_2O at the ocean surface and steadily decreased from the ocean surface to minimum values at about 100–300m depth at the base of the euphotic zone, followed by an increase to maximum values at 800m. The depletion of both ^{15}N and ^{18}O was attributed to nitrification (Dore et al, 1998; Ostrom et al, 2000; Popp et al, 2002). A more detailed study at ALOHA that included measurements of $\delta^{18}O$ in dissolved O_2 and H_2O, revealed that N_2O might be formed by two different pathways: first, by nitrification via NH_2OH or NO at most depths and, second, by nitrifier-denitrification via reduction of NO_2^- (between 350 and 500m) (Ostrom et al, 2000).

The situations in the central Arabian Sea and the eastern tropical North Pacific Ocean are more complex. N_2O was found to be strongly enriched in both ^{15}N and ^{18}O in the denitrifying oxygen minimum zone, whereas N_2O in

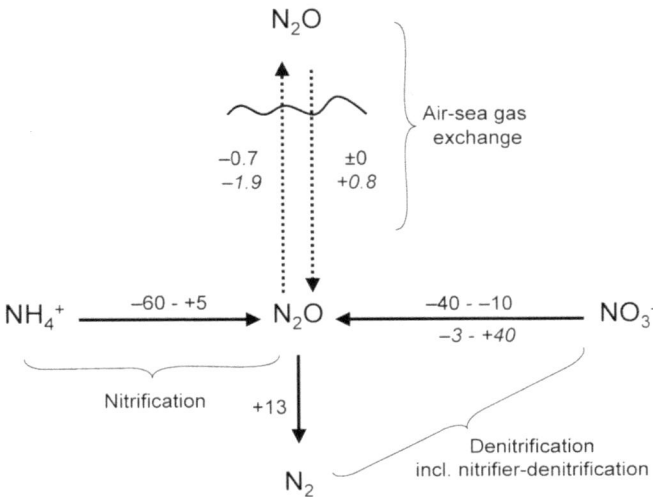

Figure 3.6 Isotopic depletion/enrichment for nitrogen and oxygen in N_2O relative to the substrates NO_3^- and NH_4^+, and the product N_2

Note: Values are given in ‰. Negative values depict isotopic depletion in N_2O and positive values depict isotopic enrichment in N_2O. Data for oxygen depletion/enrichment in N_2O are given in italics.

Source: Bange (2008)

the surface layer was depleted in ^{15}N but slightly enriched in [18]O compared to tropospheric N_2O (Yoshinari et al, 1997; Naqvi et al, 1998a, 1998b). N_2O in the core of the oxygen minimum zone was obviously formed by denitrification, since the final reduction step to N_2 should result in enriched N_2O. However, the 'light' N_2O found above the OMZ might be explained by a coupled nitrification–denitrification pathway where NO is formed during nitrification which is then reduced to N_2O during denitrification (Yoshinari et al, 1997; Naqvi et al, 1998a, 1998b).

As mentioned in the introduction, N_2O is an asymmetrical molecule and therefore it is possible to distinguish so-called isotopomers according to the position of ^{15}N within the N_2O molecule (the corresponding δ notation is given in parenthesis): [14]N[15]NO ($\delta^{15}N^a$) and [15]N[14]NO ($\delta^{15}N^b$) (Toyoda and Yoshida, 1999). The ^{15}N site preference (SP_{N2O}) in N_2O is given as $\delta^{15}N^a - \delta^{15}N^b$. Measurements of SP_{N2O} should allow for the identification of the mechanisms of N_2O formation according to the different microbial pathways (Sutka et al, 2003, 2004). Based on the results of a study with cultures of AOB, nitrifier–denitrifiers and denitrifiers, Sutka et al (2006) concluded that the characteristic SP_{N2O} of nitrification and denitrification (including nitrifier denitrification) are generally ~33 per mil and ~0 per mil, respectively. Thus, isotopomers might be used to distinguish between N_2O produced during oxidation (nitrification) and reduction (denitrification and nitrifier-denitrification) processes, however, it seems that isotopomers cannot be used to reveal subtle differences such as that between nitrifier-denitrification and denitrification (Schmidt et al, 2004; Sutka et al, 2006). So far, the oceanic distributions of N_2O isotopomers have been determined at a few stations in the North and South Pacific Oceans (Popp et al, 2002; Toyoda et al, 2002; Yamagishi et al, 2005; Charpentier et al, 2007), in the sub-oxic eastern tropical North Pacific Ocean and Gulf of California (Yamagishi et al, 2005, 2007), and in the anoxic Black Sea (Westley et al, 2006). In general, the conclusions from the SP_{N2O} distribution are in overall agreement with the $\Delta N_2O/AOU$ relationships and bulk isotopic signatures of N_2O as described above. In the North Pacific Ocean SP_{N2O} values of up to 35 per mil were determined, indicating that nitrification is the main N_2O formation process throughout the water column (Popp et al, 2002; Toyoda et al, 2002).

An additional significant contribution by nitrifier denitrification within particles was suggested for the pycnocline at 250–350m in the central South Pacific subtropical gyre because a SP_{N2O} minimum of only 10–12 per mil was found at that depth range (Charpentier et al, 2007). In contrast, Yamagishi et al (2005) suggested a net N_2O formation in the oxygen minimum zone of the western North Pacific Ocean, which mainly results from both formation and consumption during denitrification, with only a minor contribution by nitrification. SP_{N2O} data from sub-oxic (eastern tropical North Pacific Ocean, Gulf of California) and anoxic (Black Sea) environments indicate that N_2O production via nitrification and N_2O production/consumption via denitrification can be coupled and might even be concurrent at the oxic to sub-oxic/anoxic interfaces (Westley et al, 2006; Yamagishi et al, 2007).

Marine pathways to nitrous oxide and climate change

The lesson from the past

The atmospheric history of N_2O is illustrated by the ice core records which now reach back to 650,000 years before the present (yr BP) (Spahni et al, 2005). It seems that atmospheric N_2O concentrations followed glacial climate changes but in a complex way (Spahni et al, 2005). The significant variability of the atmospheric N_2O concentrations have been attributed to concurrent changes in both the terrestrial source and in the oceanic source (Sowers et al, 2003; Flückiger et al, 2004). However, the ice core data do not allow for identification of the key parameters responsible for the abrupt changes of the N_2O sources. More recently, coupled climate/biogeochemistry models were applied to investigate the role of the oceanic N_2O production during fast climate changes such as the Younger Dryas cold period (~12,000 yr BP) (Goldstein et al, 2003) and the Heinrich event H5 (~48,000 yr BP) (Schmittner and Galbraith, 2008). In both models the oceanic N_2O production was parameterized as a function of AOU. The model results of Goldstein et al (2003) suggested that the variability of the oceanic N_2O source was the main but not the sole contributor to the observed changes of atmospheric N_2O. The model results of Schmittner and Galbraith (2008) showed that the abrupt changes of atmospheric N_2O during the Heinrich event H5 could have been caused by variabilities of the oceanic sources alone. They proposed that changes of the ocean circulation results in fast adjustments of the oxygen concentrations in the thermocline, which in turn drives the oceanic N_2O production via nitrification (Schmittner and Galbraith, 2008).

Another line of argument is derived from $\delta^{15}N$ records from sediments underlying sub-oxic denitrification zones in the open ocean: several studies showed that the temporal variations of the denitrification signal in both the Arabian Sea and the eastern tropical Pacific Ocean during the last 23,000 years is paralleled by the reconstructed atmospheric N_2O concentration from ice core records (Figure 3.7) (Suthhof et al, 2001; Thunell and Kepple, 2004; Agnihotri et al, 2006; Pichevin et al, 2007). These results imply that variations in the amount of the water column denitrification might have led to changes in the magnitude of N_2O formation and its subsequent release to the atmosphere.

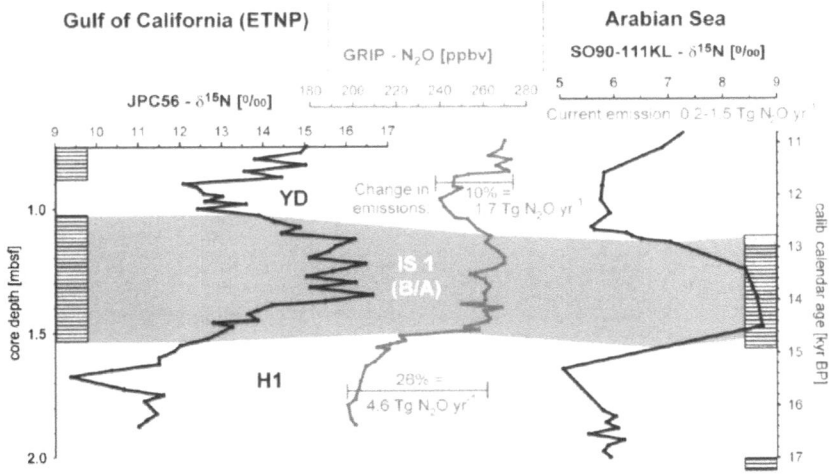

Figure 3.7 $\delta^{15}N$ profiles from sediment cores in the Gulf of California/eastern tropical North Pacific, ETNP (core no. JPC56) and Arabian Sea (core no. SO90-111KL) compared to reconstructed atmospheric N_2O data from the GRIP (Greenland Ice Core Project) ice core

Note: YD = Younger Dryas; IS1 = Interstadial 1; H1 = Heinrich event 1; B/A = Bølling/Allerød event.
Source: Suthhof et al (2001)

On the basis of the model results and sedimentary $\delta^{15}N$ records introduced above we can conclude that the rapid changes observed in the palaeorecord of N_2O concentration might be dominated by changes in the oceanic N_2O production (nitrification and/or denitrification) via pronounced changes of the dissolved oceanic oxygen concentrations.

The ongoing rapid increase in atmospheric N_2O, which started during the 19th century, is mainly attributed to the increase of agricultural activities (Kroeze et al, 1999; Ishijima et al, 2007), which in turn was caused by the expansion of agricultural land and industrialization that came along with the increasing availability of agricultural fertilizers triggered by the development of the Haber-Bosch process (see Chapters 4 and 5). A potential indirect contribution by oceanic sources (for example increased N_2O emissions as a result of eutrophication of coastal areas) has not been quantified yet.

Present day and future scenarios

Coastal ocean

As mentioned above, oceanic N_2O is exclusively produced by biological processes, thus, its production is indirectly linked to the biological productivity of the coastal and open oceans (Figure 3.8). This, in turn, implies that we have to understand how eutrophication of the coastal areas and fertilization of the open ocean influences the productivity and the resulting O_2 depletion during the remineralization (oxidation) of organic material. Both nitrification and denitrification are involved in the remineralization process and the N_2O yield of both processes depends on the prevailing O_2 concentrations (see above).

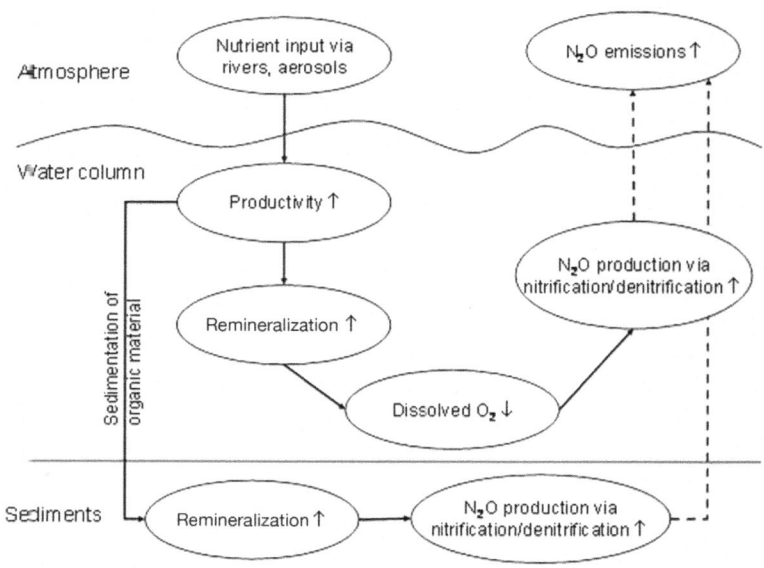

Figure 3.8 Simplified scheme of processes leading to enhanced N_2O formation in coastal areas.

Note: Up and down arrows within the ovals stand for increase or decrease, respectively.

On the basis of N_2O measurements on the shelf of the west coast of India, Naqvi et al (2000) cautioned that the N_2O emissions from shallow sub-oxic/anoxic coastal systems might increase in the future, due to the fact that the number of coastal regions with severely depleted dissolved oxygen concentrations is currently increasing worldwide (UNEP, 2004; Diaz and Rosenberg, 2008). Eutrophication can also significantly stimulate sedimentary N_2O formation by denitrification, which was demonstrated by Seitzinger and Nixon (1985) in microcosm experiments. This is in line with observations that N_2O release from mangrove ecosystems appear to be very sensitive to eutrophication: N_2O release across the mangrove sediment/atmosphere interface can be enhanced up to 2800 times when adding NH_4^+ and/or NO_3^- (Bauza et al, 2002; Muñoz-Hincapié et al, 2002; Kreuzwieser et al, 2003). Therefore, it seems realistic to expect that the N_2O emissions from shallow sub-oxic/anoxic coastal systems (including mangrove ecosystems) will increase in the near future due to increasing nutrient inputs caused by the ongoing industrialization and intensification of agricultural activities.

Open ocean

Only recently, Stramma et al (2008) showed that the oxygen minimum zones of the intermediate layers (300 to 700m water depth) in various regions of the ocean are expanding and have been losing oxygen with rates ranging from $0.09 \pm 0.21 \mu mol\ kg^{-1}\ yr^{-1}$ (in the eastern equatorial Indian Ocean) up to $0.34 \pm 0.13 \mu mol\ kg^{-1}\ yr^{-1}$ (in the eastern tropical Atlantic Ocean) during the last 50 years. In order to assess the maximum associated N_2O formation, we may roughly estimate the additional long-term N_2O formation in the tropical Atlantic Ocean: assuming a mean $\Delta N_2O/AOU$ ratio of 10^{-4} (Walter et al, 2006a) we compute an additional N_2O concentration of $1.7nmol\ kg^{-1}$. This translates into a contribution of 6 per cent of the actual mean N_2O background concentration of about $30nmol\ kg^{-1}$ at 500m depth in the tropical North Atlantic Ocean (Walter et al, 2006a). However, an N_2O accumulation at intermediate water depths in the open ocean will not lead to an immediate release of N_2O to the atmosphere because these waters are not in direct contact with the atmosphere. A major fraction of the accumulated N_2O will be subsequently released to the atmosphere from other oceanic regions when the water masses are brought back to the ocean surface.

A future increase in N_2O emissions has been suggested as an indirect result of enhanced productivity via increases in nitrogen or iron (Fe) aerosol deposition (Fuhrman and Capone, 1991; Jin and Gruber, 2003; Duce et al, 2008), N_2 fixation (Karl, 1999) and riverine nutrient inputs (Naqvi et al, 2000) (Figure 3.9).

Results from a coupled global climate/ocean-circulation/biogeochemistry model applying a business-as-usual CO_2 emission scenario indicate that the sub-oxic areas and N_2O production in the open ocean are not likely to change significantly during the next 100 years (Schmittner et al, 2008). The picture changes when the model is run for the next 4000 years: then, indeed, a 64 per

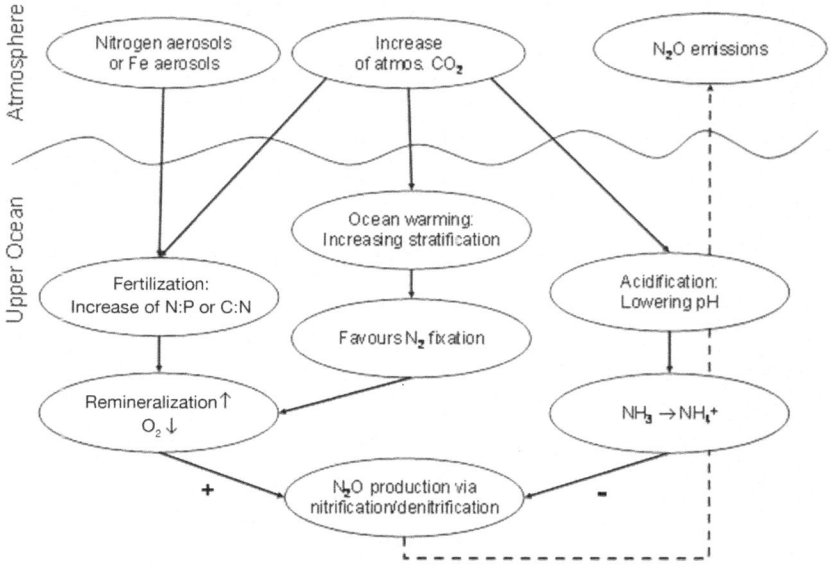

Figure 3.9 Simplified scheme of processes leading to future N_2O formation
and/or consumption.

Note: Up and down arrows within the ovals stand for increase or decrease, respectively. The plus sign (+) stands for a potential enhancement of N_2O production, whereas the minus sign (−) stands for a decrease in N_2O production.

cent increase in the oceanic N_2O production because of decreasing O_2 concentration in the open ocean is predicted (Schmittner et al, 2008). A more dramatic short-term expansion of the open ocean sub-oxic areas ($[O_2]$ <5μmol litre^{-1}) during the next 90 years was predicted with the same model when using unusually high C:N ratios (for the remineralized organic material) that were derived from mesocosm experiments simulating future high atmospheric CO_2 concentrations (Oschlies et al, 2008). One would expect that this also changes the near-future N_2O production and emissions, however, the effect of high C:N ratios on the N_2O production was not investigated in the study by Oschlies et al (2008).

Ocean acidification (Raven et al, 2005), caused by increasing atmospheric CO_2 ($CO_2 + H_2O \rightarrow HCO_3^- + H^+$), could lead to a counteracting effect because it shifts the oceanic NH_3/NH_4^+ equilibrium towards NH_4^+ with consequences for AOB because they preferably take up NH_3 and not NH_4^+ (Ward, 2008). Therefore, an overall decrease of the oceanic NH_3 concentrations might lead to a decrease in nitrification (Huesemann et al, 2002) and has the potential to decrease N_2O production via nitrification as well.

Nitrous oxide emissions and winds

The final release or uptake of N_2O across the ocean/atmosphere interface depends on physical processes such as wind-driven air–sea gas exchange

(Wanninkhof, 2007) and wind-driven oceanic circulation/mixing processes (coastal upwelling, storm events). For example, Naik et al (2008) showed that a storm can deepen the mixed layer considerably, thereby entraining N_2O from the subsurface maximum to the surface layer, where it easily escapes to the atmosphere. Therefore, N_2O emissions triggered by strong wind events may contribute significantly to both the regional and global oceanic N_2O emissions (Patra et al, 2004; Bange et al, 2008). In view of the fact that the strongest tropical cyclones, especially in the northern Indian Ocean, have been increasing in intensity in recent years (Elsner et al, 2008), we might expect an increase of N_2O emissions triggered by storm events. Therefore, any long-term changes of the atmospheric circulation that alter the wind fields with respect to the wind speeds and the wind field patterns might lead to changes in the N_2O emissions.

Possible alternative pathways to nitrous oxide

Nitrous oxide from anammox?

During the last years, anaerobic ammonium oxidation (anammox, $NO_2^- + NH_4^+ \rightarrow N_2$) has received increasing attention as an additional, previously not recognized, significant loss process of fixed nitrogen in the ocean (see for example Francis et al, 2007; Devol, 2008). N_2O has been found to be formed only in small amounts during nitric oxide detoxification ($NO_2^- \rightarrow NO \rightarrow N_2O$) that seems to be performed by the anammox bacterium *Kuenenia stuttgartiensis* as a side reaction of the anammox reaction (Kartal et al, 2007). Anammox has been found in the sub-oxic zones of eastern tropical South Pacific Ocean, in the upwelling off Namibia (Kuypers et al, 2005; Thamdrup et al, 2006; Hamersley et al, 2007) as well as in the central Baltic Sea (Hannig et al, 2007), but not in the sub-oxic zones of the central Arabian Sea (Nicholls et al, 2007). The question of whether anammox is involved in the production and/or consumption of N_2O in the ocean should be the subject of further investigation.

Nitrous oxide from nitrification and denitrification by archaea?

Archaea have been detected in almost all oceanic regions throughout the water column and in the sediments (see for example Karner et al, 2001; Sinninghe Damsté et al, 2002; Francis et al, 2005; Teira et al, 2006; Varela et al, 2008). The successful isolation of an NH_4^+-oxidizing archaeon (Könneke et al, 2005) raised the question of whether we have overlooked the role of ammonium-oxidizing archaea (AOA) in the oceanic nitrogen cycle. Meanwhile the gene amoA, which is commonly used as a marker gene for the ammonium-oxidizing enzyme ammonia mono-oxygenase in *Crenarchaeota* (the dominant group of mesophilic archaea in the ocean), has been detected in the North Atlantic Ocean, the North Sea, the Black Sea and in sediments (see for example Francis et al, 2005; Wuchter et al, 2006; Lam et al, 2007). On the basis of the dominant abundance of the AOA amoA compared to the bacterial amoA, it has been suggested that *Crenarchaeota* in the uppermost 1000m of the North Atlantic Ocean were mainly responsible for NH_4^+ oxidation (i.e. the first step of nitrification), whereas nitrifying bacteria seem to play only a minor role (Wuchter et al, 2006). Similar results were found in estuarine sediments where AOA seem to play a major role as NH_4^+ oxidizers (Caffrey et al, 2007). Archaea are also capable of performing the classical denitrification pathway, including N_2O formation and its subsequent reduction to N_2 (see for example the overview article by Cabello et al, 2004, and references therein). Despite the fact that archaea perform the same nitrogen transformation processes as bacteria, there are 'significant differences in the structure and regulation of some enzymes involved in the nitrogen metabolism in archaea' as stated by Cabello et al (2004). This might be especially important for interpreting N_2O isotopic signatures. However, N_2O production and/or consumption by archaea have not been studied yet.

Concluding remarks

Based on the points discussed in the sections above, it is clear that the marine pathways to N_2O (which are exclusively biological processes) have been and will be sensitive to ongoing environmental changes. We do not know, however, how the oceanic N_2O pathways will be altered and it is even more difficult to predict how the future oceanic N_2O emissions will be affected. Our rather poor ability to predict the future oceanic N_2O cycling (and to explain the N_2O palaeorecord) is due to the fact that we still have an only rudimentary knowledge about both the oceanic distribution of N_2O and the mechanisms of its major production processes. Emerging new aspects such as possible N_2O formation during anammox or by archaea might have the potential to change our traditional view of the oceanic N_2O pathways in the near future.

Acknowledgements

We are very grateful to Keith Smith for inviting us to contribute to this book and for his editorial comments. The writing of this review was supported by the Forschungs-bereich Marine Biogeochemie of IFM-GEOMAR.

References

Agnihotri, R., Altabet, M. A. and Herbert, T. D. (2006) 'Influence of marine denitrification on atmospheric N_2O variability during the Holocene', *Geophysical Research Letters*, vol 33, L13704, doi:10.1029/2006GL025864

Bange, H. W. (2006a) 'New Directions: The importance of the oceanic nitrous oxide emissions', *Atmospheric Environment*, vol 40, no 1, pp198–199

Bange, H. W. (2006b) 'Nitrous oxide and methane in European coastal waters', *Estuarine, Coastal and Shelf Science*, vol 70, pp361–374

Bange, H. W. (2008) 'Gaseous nitrogen compounds (NO, N_2O, N_2, NH_3) in the ocean', in D. G. Capone, D. A. Bronk, M. R. Mulholland and E. J. Carpenter (eds) *Nitrogen in the Marine Environment*, 2nd Edition, Elsevier, Amsterdam, pp51–94

Bange, H. W. and Andreae, M. O. (1999) 'Nitrous oxide in the deep waters of the world's oceans', *Global Biogeochemical Cycles*, vol 13, no 4, pp1127–1135

Bange, H. W. and Walter, S. (2007) 'Nitrous oxide in the Costa Rica Dome area (eastern tropical North Pacific Ocean)', *Geophysical Research Abstracts*, vol 9, 08615

Bange, H. W., Rapsomanikis, S. and Andreae, M. O. (2001) 'Nitrous oxide cycling in the Arabian Sea', *Journal of Geophysical Research*, vol 106, no C1, pp1053–1065

Bange, H. W., Naqvi, S. W. A. and Codispoti, L. A. (2005) 'The nitrogen cycle in the Arabian Sea', *Progress in Oceanography*, vol 65, pp145–158

Bange, H. W., Naik, H. and Naqvi, S. W. A. (2008) 'Enhancement of oceanic nitrous oxide emissions by storms', *SOLAS News*, no 8, p9

Bauza, J. F., Morell, J. M. and Corredor, J. E. (2002) 'Biogeochemistry of nitrous oxide production in Red Mangrove (*Rhizophora mangle*) forest sediments', *Estuarine, Coastal and Shelf Science*, vol 55, pp697–704

Cabello, P., Roldán, M. D. and Moreno-Vivián, C. (2004) 'Nitrate reduction and the nitrogen cycle in archaea', *Microbiology*, vol 150, pp3527–3546

Caffrey, J. M., Bano, N., Kalanetra, K. and Hollibaugh, J. T. (2007) 'Ammonia oxidation and ammonia-oxidizing bacteria and archaea from estuaries with different histories of hypoxia', *The ISME Journal*, vol 1, pp660–662

Charpentier, J., Farías, L., Yoshida, N., Boontanon, N. and Raimbault, P. (2007) 'Nitrous oxide distribution and its origin in the central and eastern South Pacific Subtropical Gyre', *Biogeosciences*, vol 4, pp729–741

Codispoti, L. A. Elkins, J. W., Yoshinari, T., Friederich, G. E., Sakamoto, C. M. and Packard, T. T. (1992) 'On the nitrous oxide flux from productive regions that contain low oxygen waters', in B. N. Desai (ed) *Oceanography of the Indian Ocean*, A. A. Balkema, Rotterdam, pp271–284

Codispoti, L. A., Yoshinari, T. and Devol, A. H. (2005) 'Suboxic respiration in the oceanic water column', in P. A. del Giorgio and P. J. le B. Williams (eds) *Respiration in Aquatic Ecosystems*, Oxford University Press, Oxford, pp225–247

Cohen, Y. (1978) 'Consumption of dissolved nitrous oxide in an anoxic basin, Saanich Inlet, British Columbia', *Nature*, vol 272, pp235–237

Cohen, Y. and Gordon, L. I. (1978) 'Nitrous oxide in the oxygen minimum of the eastern tropical North Pacific: Evidence for its consumption during denitrification and possible mechanisms for its production', *Deep-Sea Research*, vol 25, pp509–524

Cornejo, M., Farías, L. and Gallegos, M. (2007) 'Seasonal cycle of N_2O vertical distribution and air-sea fluxes over the continental shelf waters off central Chile (~36°S)', *Progress in Oceanography*, vol 75, pp383–395

Craig, H. and Gordon, L. I. (1963) 'Nitrous oxide in the ocean and the marine atmosphere', *Geochimica et Cosmochimica Acta*, vol 27, pp949–955

Devol, A. H. (2008) 'Denitrification including anammox', in D. G. Capone, D. A. Bronk, M. R. Mulholland and E. J. Carpenter (eds) *Nitrogen in the Marine Environment*, 2nd Edition, Elsevier, Amsterdam, pp263–302

Diaz, R. J. and Rosenberg, R. (2008) 'Spreading dead zones and consequences for marine ecosystems', *Science*, vol 321, pp926–929

Dore, J. E., Popp, B. N., Karl, D. M. and Sansone, F. J. (1998) 'A large source of atmospheric nitrous oxide from subtropical North Pacific surface waters', *Nature*, vol 396, pp63–66

Duce, R. A., LaRoche, J., Altieri, K., Arrigo, K. R., Baker, A. R., Capone, D. G., Cornell, S., Dentener, F., Galloway, J., Ganeshram, R. S., Geider, R. J., Jickells, T., Kuypers, M. M., Langlois, R., Liss, P. S., Liu, S. M., Middelburg, J. J., Moore, C. M., Nickovic, S., Oschlies, A., Pedersen, T., Prospero, J., Schlitzer, R., Seitzinger, S., Sorensen, L. L., Uematsu, M., Ulloa, O., Voss, M., Ward, B. and Zamora, L. (2008) 'Impacts of atmospheric anthropogenic nitrogen on the open ocean', *Science*, vol 320, pp893–897

Elkins, J. W., Wofsy, S. C., McElroy, M. B., Kolb, C. E. and Kaplan, W. A. (1978) 'Aquatic sources and sinks for nitrous oxide', *Nature*, vol 275, pp602–606

Elsner, J. B., Kossin, J. P. and Jagger, T. H. (2008) 'The increasing intensity of the strongest tropical cyclones', *Nature*, vol 455, pp92–95

Falkowski, P. G., Fenchel, T. and DeLong, E. F. (2008) 'The microbial engines that drive the Earth's biogeochemical cycles', *Science*, vol 320, pp1034–1039

Farías, L., Paulmier, A. and Gallegos, M. (2007) 'Nitrous oxide and N-nutrient cycling in the oxygen minimum zone off northern Chile', *Deep-Sea Research Part I*, vol 54, pp164–180

Firestone, M. K. and Tiedje, J. M. (1979) 'Temporal change in nitrous oxide and dinitrogen from denitrification following onset of anaerobiosis', *Applied and Environmental Microbiology*, vol 38, pp673–679

Flückiger, J., Blunier, T., Stauffer, B., Chappellaz, J., Spahni, R., Kawamura, K., Schwander, J., Stocker, T. F. and Dahl-Jensen, D. (2004) 'N_2O and CH_4 variations during the last glacial epoch: Insights into the global processes', *Global Biogeochemical Cycles*, vol 18, GB1020, doi:10.1029/2003GB002122

Francis, C. A., Roberts, K. J., Beman, J. M., Santoro, A. E. and Oakley, B. B. (2005) 'Ubiquity and diversity of ammonia-oxidizing archaea in water columns and sediments of the ocean', *Proceedings of the National Academy of Sciences of the USA*, vol 102, no 41, pp14683–14688

Francis, C. A., Beman, J. M. and Kuypers, M. M. M. (2007) 'New processes and players in the nitrogen cycle: The microbial ecology of anaerobic and archeal ammonia oxidation', *The ISME Journal*, vol 1, pp19–17

Fuhrman, J. A. and Capone, D. G. (1991) 'Possible biogeochemical consequences of ocean fertilization', *Limnology and Oceanography*, vol 36, no 8, pp1951–1959

Goldstein, B., Joos, F. and Stocker, T. F. (2003) 'A modeling study of oceanic nitrous oxide during the Younger Dryas cold period', *Geophysical Research Letters*, vol 30, no 2, doi:10.1029/2002GL016418

Goreau, T. J., Kaplan, W. A., Wofsy, S. C., McElroy, M. B., Valois, F. W. and Watson, S. W. (1980) 'Production of NO_2^- and N_2O by nitrifying bacteria at reduced concentrations of oxygen', *Applied and Environmental Microbiology*, vol 40, no 3, pp526–532

Hahn, J. (1974) 'The North Atlantic Ocean as a source of atmospheric N_2O', *Tellus*, vol 26, pp160–168

Hamersley, M. R., Lavik, G., Woebken, D., Rattray, J. E., Lam, P., Hopmans, E. C., Damste, J. S. S., Kruger, S., Graco, M., Gutierrez, D. and Kuypers, M. M. M. (2007) 'Anaerobic ammonium oxidation in the Peruvian oxygen minimum zone', *Limnology and Oceanography*, vol 52, no 3, pp923–933

Hannig, M., Lavik, G., Kuypers, M. M. M., Woebken, D., Martens-Habbena, W. and Jurgens, K. (2007) 'Shift from denitrification to anammox after inflow events in the central Baltic Sea', *Limnology and Oceanography*, vol 52, no 4, pp1336–1345

Hashimoto, L. K., Kaplan, W. A., Wofsy, S. C. and McElroy, M. B. (1983) 'Transformation of fixed nitrogen and N_2O in the Cariaco Trench', *Deep-Sea Research*, vol 30, no 6A, pp575–590

Huesemann, M. H., Skillman, A. D. and Crecelius, E. A. (2002) 'The inhibition of nitrification by ocean disposal of carbon dioxide', *Marine Pollution Bulletin*, vol 44, pp142–148

IPCC (Intergovernmental Panel on Climate Change) (2007) *Climate Change 2007: The Physical Science Basis. Contribution of Working Group I to the Fourth Assessment Report of the Intergovernmental Panel on Climate Change*, Cambridge University Press, Cambridge

Ishijima, K., Sugawara, S., Kawamura, K., Hashida, G., Morimoto, S., Murayama, S., Aoki, S. and Nakazawa, T. (2007) 'Temporal variations of atmospheric nitrous oxide concentration and its ^{15}N and [18]O for the latter half of the 20th century reconstructed from firn air analyses', *Journal of Geophysical Research*, vol 112, D03305, doi:10.1029/2006JD007208

Jin, X. and Gruber, N. (2003) 'Offsetting the radiative benefit of ocean iron fertilization by enhancing N_2O emissions', *Geophysical Research Letters*, vol 30, no 24, 2249, doi:10.1029/2003GL018458

Junge, C. and Hahn, J. (1971) 'N_2O measurements in the North Atlantic', *Journal of Geophysical Research*, vol 76, pp8143–8146

Kaiser, J., Röckmann, T. and Brenninkmeijer, C. A. M. (2003) 'Complete and accurate mass spectrometric isotope analysis of tropospheric nitrous oxide', *Journal of Geophysical Research*, vol 108, no D15, 4476, doi:10.1029/2003JD003613

Karl, D., Michaels, A., Bergman, B., Capone, D., Carpenter, E., Letelier, R., Lipschultz, F., Paerl, H., Sigman, D. and Stal, L. (2002) 'Dinitrogen fixation in the world's oceans', *Biogeochemistry*, vol 57/58, pp47–98

Karl, D. M. (1999) 'A sea of change: Biogeochemical variability in the North Pacific subtropical gyre', *Ecosystems*, vol 2, pp181–214

Karner, M. B., DeLong, E. F. and Karl, D. M. (2001) 'Archaeal dominance in the mesopelagic zone of the Pacific Ocean', *Nature*, vol 409, pp507–510

Kartal, B., Kuypers, M. M. M. , Lavik, G., Schalk, J., den Camp, H. J. M. O., Jetten, M. S. M. and Strous, M. (2007) 'Anammox bacteria disguised as denitrifiers: Nitrate reduction to dinitrogen gas via nitrite and ammonium', *Environmental Microbiology*, vol 9, no 3, pp635–642

Kim, K.-R. and Craig, H. (1993) 'Nitrogen-15 and oxygen-18 characteristics of nitrous oxide: A global perspective', *Science*, vol 262, pp1855–1857

Könneke, M., Bernhard, A. E., de la Torre, J. R., Walker, C. B., Waterbury, J. B. and Stahl, D. A. (2005) 'Isolation of an autotrophic ammonia oxidizing marine archaeon', *Nature*, vol 437, pp543–546

Kreuzwieser, J., Buchholz, J. and Rennenberg, H. (2003) 'Emission of methane and nitrous oxide by Australian mangrove ecosystems', *Plant Biology*, vol 5, pp423–431

Kroeze, C., Mosier, A. and Bouwman, L. (1999) 'Closing the global N_2O budget: A retrospective analysis 1500–1994', *Global Biogeochemical Cycles*, vol 13, no 1, pp1–8

Kuypers, M. M. M., Lavik, G., Woebken, D., Schmid, M., Fuchs, B. M., Amann, R., Jorgensen, B. B. and Jetten, M. S. M. (2005) 'Massive nitrogen loss from Benguela upwelling system through anaerobic ammonium oxidation', *Proceedings of the National Academy of Sciences of the USA*, vol 102, no 18, pp6478–6483

Lam, P., Jensen, M. M., Lavik, G., McGinnis, D. F., Muller, B., Schubert, C. J., Amann R., Thamdrup, B. and Kuypers, M. M. M. (2007) 'Linking crenarchael and bacterial nitrification to anammox in the Black Sea', *Proceedings of the National Academy of Sciences of the USA*, vol 104, no 17, pp7104–7109

Michotey, V. and Bonin, P. (1997) 'Evidence for anaerobic bacterial processes in the water column: Denitrification and dissimilatory nitrate ammonification in the northwestern Mediterranean Sea', *Marine Ecology Progress Series*, vol 160, pp47–56

Muñoz-Hincapié, M., Morell, J. M. and Corredor, J. E. (2002) 'Increase of nitrous oxide flux to the atmosphere upon nitrogen addition to red mangrove sediments', *Marine Pollution Bulletin*, vol 44, pp992–996

Naik, H., Naqvi, S. W. A., Suresh, T. and Narvekar, P. V. (2008) 'Impact of a tropical cyclone on biogeochemistry of the central Arabian Sea', *Global Biogeochemical Cycles*, vol 22, GB3020, doi:10.1029/2007GB003028

Naqvi, S. W. A., Yoshinari, T., Brandes, J. A., Devol, A. H., Jayakumar, D. A., Narvekar, P. V., Altabet, M. A. and Codispoti, L. A. (1998a) 'Nitrogen isotopic studies in the sub-oxic Arabian Sea', *Proceedings of the Indian Academy of Sciences (Earth and Planetary Sciences)*, vol 107, no 4, pp367–378

Naqvi, S. W. A., Yoshinari, T., Jayakumar, D. A., Altabet, M. A., Narvekar, P. V., Devol, A. H., Brandes, J. A. and Codispoti, L. A. (1998b) 'Budgetary and biogeochemical implications of N_2O isotope signatures in the Arabian Sea', *Nature*, vol 394, pp462–464

Naqvi, S. W. A., Jayakumar, D. A., Narvekar, P. V., Naik, H., Sarma, V. V. S. S., D'Souza, W., Joseph, S. and George, M. D. (2000) 'Increased marine production of N_2O due to intensifying anoxia on the Indian continental shelf', *Nature*, vol 408, pp346–349

Naqvi, S. W. A., Yoshinari ,T., Jayakumar, D. A., Altabet, M. A., Narvekar, P. V., Devol, A. H., Brandes, J. A. and Codispoti, L. A. (2005) 'Biogeochemical ocean-atmosphere transfers in the Arabian Sea', *Progress in Oceanography*, vol 65, pp116–144

Naqvi, S. W. A., Naik, H., Pratihary, A., D'Souza, W., Narvekar, P. V., Jayakumar, D. A., Devol, A. H., Yoshinari, T. and Saino, T. (2006) 'Coastal versus open-ocean denitrification in the Arabian Sea', *Biogeosciences*, vol 3, pp621–633

Nevison, C. D., Weiss, R. F. and Erickson III, D. J. (1995) 'Global oceanic emissions of nitrous oxide', *Journal of Geophysical Research*, vol 100, no C8, pp15809–15820

Nevison, C., Butler, J. H. and Elkins, J. W. (2003) 'Global distribution of N_2O and ΔN_2O-AOU yield in the subsurface ocean', *Global Biogeochemical Cycles*, vol 17, no 4, 1119, doi:10.1029/2003GB002068

Nevison, C., Lueker, T. and Weiss, R. F. (2004) 'Quantifying the nitrous oxide source from coastal upwelling', *Global Biogeochemical Cycles*, vol 18, GB1018, doi:10.1029/2003GB002110

Nicholls, J. C., Davis, C. A. and Trimmer, M. (2007) 'High-resolution profiles and nitrogen isotope tracing reveal a dominant source of nitrous oxide and multiple pathways of nitrogen gas formation in the central Arabian Sea', *Limnology and Oceanography*, vol 52, no 1, pp156–168

Oschlies, A., Schulz, K. G., Riebesell, U. and Schmitter, A. (2008) 'Simulated 21st century's increase in oceanic suboxia by CO_2-enhanced biotic carbon export', *Global Biogeochemical Cycles*, vol 22, GB4008, doi:10.1029/2007GB003147

Ostrom, N. E., Russ, M. E., Popp, B., Rust, T. M. and Karl, D. M. (2000) 'Mechanisms of nitrous oxide production in the subtropical North Pacific based on determinations of the isotopic abundances of nitrous oxide and di-nitrogen', *Chemosphere: Global Change Science*, vol 2, no 3–4, pp281–290

Patra, P. K., Maksyutov, S. and Nakazawa, T. (2004) 'Severe weather conditions in the Arabian Sea and their impact on atmospheric N_2O budget', *Indian Journal of Marine Science*, vol 33, no 1, pp84–94

Pichevin, L., Bard, E., Martinez, P. and Billy, I. (2007) 'Evidence of ventilation changes in the Arabian Sea during the late Quaternary: Implications for denitrification and nitrous oxide emission', *Global Biogeochemical Cycles*, vol 21, GB4008, doi:10.1029/2006GB002852

Popp, B. N., Westley, M. B., Toyoda, S., Miwa, T., Dore, J. E., Yoshida, N., Rust, T. M., Sansone, F. J., Russ, M. E., Ostrom, N. E. and Ostrom P. H. (2002) 'Nitrogen and oxygen isotopomeric constraints on the origins and sea-to-air flux of N_2O in the oligotrophic subtropical North Pacific gyre', *Global Biogeochemical Cycles*, vol 16, no 4, 1064, doi:10.1029/2001GB001806

Raven, J., Caldeira, K., Elderfield, H., Hoegh-Guldberg, O., Liss, P., Riebesell, U., Shepherd, J., Turley, C. and Watson, A. (2005) *Ocean Acidification due to Increasing Atmospheric Carbon Dioxide*, Policy document 12/05, The Royal Society, London

Schmidt, H. L., Werner, R. A., Yoshida, N. and Well, R. (2004) 'Is the isotopic composition of nitrous oxide an indicator for its origin from nitrification or denitrification? A theoretical approach from referred data and microbiological and enzyme kinetic aspects', *Rapid Communications in Mass Spectrometry*, vol 18, no 18, pp2036–2040

Schmittner, A. and Galbraith, E. D. (2008) 'Glacial greenhouse-gas fluctuations controlled by ocean circulation changes', *Nature*, vol 456, pp373–376

Schmittner, A., Oschlies, A., Matthews, H. D. and Galbraith, E. D. (2008) 'Future changes in climate, ocean circulation, ecosystems, and biogeochemial cycling simulated for a business-as-usual CO_2 emission scenario until year 4000 AD', *Global Biogeochemical Cycles*, vol 22, GB1013, doi:10.1029/2007GB002953

Schropp, S. J. and Schwarz, J. R. (1983) 'Nitrous oxide production by denitrifying microorganisms from the eastern tropical Pacific and the Caribbean Sea', *Geomicrobiology Journal*, vol 3, no 1, pp17–31

Schweiger, B. (2006) 'Messung von NH$_2$OH in ausgewählten Seegebieten', Diploma thesis, Kiel University, Kiel, Germany

Seitzinger S. P. (1990) 'Denitrification in aquatic sediments', in N. P. Revsbech and J. Sörensen (eds) *Denitrification in Soil and Sediment*, Plenum Press, New York, pp207–249

Seitzinger S. P. and Kroeze, C. (1998) 'Global distribution of nitrous oxide production and N inputs in freshwater and coastal marine ecosystems', *Global Biogeochemical Cycles*, vol 12, no 1, pp93–113

Seitzinger S. P. and Nixon, S. W. (1985) 'Eutrophication and the rate of denitrification and N$_2$O production in coastal marine sediments', *Limnology and Oceanography*, vol 30, no 6, pp1332–1339

Seitzinger S. P., Kroeze, C. and Styles, R. V. (2000) 'Global distribution of N$_2$O emissions from aquatic systems: Natural emissions and anthropogenic effects', *Chemosphere: Global Change Science*, vol 2, pp267–279

Sinninghe Damsté, J. S., Rijpstra, W. I. C., Hopmans, E. C., Prahl, F. G., Wakeham, S. G. and Schouten, S. (2002) 'Distribution of membrane lipids of planktonic Crenarchaeota in the Arabian Sea', *Applied and Environmental Microbiology*, vol 68, no 2, pp2997–3002

Sowers, T., Alley, R. B. and Jubenville, J. (2003) 'Ice core records of tropospheric N$_2$O covering the last 106,000 years', *Science*, vol 301, pp945–948

Spahni, R., Chappellaz, J., Stocker, T. F., Loulergue, L., Hausammann, G., Kawamura, K., Flückiger, J., Schwander, J., Raynaud, D., Masson-Delmotte, V. and Jouzel, J. (2005) 'Atmospheric methane and nitrous oxide of the late Pleistocene from Antarctic ice cores', *Science*, vol 310, pp1317–1321

Stramma, L., Johnson, G. C., Sprintall, J. and Mohrholz, V. (2008) 'Expanding oxygen minimum zones in the tropical oceans', *Science*, vol 320, pp655–658

Suntharalingam, P. and Sarmiento, J. L. (2000) 'Factors governing the oceanic nitrous oxide distribution: Simulations with an ocean general circulation model', *Global Biogeochemical Cycles*, vol 14, no 1, pp429–454

Suthhof, A., Ittekkot, V. and Gaye-Haake, B. (2001) 'Millennial-oscillation of denitrification intensity in the Arabian Sea during the late Quaternary and its potential influence on atmospheric N$_2$O and global climate', *Global Biogeochemical Cycles*, vol 15, no 3, pp637–649

Sutka, R. L., Ostrom, N. E., Ostrom, P. H., Gandhi, H. and Breznak, J. A. (2003) 'Nitrogen isotopomer site preference of N$_2$O produced by *Nitrosomonas europaea* and *Methylococcus capsulatus* Bath', *Rapid Communications in Mass Spectrometry*, vol 17, no 7, pp738–745

Sutka, R. L., Ostrom, N. E., Ostrom, P. H., Gandhi, H. and Breznak, J. A. (2004) 'Erratum: Nitrogen istotopomer site preference of N$_2$O produced by *Nitrosomas europaea* and *Methylococcus capsulatus* Bath', *Rapid Communications in Mass Spectrometry*, vol 18, pp1411–1412

Sutka, R. L., Ostrom, N. E., Ostrom, P. H., Breznak, J. A., Gandhi, H., Pitt, A. J. and Li, F. (2006) 'Distinguishing nitrous oxide production from nitrification and denitrification on the basis of isotopomer abundances', *Applied and Environmental Microbiology*, vol 72, no 1, pp638–644

Teira, E., Lebaron, P., Van Aken, H. and Herndl, G. J. (2006) 'Distribution and activity of bacteria and archaea in the deep water masses of the North Atlantic Ocean', *Limnology and Oceanography*, vol 51, no 5, pp2131–2144

Thamdrup, B., Dalsgaard, T., Jensen, M. M., Ulloa, O., Farias, L. and Escribano, R. (2006) 'Anaerobic ammonium oxidation in the oxygen-deficient waters off northern Chile', *Limnology and Oceanography*, vol 51, no 5, pp2145–2156

Thunell, R. C. and Kepple, A. B. (2004) 'Glacial-Holocene ^{15}N record from the Gulf of Tehuantepec, Mexico: Implications from denitrification in the eastern Pacific and changes in atmospheric N_2O', *Global Biogeochemical Cycles*, vol 18, GB1001, doi:10.1029/2002GB002028

Toyoda, S. and Yoshida, N. (1999) 'Determination of nitrogen isotopomers of nitrous oxide on a modified isotope ratio mass spectrometer' *Analytical Chemistry*, vol 71, no 20, pp4711–4718

Toyoda, S., Yoshida, N., Miwa, T., Matsui, Y., Yamagishi, H., Tsunogai, U., Nojiri, Y. and Tsurushima, N. (2002) 'Production mechanism and global budget of N_2O inferred from its isotopomers in the western North Pacific', *Geophysical Research Letters*, vol 29, no 3, doi:10.1029/2001GL014311

UNEP (United Nations Environment Programme) (2004) *Global Environment Outlook Year Book 2003*, UNEP, Nairobi

Varela, M. M., Van Aken, H. M., Sintes, E. and Herndl, G. J. (2008) 'Latitudinal trends of Crenarchaeota and bacteria in the meso- and bathypelagic water masses of the eastern North Atlantic', *Environmental Microbiology*, vol 10, no 1 pp110–124

Walter, S., Bange, H. W., Breitenbach, U. and Wallace, D. W. R. (2006a) 'Nitrous oxide in the North Atlantic Ocean', *Biogeosciences*, vol 3, pp607–619

Walter, S., Breitenbach, U., Bange, H. W., Nausch, G. and Wallace, D. W. R. (2006b) 'Distribution of N_2O in the Baltic Sea during transition from anoxic to oxic conditions', *Biogeosciences*, vol 3, pp557–570

Wanninkhof, R. (2007) 'The impact of different gas exchange fomulations and wind speed products on global air-sea CO_2 fluxes', in C. S. Garbe, R. A. Handler and B. Jähne (eds) *Transport at the Air-Sea Interface*, Springer, Berlin, pp1–23

Ward, B. B. (2008) 'Nitrification in marine systems', in D. G. Capone, D. A. Bronk, M. R. Mulholland and E. J. Carpenter (eds) *Nitrogen in the Marine Environment*, 2nd Edition, Elsevier, Amsterdam, pp199–262

Weiss, R. F. and Price, B. A. (1980) 'Nitrous oxide solubility in water and seawater', *Marine Chemistry*, vol 8, pp347–359

Westley, M. B., Yamagishi, H., Popp, B. N. and Yoshida, N. (2006) 'Nitrous oxide cycling in the Black Sea inferred from stable isotope and isotopomer distributions', *Deep-Sea Research Part II*, vol 53, pp1802–1816

Wuchter, C., Abbas, B., Coolen, M. J. L., Herfort, L., van Bleijswijk, J., Timmers, P., Strous, M., Teira, E., Herndl, G. J., Middelburg, J. J., Schouten, S. and Damste, J. S. S. (2006) 'Archaeal nitrification in the ocean', *Proceedings of the National Academy of Sciences of the USA*, vol 103, no 33, pp12317–12322

Yamagishi, H., Yoshida, N., Toyoda, S., Popp, B. N., Westley, M. B. and Watanabe, S. (2005) 'Contributions of denitrification and mixing on the distribution of nitrous oxide in the North Pacific', *Geophysical Research Letters*, vol 32, L04603, doi:10.1029/2004GL021458

Yamagish., H., Westley, M. B., Popp, B. N., Toyoda, S., Yoshida, N., Watanabe, S., Koba, K. and Yamanaka, Y. (2007) 'Role of nitrification and denitrification on the nitrous oxide cycle in the eastern tropical North Pacific and Gulf of California', *Journal of Geophysical Research*, vol 112, G02015, doi:10.1029/2006JG000227

Yoshida, N., Hattori, A., Saino, T., Matsuo, S. and Wada, E. (1984) '^{15}N/^{14}N ratio of dissolved N_2O in the eastern tropical Pacific Ocean', *Nature*, vol 307, pp442–444

Yoshinari T. (1976) 'Nitrous oxide in the sea', *Marine Chemistry*, vol 4, pp189–202

Yoshinari T., Altabet, M. A., Naqvi, S. W. A., Codispoti, L., Jayakumar, A., Kuhland, M. and Devol, A. (1997) 'Nitrogen and oxygen isotopic composition of N_2O from suboxic waters of the eastern tropical North Pacific and the Arabian Sea – Measurements by continuous-flow isotope-ratio monitoring', *Marine Chemistry*, vol 56, no 3–4, pp253–264

4

The Global Nitrous Oxide Budget: A Reassessment

Keith Smith, Paul Crutzen, Arvin Mosier and Wilfried Winiwarter

Introduction

An understanding of the global cycle of N_2O and the capacity to construct a balanced budget of global sources and sinks, requires knowledge of the concentration of the gas in earth's atmosphere and its average lifetime, how that concentration has changed with time, and the principal influences that affect the magnitude of the sources and sinks. An essential component of this knowledge is the estimation of the present N_2O concentration in the atmosphere, by high-precision gas chromatographic analysis of air samples collected at monitoring stations around the world (for example CSIRO, 2009), and the determination of how the atmospheric concentration has varied over time.

This latter information comes from the analysis of air trapped in firn (the ice formed from relatively recent snowfalls on glaciers) and in deeper glacier ice of greater age (for example Wolff and Spahni, 2007). In recent years, drilling through the Antarctic and Greenland ice caps (Plate 4.2) has yielded cores of ice up to several hundred thousand years in age, and the trapped air has given the concentrations of all the major long-lived greenhouse gases – carbon dioxide, methane and N_2O – through several glacial–interglacial cycles back to 650,000 yr BP. The N_2O record is less complete than those of the other gases, because of suspected artifacts affecting the gas concentration in certain periods (Wolff and Spahni, 2007), but it is clear that the concentration varied from highs of around 270ppbv (270nmol mol^{-1}) in interglacial periods to lows around 200ppbv in glacial periods (for example Sowers, 2001; Flückiger et al, 2004). Over most of the last four millennia N_2O mixing ratios (i.e. the concentrations in dry air) have been close to 270ppbv (Flückiger et al, 2002), but in more recent times there has been a 20 per cent increase, beginning around AD1850. The mixing ratio exceeded 280ppbv for the first time in 1905, reached 300ppbv in the mid-1970s, 319ppbv in 2005 (IPCC, 2007), and now exceeds 320 ppbv. The trends, on different timescales, are shown in Figure 4.1 and Plate 4.3. This increase has no precedent over the last 650,000 years, and whereas in the past the sources and sinks were broadly in line, limiting the fluctuations in atmospheric mixing ratio to within a narrow range, the only rational explanation for the increase since 1850 is a change in the magnitude of sources, and/or of sinks, by human activity; these issues are explored in later sections. The rate of increase in atmospheric N_2O concentrations increased from approximately 0.15ppb yr^{-1} between 1900 and 1955 to the current linear increase rate of about 0.7ppb yr^{-1}.

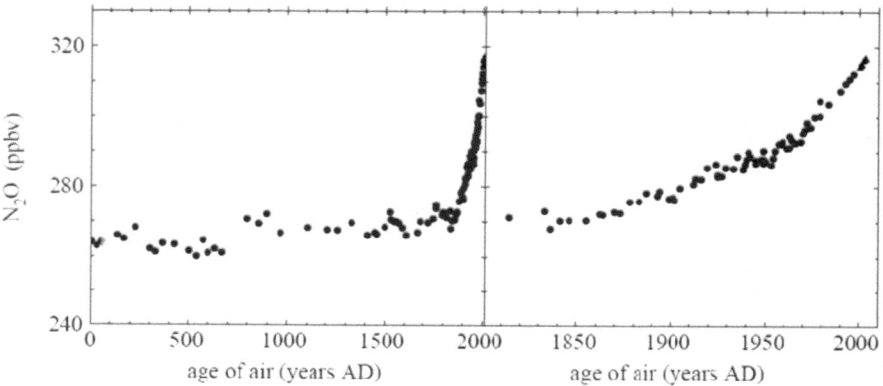

Figure 4.1 N_2O concentrations over the last 2000 years (left) and 200 years (right), based on measurements from firn air and ice at Law Dome, Antarctica (circles) and annual averages of flasks collected at Cape Grim Observatory, Tasmania (triangles)

Source: Based on data of MacFarling Meure et al (2006) and in Wolff and Spahni (2007) courtesy of The Royal Society and the authors

Stratospheric sink for nitrous oxide

In tropospheric chemistry, N_2O as a relatively stable compound is of marginal importance. It gradually diffuses into the stratosphere, where the predominant sink occurs, as a result of destruction by direct photolysis by ultraviolet radiation:

$$N_2O + h\nu \rightarrow N_2 + O(^1D) \tag{4.1}$$

and by reaction with the excited oxygen atoms, $O(^1D)$, formed in ozone photolysis:

$$N_2O + O(^1D) \rightarrow 2NO \quad (60 \text{ per cent})$$

$$N_2O + O(^1D) \rightarrow N_2 + O_2 (40 \text{ per cent}) \tag{4.2}$$

Smaller amounts of other oxygenated nitrogen compounds (including NO_2, HNO_3 and N_2O_5) are also formed (Prather, 2007). Although the relative proportions of the two reactions vary throughout the stratosphere, direct photolysis is responsible for about 90 per cent of the total stratospheric decomposition, while the reaction with excited oxygen accounts for the remaining 10 per cent (Garcia and Solomon, 1994). The destruction of N_2O, while removing a potent greenhouse gas from the atmosphere, results in a contribution to a different environmental problem: the NO formed by the reaction of N_2O with $O(^1D)$ causes the decomposition of stratospheric ozone, O_3 (Crutzen, 1970), thus diminishing the protective role of the ozone layer against harmful effects of UV radiation on organisms at the surface of earth. It should be remembered that the recognition of the environmental importance of this process preceded any focus on the role of N_2O as a greenhouse gas, and it is now again recognized as important, as other ozone-depleting substances regulated by the Montreal Protocol (unlike N_2O) are gradually removed from the stratosphere (Ravishankara et al, 2009).

Lifetime of nitrous oxide in the atmosphere

The rate of N_2O photolysis varies spatially and temporally throughout the stratosphere, peaking near the equator at altitudes between 30 and 35km and at noon when the solar UV radiation is most intense (Butenhoff and Khalil, 2007). N_2O abundances are about 0.8ppbv greater in the Northern Hemisphere than in the Southern Hemisphere, consistent with about 60 per cent of emissions occurring in the Northern Hemisphere (IPCC, 2001). Almost no vertical gradient is observed in the troposphere, but N_2O abundances decrease in the stratosphere as a result of the UV-driven decomposition, for example falling to about 120ppbv at 30km altitude at mid-latitudes (IPCC, 2001). The average lifetime of N_2O is in excess of 100 years; the work of Volk et al (1997) and Prinn and Zander (1999) gives a value of 120 years, but there is a feedback whereby the N_2O mixing ratio has an impact on its own lifetime (Prather, 1998), and the application of 2-D stratospheric chemical models indicates that the global mean atmospheric lifetime of N_2O decreases about 0.5 per cent for every 10 per cent increase in the amount of N_2O present (IPCC, 2001). The globally averaged surface abundance of N_2O was 314ppbv in 1998, corresponding to a global burden of 1510Tg N. Division of this value by the lifetime, 120 years, gives an annual stratospheric sink rate of 12.6Tg N (Table 4.1). The most recent lifetime assessment listed by IPCC (2007) is the one by Montzka et al (2003) that recommended a lifetime of 114 years. Taking the atmospheric mixing ratio in the middle of the present decade as 319ppbv, the corresponding atmospheric burden is 1534Tg N, and the stratospheric sink becomes $1534/114 = 13.5$Tg N yr^{-1}.

Table 4.1 Estimates of the global nitrous oxide budget (in Tg N yr^{-1}) from different sources

Reference	Mosier et al (1998)/ Kroeze et al (1999)		Olivier et al (1998)		IPCC (1996)	IPCC (2001)
Base year	1994	range	1990	range	1980s	1990s
Sources						
Ocean	3.0	1–5	3.6	2.8–5.7	3	
Atmosphere (NH$_3$ oxidation)	0.6	0.3–1.2	0.6	0.3–1.2		
Soils:						
Tropical wet forest	3.0	2.2–3.7			3	
Tropical dry savannas	1.0	0.5–2.0			1	
Temperate forest	1.0	0.1–2.0			1	
Temperate grassland	1.0	0.5–2.0			1	
All natural soils			6.6	3.3–9.9		
Natural sub-total	9.6	4.6–15.9	10.8	6.4–16.8	9	
Agricultural soils	4.2	0.6–14.8	1.9	0.7–4.3	3.5	
Biomass burning	0.5	0.2–1.0	0.5	0.2–0.8	0.5	
Industrial sources	1.3	0.7–1.8	0.7	0.2–1.1	1.3	
Cattle and feedlots	2.1	0.6–3.1	1.0	0.2–2.0	0.4	
Anthropogenic sub-total	8.1	2.1–20.7	4.1	1.3–7.7	5.7	6.9
Total sources	**17.7**	**6.7–36.6**	**14.9**	**7.7–24.5**	**14.7**	
Imbalance (trend)	3.9	3.1–4.7			3.9	3.8
Total sinks (stratospheric)	**12.3**	**9–16**		**2.3**	**12.6**	
Implied total source	16.2			16.2	16.4	

Source: IPCC (2001)

Global warming potential

The global warming potential (GWP) of a greenhouse gas is a measure of its contribution to global warming relative to that of a corresponding amount of a reference gas, carbon dioxide. The concept was developed as a tool, following the adoption of the Kyoto Protocol, so that nations and industries could compare the cost effectiveness of different mitigation measures (IPCC, 2007). The calculation of the GWP of N$_2$O or any other greenhouse gas requires knowledge of its contribution to climate change from emissions over time. There are a number of complications associated with this calculation that cannot be covered here (the reader is referred to IPCC, 2007, for details), and the simplified method commonly used is to integrate the global mean radiative forcing over time of a pulse emission of 1kg of a compound (i) relative to that of 1kg of CO_2 (IPCC, 1990). Thus the GWP is defined by:

$$GWP_i = \frac{\int_0^{TH} RF_i\,(t)dt}{\int_0^{TH} RF_r\,(t)dt} = \frac{\int_0^{TH} a_i \cdot [C_i(t)]dt}{\int_0^{TH} a_r \cdot [C_r(t)]dt} \tag{4.3}$$

where TH is the time horizon, RF_i is the global mean RF (radiative forcing) of gas i, ai is the RF per unit mass increase in atmospheric abundance of gas i (radiative efficiency), $[C_i(t)]$ is the time-dependent abundance of i, and the corresponding quantities for the reference gas (r) in the denominator. The numerator and denominator are called the absolute global warming potential (AGWP) of i and r, respectively (IPCC, 2007).

The use of a pulse emission in the GWP calculation has been the subject of considerable debate (for example O'Neill, 2000), but the concept, even with its limitations, makes possible a useful evaluation of the relative climate change impacts of the long-lived greenhouse gases. Table 4.2 shows the GWPs of CO_2, methane and N_2O, calculated over three timescales: 20, 100 and 500 years. The 100-year time horizon is the one most commonly used in climate change assessments, and over this period N_2O is almost 300 times as potent as CO_2 as a global warming agent.

Table 4.2 Global warming potentials of the long-lived greenhouse gases CO_2, CH_4 and N_2O

Greenhouse gas	GWP for each time horizon		
	20 years	100 years	500 years
CO_2	1	1	1
CH_4	72	25	7.6
N_2O	289	298	153

Source: IPCC (2007)

Global emissions estimates and source uncertainties

As indicated above, the amount of N_2O in the atmosphere has been increasing approximately linearly, at around 0.7ppb, or 0.26 per cent, per year for the past few decades (Plate 4.3). The IPCC Third Assessment Report (IPCC, 2001) concluded that the primary driver of this increase was enhanced microbial production of N_2O in expanding and fertilized agricultural lands. In the report, as in the earlier evaluations (IPCC, 1990, 1992, 1996) and the later Fourth Assessment Report (IPCC, 2007), the contribution from agriculture was based on 'bottom-up' extrapolations to the global scale from measurements at a limited number of experimental sites. The same was true for the contributions from natural terrestrial ecosystems. (In a later section we compare this bottom-up approach to assessing agricultural emissions with a top-down calculation.)

In IPCC (1990) and IPCC (1992), the uncertainty concerning agricultural N_2O emissions was so great that the range of estimates covered two orders of magnitude, 0.01–2.2 and 0.03–3.3 Tg N yr^{-1}, respectively. The widely varying conditions in agroecosystems, particularly of N supply and water availability, inevitably led to measured annual emissions covering a huge range, from a few grams of N_2O per hectare to tens of kilograms per hectare (see Chapter 5). This contributed to the huge uncertainties in the early global assessments. However, many more detailed emission measurements at agricultural sites, and their relationships to N inputs, were carried out from the early 1990s onwards, and provided convincing evidence that emissions of N_2O from agricultural systems make up the largest single global source. Thus in the next assessment (IPCC, 1996) the mean value for the N_2O agricultural source was given as 3.5 Tg N yr^{-1}, which was above the previous upper limit, and the range had narrowed from two orders of magnitude to ± 50 per cent of the mean (Table 4.3). Since then there has been a consensus that the emissions from agriculture – direct and indirect – are of this order, i.e. several Tg N yr^{-1} (Table 4.3), and constitute the largest single source category (Table 4.1), while uncertainty margins have been extended again.

Table 4.3 Changing assessment since 1990 of agricultural contribution to global N_2O emissions

Assessment	Estimated global N_2O emission from agriculture (Tg N_2O-N yr^{-1})	Description of source
IPCC (1990)	0.01–2.2	Fertilizer, including emission from groundwater
IPCC (1992)	0.03–3.0	Cultivated soils
IPCC (1996)	3.5 (1.8–5.3)	Mineral N (fertilizer) + animal waste + N-fixation
Mosier et al (1998) and IPCC (200)	4.2 (0.6–14.8) 2.1 (0.6–3.1) 0.5 (0.2–1.0) Total: 6.8 (1.0–18.9)	N added to soils + indirect emissions Manure management Biomass burning

By contrast, the trend in the estimates of N_2O emissions from the largest natural terrestrial source, wet tropical forests, has gone in the opposite direction. McElroy and Wofsy (1986) estimated an annual global emission from this biome of 7.4 ± 4 Tg N_2O–N, nearly ten times their estimate for fertilized agricultural land (Table 4.1). The forest estimate was based on measurements at only five sites, and not long afterwards Matson and Vitousek (1990), using data from about 30 sites, reduced the estimate to 2.4 Tg N_2O-N

yr^{-1}, and the assessments of Mosier et al (1998) and Kroeze et al (1999) were similar. The scaling of observed fluxes in wet tropical forests to annual emissions was aided by the relatively small diurnal and month-to-month variations in N_2O emissions (Matson and Vitousek, 1990) because of the rapid and continuous N mineralization that takes place (Keller and Reiners, 1994), providing a steady supply of mineral N (NH_4^+ and NO_3^-) as substrates for the N_2O-producing microorganisms (see Chapter 2). Nonetheless, the global upscaling has large associated uncertainties.

The global problem: Too much or too little nitrogen

Nitrogen, contained in amino acids, proteins and DNA, is necessary for life. While there is an abundance of nitrogen in nature, almost all is in an unreactive form (gaseous nitrogen, N_2) that is not usable by most organisms. N compounds fall into two groups – non-reactive and reactive. Non-reactive N is N_2. Reactive N (Nr) includes all biologically active, photochemically reactive, and radiatively active N compounds in the atmosphere and biosphere of the earth. Thus Nr includes inorganic reduced forms of N (for example, NH_3, NH_4^+), inorganic oxidized forms (for example, NO_x, HNO_3, N_2O, NO_3^-), and organic N compounds (for example, urea, amines, proteins, nucleic acids) that can undergo biochemical transformations. In the absence of human intervention, the supply of reactive nitrogen in the environment is not sufficient to sustain the current abundance of human life. Thus humans learned in the early 20th century how to convert gaseous N_2 into forms that could sustain food production. Over 40 per cent of the world's population is here today because of that capability (Galloway et al, 2004, 2008).

Reactive N is introduced to the natural terrestrial environment primarily by biological nitrogen fixation in forests and grasslands, particularly in the tropics. Human activity introduces reactive N inadvertently by fossil fuel combustion and purposefully through biological nitrogen fixation associated with agricultural crops and through the Haber-Bosch process that allows the manufacture of synthetic fertilizer. Human introduction of Nr has changed with time relative to natural sources (Figure 4.2). In 1860, natural terrestrial nitrogen fixation introduced between 100 and 200Tg per year of Nr (Galloway et al, 2004). Within the last few decades, human activities have roughly doubled this supply. Nr creation continues to increase every year. It is dominated by agricultural activities, but fossil fuel energy plays an important role, and the increasing use of biofuels is adding a new and rapidly changing dimension. From 1860 to 1995, energy and food production increased steadily on both an absolute and per capita basis; Nr creation also increased from around 15Tg N yr^{-1} in 1860 to 156Tg N yr^{-1} in 1995, and increased further to 187Tg N yr^{-1} in 2005, in large part because cereal production increased by about 20 per cent and meat production by 26 per cent over about 10 years. These rising agricultural demands were sustained by a matching rise in Nr creation by the Haber-Bosch process from 100 to 121Tg N yr^{-1} (FAO, 2006).

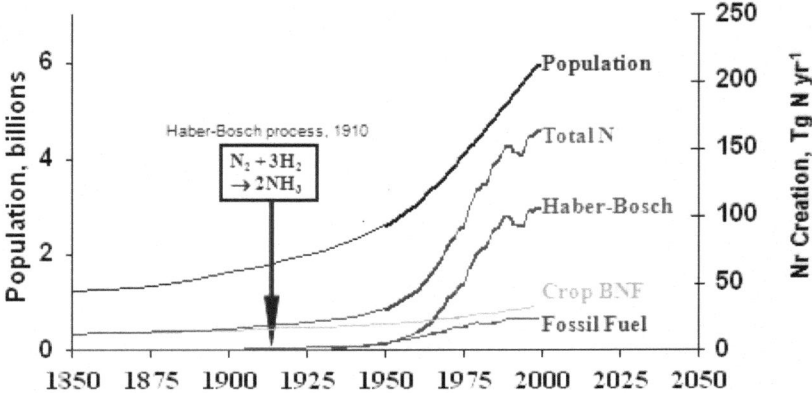

Figure 4.2 Global anthropogenic creation rates of reactive nitrogen by the Haber-Bosch process, biological nitrogen fixation associated with agricultural crops, and combustion of fossil fuels

Source: Adapted from Galloway et al (2003)

Cultivation-induced biological nitrogen fixation (C-BNF) occurs in several agricultural systems, with crop, pasture and fodder legumes being the most important. The C-BNF estimate for 1995 was 31.5Tg N and, because of the increase in soya bean and meat production over the past decade, Galloway et al (2008) estimated that in 2005 C-BNF was 40Tg N. In parallel, primary commercial energy production by coal, natural gas and petroleum combustion increased from 8543 million tons of oil equivalents (Mtoe) to 10,600 Mtoe (24 per cent) (BP Global, 2007).

The proportion of Nr created by the Haber-Bosch process that is not used for fertilizer-N production is used as a raw material in various industries, to make products such as nylon and other plastics, resins, glues, melamine, animal/fish/shrimp feed supplements, and explosives (Galloway et al, 2008). In 2005, around 23Tg N, accounting for 20 per cent of Haber-Bosch Nr, was used in this way (Prud'homme, 2007), but little is known about the ultimate fate of the Nr.

The nitrogen cascade

Nitrogen is now known to be unusual among the elements that have had their cycles significantly perturbed by human action. The great significance of nitrogen is that it is linked to so many of the major global and regional environmental challenges that we face today: ozone layer depletion, acidification of soils and surface waters, global warming, surface and groundwater pollution, biodiversity loss, and human health and vulnerability. As nitrogen moves along its biogeochemical pathway, the same atom can contribute to many different impacts. This sequence of effects has been termed the 'nitrogen cascade' (Galloway and Cowling, 2002). It is depicted in Figure 4.3.

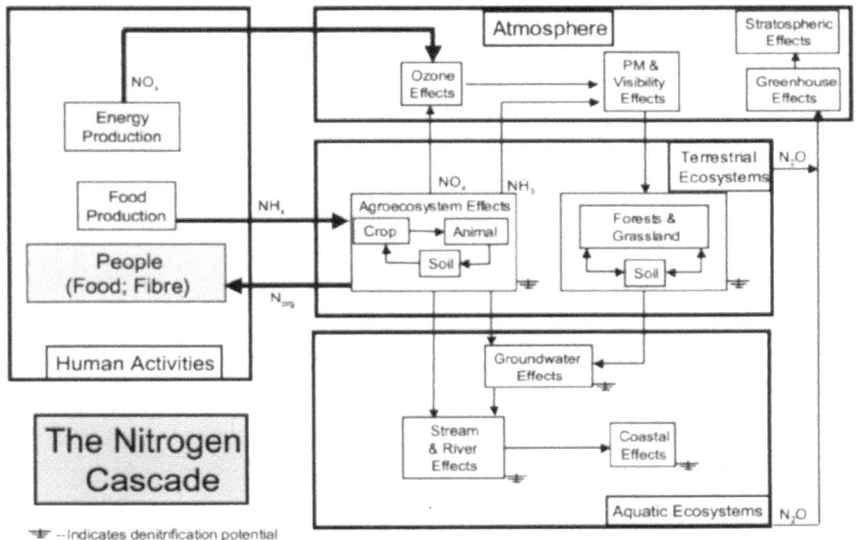

Figure 4.3 The nitrogen cascade, illustrating the movement of human-produced reactive nitrogen (Nr) as it cycles through the atmosphere, terrestrial ecosystems and aquatic ecosystems

Source: Galloway and Cowling (2002), with permission from the Royal Swedish Academy of Sciences

The concept of the cascade, and the extensive research that underlies it, has allowed us not only to determine the linkages among the various aspects of the nitrogen cycle, but also to begin to assess how changes in one part of the cycle can delay or enhance the transfer of nitrogen to other parts of the cycle. The cascade continues as long as the nitrogen remains active in the environment, and it ceases only when reactive nitrogen is stored for a very long time, or is converted back to non-reactive N_2.

Reactive nitrogen released into the environment in gaseous forms (NH_3 and NO_x) can have remote as well as local impacts; the gases may be transported thousands of kilometres from the point of emission before being deposited to ecosystems. As an example, Plate 4.1 shows the change in the rates of deposition from the atmosphere of reactive nitrogen, relative to the predominantly natural conditions prevailing at the start of the industrial era. In addition to atmospheric emission and transport, much nitrogen is transported via river systems to coastal marine waters. Human activity has had great influence on nitrogen fluxes to coastal oceans in some parts of the world, but very little effect elsewhere. The nitrogen cycle is most altered where farming and industrial activity in the river basins and watersheds is most intense (Howarth et al, 1996, 2002; Boyer and Howarth, 2002).

The sequence of events linking the addition of new reactive N into the environment specifically to N_2O emissions is depicted in Figure 4.4. Here, the dominant sources are nitrogen fertilizers and manures, and the release of N from the mineralization of soil organic matter following cultivation, especially as a result of land-use change (see Chapters 5–7).

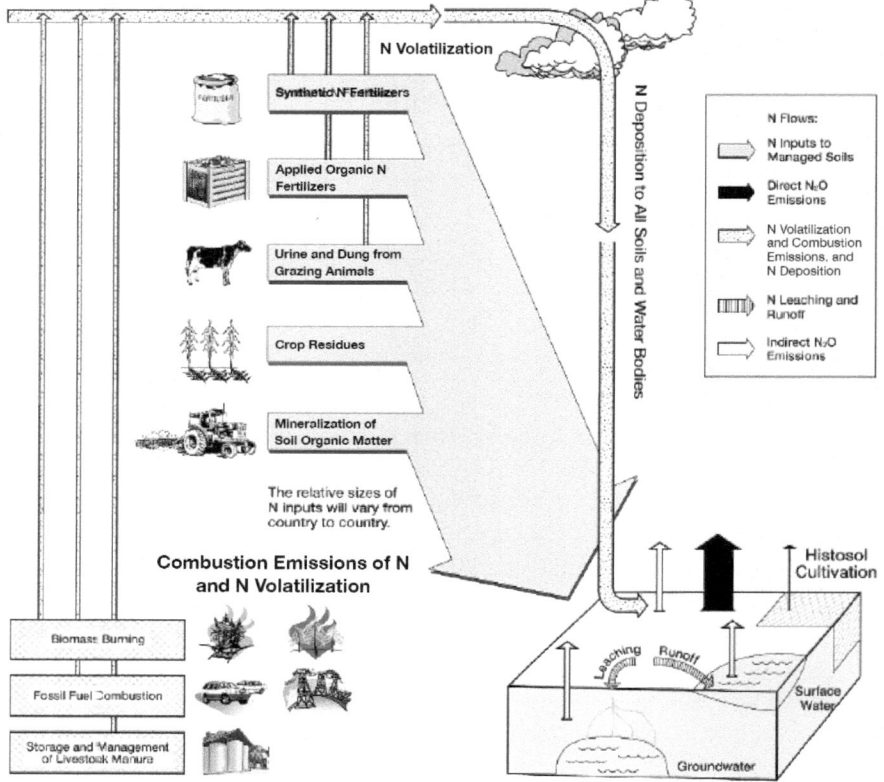

Figure 4.4 Schematic diagram illustrating the sources and pathways of reactive nitrogen that result in direct and indirect N_2O emissions from soils and waters

Source: IPCC (2006)

Top-down and bottom-up estimates of global nitrous oxide emissions to the atmosphere

Nitrous oxide is produced in 'natural' and agricultural soils almost exclusively as a result of microbial processes (see Chapter 2). The main microbial reactions involved in the production of N_2O are nitrification (oxidation of ammonium to nitrite) and denitrification (reduction of nitrate to dinitrogen; N_2O is an obligate intermediate in this latter process). Although nitrification is basically an aerobic process and denitrification is essentially an anaerobic one, both can take place in the soil in close proximity under soil conditions that are amenable to upland crop and forage production. Because agricultural systems are typically 'leaky,' a considerable portion of new N additions is lost to the environment through leaching and runoff, ammonia volatilization and denitrification. Van der Hoek (1998) estimated that more than 60 per cent of the annual N input into food production was not converted into usable products. Globally, cereal crop N use efficiency remains near 40 per cent (Cassman et al, 2002; Balasubramanian et al, 2004).

Two general methods used to estimate soil N_2O emissions can be broadly considered as either (1) bottom-up approaches based on soil surface gas flux measurements or models based on soil application of N, or (2) top-down approaches based on changes in atmospheric concentration of N_2O and estimates of sink strength (Del Grosso et al, 2008). The bottom-up approaches considered in this chapter are (1) the IPCC (2006) methodology, (2) DAYCENT (daily version of Century) ecosystem model estimates of direct N_2O emissions (Del Grosso et al, 2006), and (3) field-scale estimates based on soil surface gas flux measurements.

Top-down estimates

Nitrous oxide and crop production

Using a global top-down approach, Crutzen et al (2008) estimated that at the global scale 3–5 per cent of all new reactive N input into terrestrial systems is converted to N_2O. This N_2O conversion range was based on data compiled by Prather et al (2001) and Galloway et al (2004). New N input includes the N that is produced by chemical, biological and atmospheric processes (i.e. fertilizer nitrogen produced by the Haber-Bosch process, N oxides produced by fossil fuel combustion, and biological N fixation). The pre-industrial, natural N_2O sink and source at an atmospheric mixing ratio of 270ppb was calculated to be equal to 10.2Tg N_2O-N yr^{-1} (Prather et al, 2001), which includes marine emissions. By the start of the present century, when the atmospheric volume mixing ratio was 315ppb, the stratospheric photochemical sink of N_2O was about 11.9Tg N_2O-N yr^{-1}. The total N_2O source at that time was equal to the photochemical sink (11.9Tg N_2O-N yr^{-1}) plus the atmospheric growth rate (3.9Tg N_2O-N yr^{-1}), together totalling 15.8Tg N_2O-N yr^{-1} (Prather et al, 2001). The anthropogenic N_2O source is the difference between the total source strength, 15.8Tg N_2O-Nr yr^{-1}, and the current natural source, which is equal to the pre-industrial source of 10.2Tg N_2O-N yr^{-1} minus an uncertain 0–0.9Tg N_2O-N, with the last number taking into account a decreased natural N_2O source due to 30 per cent global deforestation (Klein Goldewijk, 2001). Thus an anthropogenic N_2O source of 5.6–6.5Tg N_2O-N yr^{-1} was derived. To obtain the agricultural contribution, the estimated industrial source of 0.7–

1.3Tg N_2O-N yr^{-1} (Prather et al, 2001) was subtracted, giving a range of 4.3–5.8Tg N_2O-N yr^{-1}. This is 3.8–5.1 per cent of the anthropogenic 'new' reactive nitrogen input of 114Tg N yr^{-1} for the early 1990s. This input value is derived from the 100Tg of N fixed by the Haber-Bosch process, plus 24Tg of reactive N released by fossil fuel combustion and a 3.5Tg N increase through biological nitrogen fixation (BNF), between current and pre-industrial times (Galloway et al, 2004), minus the 14Tg of Haber-Bosch N not used as fertilizer (Smeets et al, 2007). In comparison, the N_2O-N emission estimated by Prather et al (2001) is 2.9–6.3Tg N_2O-N yr^{-1}, or 3.4–6.8Tg N_2O-N yr^{-1} if we also include biomass and biofuel burning (which we consider an agricultural source), leading to N_2O-N yields of 2.6–5.5 per cent or 3.0–6.0 per cent, respectively.

The global source and sink of N_2O in pre-industrial, natural conditions was 10.2Tg N_2O-N yr^{-1} (see above). Of that, 6.2–7.2Tg N_2O-N yr^{-1} came from the land and coastal zones (Prather et al, 2001), and was derived from an estimated fresh reactive N input of 141Tg N yr^{-1} (Galloway et al, 2004). This gives an N_2O-N yield of 4.4–5.1 per cent. Thus, both for the pre-Haber-Bosch natural terrestrial emissions and the agricultural emissions in the Haber-Bosch era, we find that the N_2O output (or emission factor, EF) is 3–5 per cent of the fresh reactive N input. This is a parametric relationship, based on the global budgets of N_2O and reactive N input, and on atmospheric concentrations and known lifetime of N_2O, and thus is not dependent on detailed knowledge of the terrestrial N cycle. When the relationship was applied to the fertilizer-N input to biofuel crop production – for example, production of maize for ethanol and rapeseed oil for biodiesel – the global warming impact of the resulting N_2O emissions was found to match or even exceed the corresponding 'cooling' achieved by replacement of fossil fuels by the biofuels (Crutzen et al, 2008). Subsequent work in which the global EF was introduced into existing life-cycle analysis models gave broadly similar outcomes (Mosier et al, 2009).

We make the assumption here that this top-down approach for estimating the impact of Nr on N_2O emissions is applicable to all crop production and is not limited solely to biofuel crop production, and in the next section we elaborate on how the underlying concepts compare with those that are the basis for bottom-up estimation procedures, and compare the results given by the different procedures.

Comparison of top-down and bottom-up estimates

The Crutzen et al (2008) and IPCC (2006) methods to estimate global nitrous oxide yields

The basis of the Crutzen et al (2008) methodology is that the newly fixed N entering agricultural systems (synthetic fertilizer-N and N from BNF) is regarded as the source of all agriculture-related N_2O emissions. These emissions will not all happen in the season when fixation takes place, but will involve longer cycling times (which are nonetheless short compared with the >100 year lifetime of N_2O in the environment). We may consider three categories:

1   direct emissions from N-fertilized soils;
2   'secondary' emissions resulting from the complex transformations of N compounds in the various flows within agricultural systems;
3   indirect emissions (in the IPCC meaning of the phrase) arising from leached N leaving agricultural fields and entering water systems, and from volatilized N deposited onto natural ecosystems.

Examples of the 'secondary' emission sources are:

- crop residues ploughed in as fertilizer for a successor crop;
- dung and urine from livestock (both grazing and housed) fed variously on N-fertilized grain crops, feeds containing BNF-N (for example soya bean meal, alfalfa, clover-rich pasture and silage in Europe, and tropical grasses with *Azospirillum* associations in Brazil);
- N mineralized from soil organic matter and root residues following cultivation or grassland renewal.

In contrast, in the IPCC approach, emissions from crop residues and mineralization are included in the 'direct' emissions and have the same EF; separate EFs are used for emissions from grazing animals, and the N source here is quantified on the basis of the N excreted, and essentially is treated as a 'new' N source, not as fertilizer- or BNF-derived N (see also Chapters 5 and 6). The fractions of the N applied to fields that are lost by leaching, runoff and volatilization have additional EFs applied to them to describe the resulting 'indirect' emissions. The aggregate emissions from agriculture are arrived at by summing all these individual sources. The IPCC's 1 per cent EF for direct N_2O emissions contains an uncertainty of one third to three times the default value (IPCC, 2006). The default EF for emissions from cattle, poultry and pigs is 2 per cent of the N excreted, with a range of 0.7 per cent to 6 per cent – again, from one third to three times the default value. The EFs for N derived from N volatilization and re-deposition and N derived from leaching and runoff are 1 per cent (uncertainty range 0.2–5 per cent) and 0.75 per cent (0.05–2.5 per cent), respectively. At default volatilization fractions of 10 per cent (mineral fertilizer) or 20 per cent (animal manure), and the default leaching fraction of 30 per cent, indirect emissions amount to 0.35–0.45 per cent of the N applied to the land. Each of the source terms in the bottom-up IPCC method is very uncertain. However, their sum is not inconsistent with the total derived by the top-down methodology.

Modelling-based comparisons

As noted above, the IPCC (2006) methodology is based on soil surface gas flux measurements from numerous global sites. Emissions are assumed to be proportional to soil N inputs from various sources (Table 4.1). This method also accounts for emissions from burning crop biomass and from N transformations occurring in manure management systems. The DAYCENT model (Del Grosso et al, 2006) is an example of a more sophisticated bottom-up approach. In addition to N inputs, DAYCENT accounts for the influence of other factors (water, temperature, O_2 and labile C availability, and plant N demand) that influence direct soil N_2O emissions. Model predictions are evaluated on the basis of soil surface flux measurements. In contrast, the top-down approach infers anthropogenic N_2O emissions from changes in atmospheric N_2O concentration and N_2O removal rates (Crutzen et al, 2008). Using DAYCENT, Del Grosso et al (2008 calculated N_2O emissions from agricultural systems in the US and for the entire globe using bottom-up approaches, and compared these results with the range of N_2O emissions estimated using the top-down approach of Crutzen et al (2008) that, as explained above, calculates the N_2O emissions range as 3–5 per cent of the combined N inputs from symbiotic N fixation and synthetic fertilizer application. To obtain a value for the US national greenhouse gas inventory, Del Grosso et al (2008) calculated N_2O emissions for major crops and grasslands from DAYCENT model simulations, and for other crops and manure management systems using the IPCC (2006) methodology (US EPA, 2008). DAYCENT results for N volatilization and leached/runoff N were combined with IPCC (2006) methodology to estimate indirect N_2O emissions. The N_2O emission from agricultural systems in the US for 2005, obtained using the bottom-up approaches, was 0.6Tg N yr^{-1} (Table 4.4), somewhat below the range of 0.8–1.4Tg N yr^{-1} based on the top-down approach. However, using the IPCC (2006) methodology, Del Grosso et al (2008) also calculated that approximately 5.8Tg of N from N_2O are currently emitted annually from agricultural systems at the global scale (Table 4.4). This is close to the middle of the range (4.2–7.0Tg N_2O-N yr^{-1}) given by the top-down approach. Del Grosso et al concluded that, at sufficiently large scales, the use of these top-down and bottom-up approaches to calculate N_2O emissions from agricultural systems yield similar estimates. They emphasized that the EF for the global top-down approach (3–5 per cent of N inputs from symbiotic N fixation and synthetic fertilizer production) cannot be compared directly with the emissions factors used in the IPCC (2006) method because the methods consider different sources of N inputs. N_2O emissions are highly variable in space and time, and different methodologies have not agreed closely, especially at small scales. However, as scale increases, so does the agreement between estimates based on soil surface measurements (bottom-up approach) and estimates derived from changes in atmospheric concentration of N_2O (top-down approach). Del Grosso et al (2008) concluded that:

> the convergence of top-down and bottom-up approaches increases confidence in emissions estimates because the methods are based on different assumptions, and this convergence suggests that we have at least a rudimentary understanding of the factors that control emissions at large spatial and temporal scales.

Table 4.4 Global and US N_2O sources and estimates of emissions from agricultural systems in 2005

Region	N Source (Tg N)				N_2O Emission (Tg N)	
	Synthetic fertilizer	N-fixation	Livestock manure	Crop residue	Bottom-up	Top-down
USA	10.4	16.8	6.7	5.2	0.6*	0.8–1.4
World	88.0	53.0	161	30.7	5.8	4.2–7.0

Note: * Value amended from the '0.9Tg N' in Del Grosso et al (2008), which resulted from failure to convert from Tg of N_2O to Tg of N (S. Del Grosso, personal communication to A. R. Mosier, 2009).
Source: Based on Del Grosso et al (2008), with world bottom-up calculations based on IPCC (2006) methodology; US calculations based on DAYCENT model simulations and IPCC (2006) methodology (US EPA, 2008). Top-down calculations based on Crutzen et al (2008).

Although the Del Grosso et al (2008) paper contains too large a value for US emissions modelled by DAYCENT (see note to Table 4.4), their conclusions about the agreement at the global scale are unaffected and encouraging. The numbers demonstrate that the 3–5 per cent EF relates to agriculture as a whole – i.e. it is not limited to crop-based biofuel production. Crutzen et al (2008) focused only on crop-based biofuels because of a logical inconsistency in one of the motivations for their production, i.e. the abatement of greenhouse gas emissions by replacing fossil fuel, while at the same time introducing another, much more potent, greenhouse gas. As food production for a still-growing world population will require further increases in fertilizer nitrogen application (Erisman et al, 2008), efforts to limit N_2O emissions cannot be allowed to affect production. Instead, efforts need to focus on improvement of the nitrogen use efficiency throughout the production chain. This entails improvement of nitrogen use efficiency in the field, but also directs attention towards reducing wastage of food and prioritizing production of essential proteins for human consumptions in nitrogen-efficient pathways. This may in practice mean dietary changes towards more vegetarian food and less meat consumption.

Very recently, Davidson (2009) reported that although our calculation of a global EF of 4 ± 1 per cent (Crutzen et al, 2008) fits the data well for the two reference dates used in that paper (1860 and the 1990s), it underestimates the emissions in the late 19th and early 20th centuries, and Davidson argues that other sources of N_2O were important in that period. Much of his paper is devoted to the calculation of alternative EFs for manure-N and fertilizer-N, and he achieves a good match to the rise in atmospheric concentration with EF values of 2 per cent and 2.5 per cent, respectively. However, although Davidson (2009) also discusses the 'mining' of soil N, i.e. the release of reactive N from inactive forms in old soil organic matter, under the influence of ploughing and over-grazing as agriculture expanded into new regions, and argues that this was a likely source of increasing N_2O between 1860 and 1960, he does not make any calculation of the actual contribution. We (in Crutzen et al, 2008) did not attempt to match the global EF to the gradual rise in atmospheric N_2O in the period between 1860 and the 1990s, and we readily accept that N mining would have been important, as the period coincided with the major expansion of crop and grazing land in, for example, the US, Canada, Argentina, South Africa, Australia and New Zealand, culminating with the massive Virgin Lands Programme in the former USSR in 1954–1964 (McCauley, 1976). Indeed, we hypothesize that this source is likely to have been responsible for much of the discrepancy that Davidson (2009) has noted. Mineral N arising from the decomposition of old soil organic matter (perhaps hundreds or even thousands of years old) in effect introduces additional reactive N into the terrestrial biosphere and its role in contributing to direct N_2O emissions is now recognized (see above). Likewise, mineral nitrates (saltpetre) from dry Chilean deserts introduced into agricultural systems in the late 19th century (Smil, 2001) was another source of reactive N, just like any industrially produced nitrogen fertilizer compound. If the term we used, 'newly fixed N', were to be replaced by 'new reactive N', i.e. including mined soil N and saltpetre N as well as BNF-N and synthetic fertilizer N, then the apparent differences between Davidson's analysis and ours would be largely reconciled. Ojima et al (1993) estimated that 13.2–25.5Pg of carbon was released by land-use change to agriculture in the period 1800–1990. It seems reasonable to assume that at least 10Pg of this C was released between 1860 and 1960 – the period of greatest agricultural expansion – and with it around 0.8Pg of N. This is equivalent to, on average, 8Tg N yr^{-1}, which is similar to Davidson's (2009) estimate of annual fertilizer N use in the mid-1950s. Obviously, the annual rate of soil C and N release will have fluctuated considerably during the 100-year period, and with it the associated N_2O emissions. We consider that a detailed calculation of the global N release rates, on a decade-by-decade basis if not annually, would be a worthwhile project to undertake.

Soil nitrous oxide emissions and national greenhouse gas inventories

Important conclusions on a country's greenhouse gas emissions can be derived from the official submission of emission data to the United Nations Framework Convention on Climate Change (UNFCCC) as part of the national reporting obligations of the developed countries of the world (as listed in Annex I to the Kyoto Protocol). These data are available via UNFCCC's web pages, together with background information. Not only are they prepared in a standardized manner for all countries (based on IPCC guidelines: IPCC, 2006) but they also represent the emissions to which a country is officially committed. According to this information, soil N_2O is generally not of very great importance to the overall greenhouse gas emissions of a country. As an example, in the European Union, about 4 per cent of total greenhouse gas emissions derive from this source. However, in a country where agriculture is the dominant industry, for example New Zealand, N_2O is much more important.

Most countries reporting emissions of soil N_2O apply the default IPCC methodology (IPCC, 1996, 2006), as described above, which has a very large associated uncertainty range. This uncertainty also affects the overall uncertainty of greenhouse gas inventories. A number of studies (Rypdal and Winiwarter, 2001; Winiwarter and Rypdal, 2001; Monni et al, 2004; Ramírez et al, 2008) demonstrate that for national inventories of different developed countries, soil N_2O is the single most important contributor to overall uncertainty. The extent of its contribution varies and may depend on the relative importance of agriculture in the economy of a country as well as on some subjective choices of parameters by the national experts. Despite such variations, uncertainty associated with soil N_2O emissions dominates the overall inventory uncertainty in all countries investigated. Therefore, if one wishes to improve the overall reliability of a national greenhouse gas balance, the first improvement to be made is to the estimate of soil nitrous oxide emissions. For reasons related to the mathematical procedure involved in error propagation, the dominance of N_2O emissions in the overall uncertainty applies only to those assessed for a specific year. For emission trends (the difference between two years) the influence of soil N_2O is considerably smaller, as long as the quoted uncertainty in EFs applies similarly to both years of the trend.

The large uncertainty associated with N_2O emissions from soils leads to repercussions with respect to interpretations of emissions reported, not only regarding reported annual emissions, but also regarding trends. Because of the way in which the EF for direct soil N_2O emissions is defined (i.e. as a percentage of the N applied to the land), a decreasing trend in reported N_2O emissions derives not from observations but from the decrease in fertilizer application, which in reality may or may not have led to actual decreases of nitrous oxide release. In general, the relationship between emissions and nitrogen availability is well established (see Stehfest and Bouwman, 2006, and citations therein, and Chapters 5 and 6). Nevertheless, potentially changes in weather patterns as well as changes in the practice of N application may lead to deviations from this relationship that are very difficult to trace. Consequently, a global assessment of N_2O emissions is an essential tool for validation, where the source term can be estimated fully independently on the basis of the atmospheric concentration increase (Kroeze et al, 1999; Crutzen et al, 2008).

While it may be difficult, from such an independent 'top-down' estimate, to identify an individual (national) source and to conclude what useful abatement measures might be employed, at least the uncertainty margin is considerably smaller than for estimates based on plot-scale measurements. Approaches like the one of IPCC, which are aimed at source attribution and assigning responsibility, will have to continue to rely on source-based assessments. Thus alternative approaches need to be identified which make it possible to draw better connections between the release processes and the input parameters. One promising option is biophysical modelling, but more still needs to be done before models are fully adequate for the purpose (see Chapter 5).

Concluding remarks

The study of the composition of the atmosphere in the recent and the distant past has shown conclusively that the concentration of N_2O began to increase around the middle of the 19th century, after millennia when the concentration was stable, and that the rate of increase has accelerated over the last few decades. The annual global increase of 3.9Tg N_2O-N is evidently the outcome of the release of substantial quantities of reactive nitrogen into the environment, a fraction of which becomes transformed into N_2O, in both terrestrial and aquatic ecosystems. In this chapter we have sought to demonstrate, by a top-down analysis, that that fraction is of the order of 4 ± 1 per cent of the anthropogenically released reactive N, and furthermore that the dominant pathway for reactive N entering the biosphere is through agriculture.

This EF is applicable to agriculture in general. As food production is essential to feed the growing world population, an increase of the efficiency in nitrogen use is urgently needed for the overall system. Our overall factor of 4 ± 1 per cent, while not inconsistent with the IPCC estimates, points out much more strongly than their bottom-up approach how dominant the agricultural contribution to N_2O emissions is. In order to arrive at higher efficiency in agricultural nitrogen use, we need improvements in agronomy, but also require to better understand the nitrogen fluxes that follow losses of reactive N from agricultural areas to other parts of the environment, leading to remote N_2O formation in natural soils, in groundwater and in surface water, but also in the oceans (see Duce et al, 2008).

References

Balasubramanian, V., Alves, B., Aulakh, M., Bekunda, M., Cai, Z., Drinkwater, L., Mugendi, D., van Kessel, C. and Oenema, O. (2004) 'Crop, environmental, and management factors affecting nitrogen use efficiency', in A. R. Mosier, J. K. Syers and J. Freney (eds) *Agriculture and the Nitrogen Cycle*, SCOPE 65, Island Press, Washington, DC, pp19–33

BP Globa (2007) *Statistical Review of World Energy 2007*, www.bp.com/productlanding.do?categoryId=6848&contentId=7033471

Boyer, E. W. and Howarth, R. W. (eds) (2002) *The Nitrogen Cycle at Regional to Global Scales*, Kluwer Academic Publishers, Boston

Butenhoff, C. L. and Khalil, M. A. K. (2007) 'Stratospheric sinks of nitrous oxide', in D. Reay, N. Hewitt, K. A. Smith and J. Grace (eds) *Greenhouse Gas Sinks*, CABI, Wallingford, pp207–226

Cassman, K. G., Dobermann, A. and Walters, D. T. (2002) 'Agroecosystems, nitrogen-use efficiency, and nitrogen management', *Ambio*, vol 31, pp132–140

Crutzen, P. J. (1970) 'The influence of nitrogen oxides on the atmospheric ozone content , *Quarterly Journal of the Royal Meteorological Society*, vol 96, pp320–325

Crutzen, P. J., Mosier, A. R., Smith, K. A. and Winiwarter, W. (2008) 'N_2O release from agro-biofuel production negates global warming reduction by replacing fossil fuels', *Atmospheric Chemistry and Physics*, vol 8, pp389–395

CSIRO (Commonwealth Scientific and Industrial Research Organization) (2009) 'CSIRO GASLAB – N_2O (nitrous oxide) analysis datasets – global atmospheric sampling laboratory – flask sampling network', ftp://gaspublic@ftp.dar.csiro.au/data/gaslab/

Davidson, E. A. (2009), 'The contribution of manure and fertilizer nitrogen to atmospheric nitrous oxide since 1860', *Nature Geoscience*, doi:10.1038/NGEO608

Del Grosso, S. J., Parton, W. J., Mosier, A. R., Walsh, M. K., Ojima, D. S. and Thornton P. E. (2006) 'DAYCENT national scale simulations of N_2O emissions from cropped soils in the USA, *Journal of Environmental Quality*, vol 35, pp1451–1460

Del Grosso, S. J., Wirth, T., Ogle, S. M. and Parton, W. J. (2008) 'Estimating agricultural nitrous oxide emissions', *Eos (Transactions of the American Geophysical Union)*, vol 89, pp529–530

Duce, R. A., LaRoche, J., Altieri, K., Arrigo, K. R., Baker, A. R., Capone, D. G., Cornell, S., Dentener, F., Galloway, J., Ganeshram, R. S., Geider, R. J., Jickells, T., Kuypers, M. M., Langlois, R., Liss, P. S., Liu, S. M., Middelburg, J. J., Moore, C. M., Nickovic, S., Oschlies, A., Pedersen, T., Prospero, J., Schlitzer, R., Seitzinger, S., Sorensen, L. L., Uematsu, M., Ulloa, O., Voss, M., Ward, B. and Zamora L. (2008) 'Impacts of atmospheric anthropogenic nitrogen on the open ocean', *Science*, vol 320, pp893–897

Erisman, J. W., Sutton, M. A., Galloway, J., Klimont, Z. and Winiwarter, W. (2008) '100 years of ammonia synthesis: how a single patent changed the world', *Nature Geoscience*, vol 1, pp636–639

FAO (Food and Agriculture Organization) (2006) *FAO Statistical Databases*, FAO, Rome, http://faostat.fao.org/default.aspx

Flückiger, J. M., Monnin, E., Stauffer, B., Schwander, J., Stocker, T. F., Chappellaz, J., Raynaud, D. and Barnola, J. M. (2002) 'High-resolution Holocene N_2O ice core record and its relationship with CH_4 and CO_2', *Global Biogeochemical Cycles*, vol 16, article no 1010, doi:10.1029/2001GB001417

Flückiger, J. M., Blunier, T., Stauffer, B., Chappellaz, J., Spahni, R., Kawamura, K., Schwander, J., Stocker, T. F. and Dahl-Jensen, D. (2004) 'N_2O and CH_4 variations during the last glacial epoch: Insight into global processes', *Global Biogeochemical Cycles*, vol 18, GB1020, doi:10.1029/2003GB002122

Galloway, J. N and Cowling, E. B. (2002) 'Reactive nitrogen and the world: 200 years of change', *Ambio*, vol 31, pp64–71

Galloway, J. N., Aber, J. D., Erisman, J. W., Seitzinger, S. P., Howarth, R. H., Cowling, E. B. and Cosby B. J. (2003) 'The nitrogen cascade', *Bioscience*, vol 53, pp341–356

Galloway, J. N., Dentener, F. J., Capone, D. G., Boyer, E. W., Howarth, R. W., Seitzinger, S. P., Asner, G. P., Cleveland, C.C., Green, P. A., Holland, E. A., Karl, D. M., Michaels, A. F., Porter, J. H., Townsend, A. R. and Vörösmarty, C. J. (2004) 'Nitrogen cycles: Past, present, and future', *Biogeochemistry*, vol 70, pp153–226

Galloway, J. N., Townsend, A. R., Erisman, J. W., Bekunda, M., Cai, Z., Freney, J. R., Martinelli, L. A., Seitzinger, S. P. and Sutton M. A. (2008) 'Transformation of the nitrogen cycle: Recent trends, questions, and potential solutions', *Science*, vol 320, pp889–892

Garcia, R. R. and Solomon, S. (1994) 'A new numerical model of the middle atmosphere: 2. Ozone and related species', *Journal of Geophysical Research*, vol 99, pp12937–12951

Howarth, R. W., Billen, G., Swaney, D., Townsend, A., Jawonski, N, Lajha, K., Downing, J. A., Delmgren, R., Caraco, N., Jordan, T., Berendse, F., Freney, J., Kudeyarov, V., Murdoch, P. and Zhu, Z. L. (1996) 'Regional nitrogen budgets and riverine N and P fluxes for the drainages to the North Atlantic Ocean: Natural and human influences', *Biogeochemistry*, vol 35, pp75–139

Howarth, R. W., Boyer, E. W., Pabgich, W. J. and Galloway, J. N. (2002), 'Nitrogen use in the United States from 1961–2000 and potential future trends', *Ambio*, vol 31, pp88–96

IPCC (Intergovernmental Panel on Climate Change) (1990) *Climate Change: Scientific Assessment of Climate Change. Contribution of Working Group I to the First Assessment Report*, J. T. Houghton, G. J. Jenkins and J. J. Ephraums (eds), Cambridge University Press, Cambridge

IPCC (1992) *Climate Change 1992: The Supplementary Report to the IPCC Scientific Assessment*, J. T. Houghton, B. A. Callander and S. K. Varney (eds), Cambridge University Press, Cambridge

IPCC (1996) *Climate Change 1995: The Science of Climate Change. Contribution of Working Group I to the Second Assessment Report*, J. T. Houghton, L. G. Meiro Filho, E. A. Callander, N. Harris, A. Kattenberg and K. Maskell (eds), Cambridge University Press, Cambridge

IPCC (2001) *Climate Change 2001: The Scientific Basis. Contribution of Working Group I to the Third Assessment Report*, J. T. Houghton, Y. Ding, D. J. Griggs, M. Noguer, P. J. van der Linden, X. Dai, K. Maskell and C. A. Johnson (eds), Cambridge University Press, Cambridge

IPCC (2006) *2006 IPCC Guidelines for National Greenhouse Gas Inventories*, Vol. 4, Chapter 11, 'N$_2$O emissions from managed soils and CO$_2$ emissions from lime and urea application', H. S. Eggelston, L. Buenida, K. Miwa, T. Ngara, and K. Tanabe (eds), IGES, Hayama, Japan

IPCC (2007) *Climate Change 2007: The Physical Science Basis. Contribution of Working Group I to the Fourth Assessment Report*, S. Solomon, D. Qin, M. Manning, Z. Chen, M. Marquis, K. B. Averyt, M. Tignor and H. L. Miller (eds), Cambridge University Press, Cambridge

Keller, M. and Reiners, W. A. (1994) 'Soil-atmosphere exchange of nitrous oxide, nitric oxide, and methane under secondary succession of pasture to forest in the Atlantic lowlands of Costa Rica', *Global Biogeochemical Cycles*, vol 8, pp399–409

Klein Goldewijk, C. G. M. (2001) 'Estimating global land use change over the past 300 years: The HYDE data base', *Global Biogeochemical Cycles*, vol 15, pp415–434

Kroeze, C., Mosier, A. and Bouwman, L. (1999) 'Closing the global N$_2$O budget: A retrospective analysis 1500–1994', *Global Biogeochemical Cycles*, vol 13, pp1–8

MacFarling Meure, C., Etheridge, D., Trudinger, C., Steele, P., Langenfelds, R., van Ommen, T., Smith, A. and Elkins J. (2006) 'Law Dome CO$_2$, CH$_4$ and N$_2$O ice core records extended to 2000 years BP', *Geophysical Research Letters*, vol 33, L14 810, doi:10.1029/2006GL026152

Matson, P. A. and Vitousek, P. M. (1990) 'Ecosystem approach to a global nitrous oxide budget', *BioScience*, vol 40, pp667–672

McCauley, M. (1976) *Khrushchev and the Development of Soviet Agriculture: The Virgin Lands Program 1953–1964*, Holmes and Meier Publishers, Inc., New York (summary available at www.sjsu.edu/faculty/watkins/virginlands.htm)

McElroy, M. B. and Wofsy, S. C. (1986) 'Tropical forests: Interactions with the atmosphere', in G. T. Prance (ed) *Tropical Rain Forests and the World Atmosphere*, Westview Publishers, Boulder, CO, pp33–60

Monni S., Syri, S. and Savolainen, I. (2004) 'Uncertainties in the Finnish greenhouse gas emission inventory', *Environmental Science and Policy*, vol 7, pp87–98

Montzka, S. A., Fraser, P. J., Butler, J. H., Connell, P. S., Cunnold, D. M., Daniel, J. S., Derwent, R. G., Lal, S., McCulloch, A., Oram, D. E., Reeves, C. E., Sanhueza, E., Steele, L. P., Velders, G. J. M., Weiss, R. F. and Zander, R. J. (2003) 'Controlled substances and other source gases', in *Scientific Assessment of Ozone Depletion: 2002*, World Meteorological Organization, Geneva, pp1.1–1.83

Mosier, A. R. and Galloway, J. N. (2005) 'Setting the scene – the International Nitrogen Initiative', *Proceedings of the International Fertilizer Industry Association (IFA) Workshop on Enhanced-Efficiency Fertilizers*, Frankfurt, Germany, 28 June 2005 (on CD)

Mosier, A. R., Kroeze, C., Nevison, C., Oenema, O., Seitzinger, S. and van Cleemput, O. (1998) 'Closing the global N_2O budget: Nitrous oxide emissions through the agricultural nitrogen cycle – OECD/IPCC/IEA phase II development of IPCC guidelines for national greenhouse gas inventory methodology', *Nutrient Cycling in Agroecosystems*, vol 52, pp225–248

Mosier, A. R., Crutzen, P. J., Smith, K. A. and Winiwarter, W. (2009) 'Nitrous oxide's impact on net greenhouse gas savings from biofuels: Life-cycle analysis comparison', *International Journal of Biotechnology*, vol 11, pp60–74

Ojima, D. S., Dirks, B. O. M., Glenn, E. P., Owensby, C. E. and Scurlock, J. O. (1993) 'Assessment of C budget for grasslands and drylands of the world', *Water, Air, and Soil Pollution*, vol 70, pp95–109

Olivier, J. G. J., Bouwman, A. F., van der Hoek, K. W. and Berdowski, J. J. M. (1998) 'Global air emission inventories for anthropogenic sources of NO_x, NH_3 and N_2O in 1990', *Environmental Pollution*, vol 102, pp135–148

O'Neill, B. C. (2000) 'The jury is still out on global warming potentials', *Climatic Change*, vol 44, pp427–443

Prather, M. J. (1998) 'Time scales in atmospheric chemistry: Coupled perturbations to N_2O, NO_y, and O_3', *Science*, vol 279, pp1339–1341

Prather, M. J. (2007) 'Lifetimes and time scales in atmospheric chemistry', *Philosophical Transactions of the Royal Society, Series A*, vol 365, pp1705–1726

Prather, M. J., Ehhalt, D., Dentener, F., Derwent, R., Dlugokencky, E., Holland, E., Isaksen, I., Katima, J., Kirchhoff, V., Matson, P., Midgley, P., Wang, M. and many contributing authors (2001) 'Atmospheric chemistry and greenhouse gases', in J. T. Houghton, Y. Ding, D. J. Griggs, M. Noguer, P. J. van der Linden, X. Dai, K. Maskell and C. A. Johnson (eds) *Climate Change 2001: The Scientific Basis*, Cambridge University Press, Cambridge, pp239–287

Prinn, R. G. and Zander, R. (1999) 'Long-lived ozone related compounds', in *Scientific Assessment of Ozone Depletion: 1998*, Global Ozone Research and Monitoring Project, Report No. 44, World Meteorological Organization, Geneva, pp1.1–1.54

Prud'homme, M. (2007) *Global Fertilizers and Raw Materials Supply and Supply/Demand Balances*, International Fertilizer Industry Association, Istanbul

Ramírez, A., de Keizer, C., Van der Sluijs, J. P., Olivier, J. and Brandes, L. (2008) 'Monte Carlo analysis of uncertainties in the Netherlands greenhouse gas emission inventory for 1990–2004', *Atmospheric Environment*, vol 42, pp8263–8272

Ravishankara, A. R., Daniel, J. S. and Portmann, R. W. (2009) 'Nitrous oxide (N_2O): The dominant ozone-depleting substance emitted in the 21st century', *Science*, vol 326, pp123–125

Rypdal, K. and Winiwarter, W. (2001) 'Uncertainties in greenhouse gas inventories – evaluation, comparability and implications', *Environmental Science and Policy*, vol 4, pp107–116

Smeets, E., Bouwman, A. F. and Stehfest, E. (2007) 'Interactive comment on "N_2O release from agro-biofuel production negates global warming reduction by replacing fossil fuels" by P. J. Crutzen et al', *Atmospheric Chemistry and Physics Discussions*, vol 7, ppS4937–S4941

Smil, V. (2001) *Enriching the Earth: Fritz Haber, Carl Bosch, and the Transformation of Food Production*, MIT Press, Cambridge, MA

Sowers, T. (2001) 'N_2O record spanning the penultimate deglaciation from the Vostok ice core', *Journal of Geophysical Research*, vol 106, pp31903–31914

Stehfest, E. and Bouwman, A. F. (2006) 'N_2O and NO emissions from agricultural fields and soils under natural vegetation: Summarizing available measurement data and modelling of global annual emissions', *Nutrient Cycling in Agroecosystems*, vol 74, pp207–228

US EPA (United States Environmental Protection Agency) (2008) 'Inventory of US greenhouse gas emissions and sinks: 1990–2006', Office of Atmospheric Programs, Washington, DC, www.epa.gov/climatechange/emissions/usinventoryreport.html

Van der Hoek, K. W. (1998) 'Nitrogen efficiency in global animal production', *Environmental Pollution*, vol 102, pp127–132

Volk, C. M., Elkins, J. W., Fahey, D. W., Dutton, G. S., Gilligan, J. M., Loewenstein, M., Podolske, J. R., Chan, K. R. and Gunson, M. R. (1997) 'Evaluation of source gas lifetimes from stratospheric observations', *Journal of Geophysical Research*, vol 102, pp25543–25564

Winiwarter, W. and Rypdal, K. (2001) 'Assessing the uncertainty associated with national greenhouse gas emission inventories: A case study for Austria', *Atmospheric Environment*, vol 35, pp5425–5440

Wolff, E. and Spahni, R. (2007) 'Methane and nitrous oxide in the ice core record', *Philosophical Transactions of the Royal Society, Series A*, vol 365, pp1775–1792

5

Nitrous Oxide Emissions from the Nitrogen Cycle in Arable Agriculture: Estimation and Mitigation

Lex Bouwman, Elke Stehfest and Chris van Kessel

Introduction

Human activities are probably the major cause of the increase in the atmospheric N_2O concentration of 0.7ppb per year and of the increased release of NO into the atmosphere. Nitrous oxide is a greenhouse gas, which constitutes 6 per cent of the anthropogenic greenhouse effect (Forster et al, 2007), and also contributes to the depletion of stratospheric ozone (IPCC, 2001). Although the contributions of N_2O from the different sources remain less well known than for all other greenhouse gases, it is generally accepted that the use of mineral N fertilizers and animal manure management are the major anthropogenic sources of N_2O (Mosier et al, 1998; IPCC, 2001).

Many factors associated with crop, soil, water and N fertilizer management influence soil conditions and processes, and thus N_2O emissions. The bacterial processes of denitrification and nitrification are the dominant sources of N_2O and NO in most soil systems, while denitrification is also a sink for N_2O. Nitrification is an aerobic process that occurs across all ecosystems. The availability of ammonium (NH_4^+) and O_2 is the most important factor controlling soil nitrification (Firestone and Davidson, 1989). Denitrification is an anaerobic process, and rates can be highly variable across time and space. The major controls of biological denitrification include the availability of C and NO_3^- and other N oxides, and the O_2 supply (Tiedje, 1988). N_2O can also be produced during chemical decomposition of HNO_2 under limited O_2 conditions and at low soil pH (Neff et al, 1995; Bremner, 1997; McKenney and Drury, 1997; Veldkamp and Keller, 1997).

Direct anthropogenic emissions of N_2O occur in N-fertilized cropland and grassland, from grazed pastures, and from animal barns and manure storage systems. Indirect anthropogenic emissions occur through degassing of N_2O from aquifers and surface waters, stemming from N_2O dissolved in water draining through soils, or from denitrification in groundwater of nitrate leached from fertilized soils. Furthermore, NH_3 volatilization and NO emissions from agricultural systems may lead to N_2O emission, following re-deposition on land or water. These nitrogen transformation processes, known as the 'nitrogen cascade' (Galloway et al, 2003), are illustrated in Figure 4.3.

In this chapter we first discuss different methods for estimating direct N_2O emissions, including the EF approach, and empirical and process-based models. For indirect emissions we discuss different approaches to determine EFs. We then present a global inventory of fertilizer-N use and animal manure management systems and associated N_2O emissions. This inventory is used in subsequent sections to discuss the magnitude of the various sources of N_2O within the agricultural system, the various mitigation strategies for reducing N_2O emissions, and their emission reduction potential. Finally, we present a summary and conclusions.

Approaches to estimating direct and indirect nitrous oxide emissions

The estimation of N_2O emissions is still highly uncertain, due to their large variability in time and space. Large variability is caused by the variable rates at which the processes of nitrification and denitrification occur. These processes, in turn, are controlled by biophysical and chemical conditions in soil micro-sites, which often show strong non-linear relationships with emissions of N_2O. This non-linearity makes upscaling difficult. In order to estimate direct N_2O emissions from the plot scale to the global scale, various methodologies have been developed, ranging from factorial approaches and statistical models to process-based simulation models. All approaches are trying to find relationships between controlling factors and N_2O emissions, on different scales, and apply them to estimate N_2O emissions using information on the controlling factors as input parameters.

The emission factor approaches such as those used by Bouwman (1996) are based on increasing N_2O emissions following an increase in fertilizer-N input (Kaiser and Ruser, 2000; McSwiney and Robertson, 2005). IPCC (2006) estimated the proportion of N fertilizer applied that is emitted as N_2O to be 1 per cent, i.e. an EF of 0.01 (or 1 per cent). This EF is used as a 'default factor' and reflects 'direct emissions' (Table 5.1). Recently, Crutzen et al (2008) proposed that instead of using a bottom-up approach for estimating the EF in agricultural systems, a top-down approach would give a more realistic estimate of N_2O emissions in agriculture. The top-down approach includes secondary N_2O emissions within agricultural systems, for example via livestock feed and manure, and indirect emissions of N_2O from N leached from agricultural fields

Table 5.1 Default emission factors to calculate the anthropogenic N_2O emission for arable land, including direct and indirect emissions

Type	Code	Default value	Range	Description
Direct emission	EF1	0.01	0.003–0.03	Emission from N input of fertilizer, manure, crop residues, mineralization
Indirect emission	EF4	0.01	0.002–0.05	Emission from N input from volatilization and re-deposition
Indirect emission	EF5	0.0075	0.0005–0.025	Emission from nitrate leaching and N runoff

Source: IPCC (2006)

into streams and open waters. Nitrogen newly fixed by biological N_2 fixation is also included. By following the top-down avenue, Crutzen et al (2008) calculated that between 3 and 5 per cent of newly fixed N is emitted as N_2O.

Statistical methods have identified factors that control annual N_2O emissions at the field scale, and form the basis of the default emission factor adopted by IPCC to estimate direct N_2O emissions (Bouwman et al, 2002b; Freibauer and Kaltschmitt, 2003; Stehfest and Bouwman, 2006) (Figure 5.1). The most important controlling factors in these statistical approaches are the rates of N-fertilizer application, crop type, fertilizer type, soil organic C content, soil pH and texture, and climate. As these relationships are identified at a larger temporal and spatial scale by statistical methods, no causal relationships can be derived from these correlations. In general, it cannot be determined whether the emissions occur via nitrification or denitrification.

In contrast, mechanistic models make use of (sub-)daily measurements of N_2O emissions at the plot scale and the results of laboratory experiments to identify the controlling factors for N_2O emissions from nitrification and denitrification. One of the first conceptual mechanistic N_2O emissions models is the 'hole-in-the-pipe' model (Firestone and Davidson, 1989), which relates N_2O emissions to the total amount of N flowing through the soils system (the pipe), and to factors controlling the loss of N_2O during this flow (the hole). More detailed process-based models such as DNDC ('denitrification/-decomposition') (Li and Aber, 2000) and DAYCENT (Parton et al, 1996, 2001; Del Grosso et al, 2000) describe the processes of nitrification and denitrification with separate N_2O emissions (Figure 5.2). In both models, the soil water content and aeration status of the soils are key controlling factors. The DNDC model also explicitly includes gas diffusion through the soil profile.

An overview of the controlling factors for N_2O emissions from nitrification and denitrification is provided elsewhere (Farquharson and Baldock, 2008). Although DNDC and DAYCENT are the most well-known process-based

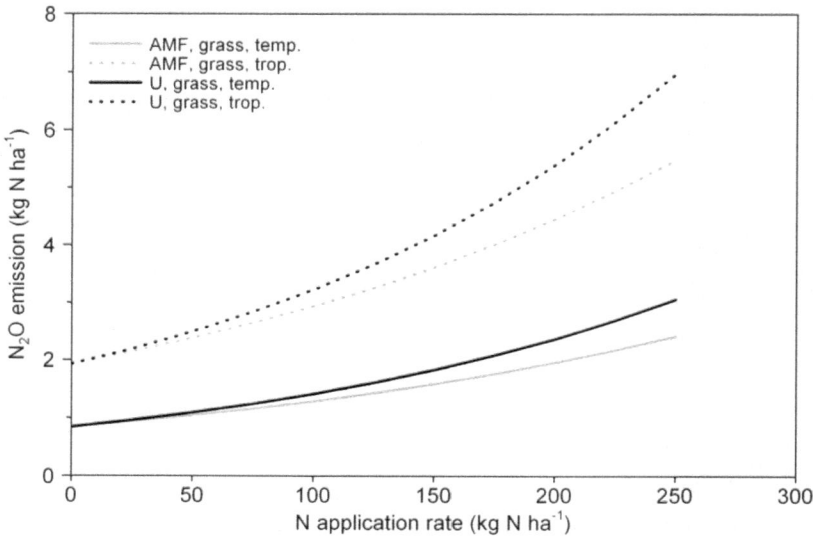

Figure 5.1 Example of calculations using the REML model for N_2O developed by Bouwman et al (2002b) for urea (U) and mixed animal manure and synthetic fertilizer (AMF) and varying N application rates for grass, fine soil texture, poor soil drainage, >3% soil organic C, pH 5.5–7.3, and temperate and tropical climates

Source: Bouwman et al (2002b)

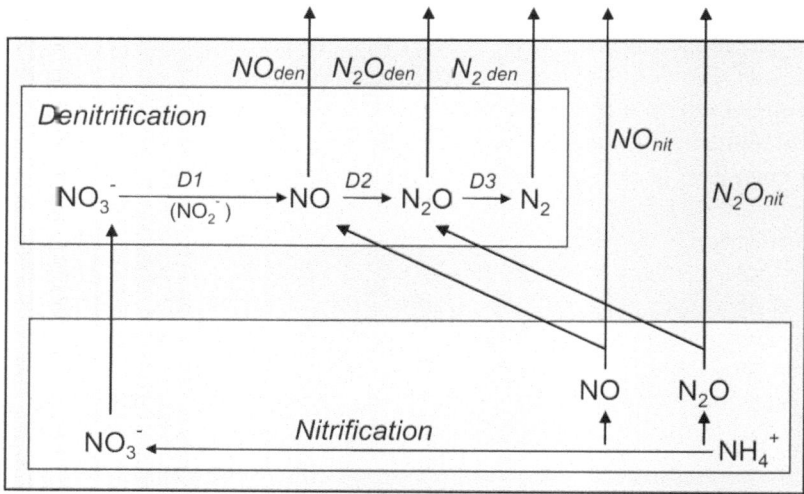

Figure 5.2 Scheme of the revised DAYCENT trace gas module

Note: D1, reduction of nitrate to NO; D2, reduction of NO to N_2O; D3, reduction of N_2O to N_2; N_{2den}, production of N_2 from denitrification; N_2O_{den}, production of N_2O from denitrification; N_2O_{nit}, production of N_2O from nitrification; NO_{den}, production of NO from denitrification; NO_{nit}, production of NO from nitrification.
Source: Stehfest (2005)

models for simulating N_2O emissions on a field to regional scale, other models have included algorithms to estimate N_2O emissions. Examples of these models are EPIC, LPJ (Xu-Ri and Prentice, 2008), DRAINMOD-N (Youssef et al, 2005), NLOSS (Riley and Matson, 2000), WNMM (Li et al, 2005, 2007), FASSET (Chatskikh et al, 2005), and CERES-NOE (Gabrielle et al, 2006). A comprehensive review of field- to regional-scale N_2O models has been carried out by Chen et al (2008).

Better performance of mechanistic models for N_2O emissions at these scales is not only hampered by insufficient understanding of the various processes controlling N_2O emissions, but also by lack of data on hydraulic conductivity and porosity, and their spatial heterogeneity, among other variables. As a result, the performance of the above mechanistic models remains rather poor, and uncertainty ranges are large, both for process-based models and for statistical approaches (Figure 5.3). The modelling efficiency or R^2 correlations with measurement data are often not reported or average 0.5 or less when data cover periods up to the crop growing season; agreement tends to be poorer when models are evaluated with daily measurements (Del Grosso et al, 2000, 2009; Li et al, 2005). For the EF approaches, the uncertainty range remains large, for example from -85 per cent to $+250$ per cent (Stehfest and Bouwman, 2006).

In addition to the direct N_2O emissions that occur in an agricultural field, all N leaching, NH_3 volatilization and NO_x emission from the field can eventually be converted into N_2O in groundwater and surface water, leading to indirect emission of N_2O. So far, there is no process-based modelling of indirect N_2O emissions, and all estimates are based on the emission factor approach as outlined by IPCC (2006) (Table 5.1). Starting from the amount of N added to the system as synthetic fertilizer input, as organic N input from manure spreading, dung and urine excretion on the field and crop residues, or from mineralization of soil organic matter following a land-use change, the fractions that are volatilized and leached and the subsequent N_2O production are given as generic factors. The N_2O emissions from groundwater and surface drainage are not based on N input, but rather on the relative concentrations of N_2O and dissolved inorganic N in the water. The EFs for indirect N_2O emissions are based on a rather small number of measurements with an uncertainty range that often exceeds one order of magnitude (IPCC, 2006).

Global nitrous oxide emission from arable systems

For estimating global N_2O emissions from arable fields, we use the 0.5 by 0.5 degree resolution global inventory of fertilizer-N use and manure management in arable systems for the year 2000 (Bouwman et al, 2006b; Beusen et al, 2008). This inventory is based on the land-cover data set described by Klein Goldewijk et al (2006), which includes agricultural areas for cropland and grassland that are consistent with statistical information from the Food and Agriculture Organization of the United Nations (FAO) (2008).

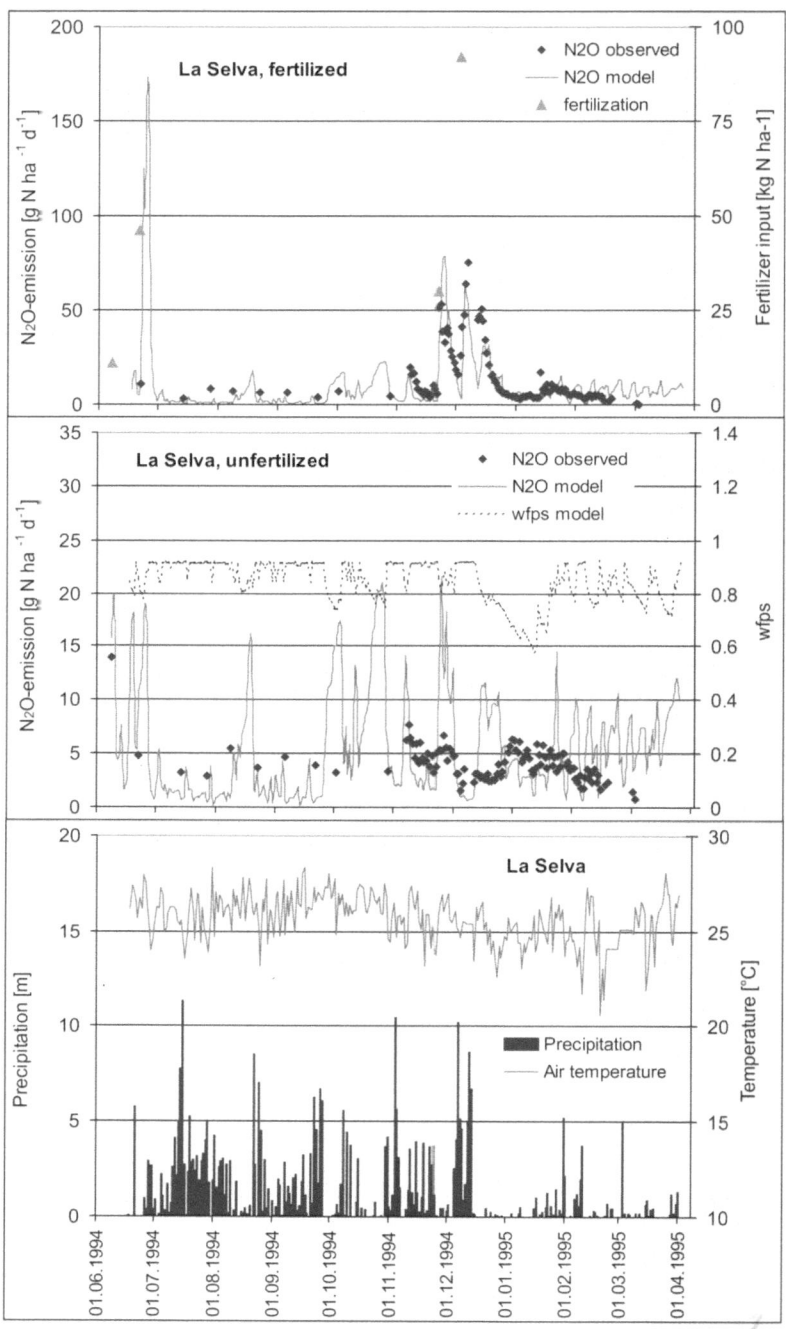

Figure 5.3 Fertilization and observed and simulated N$_2$O emission rates from fertilized and unfertilized maize cropping in La Selva, Costa Rica

Source: Based on field data from Weitz et al (2001)

We present here the N cycle in both livestock and crop production systems, because their interrelations are crucial for N input and N_2O emissions from agricultural land. Livestock production takes place in two production systems: (1) pastoral systems and (2) mixed and landless systems (Seré and Steinfeld, 1996; Bouwman et al, 2005). Mixed systems comprise both crop and livestock production, linked by animal feed and fodder, manure, etc. Landless ruminant production systems are included in mixed and landless systems, because they have the same interrelationships between crop and livestock production. Extensive grazing by ruminants is dominant in the vast areas of grassland in pastoral systems, where crop production is only a minor activity. Pork, poultry and eggs are only produced in mixed and landless systems. However, ruminant meat and milk can be produced in pastoral as well as in mixed and landless systems.

For the inventory, we used data on total fertilizer-N consumption per country for the year 2000 from FAO (2008), and data on N fertilizer use by crop from IFA/IFDC/FAO (2003). Data for animal stocks for ten animal categories (beef cattle, dairy cattle, buffalo, pigs, poultry, sheep and goats, along with the category of small ruminants, donkeys, mules, horses and camels) for the year 2000 were obtained from FAO (2008). Associated N excretion rates, specified for Western Europe, North America, the remaining industrialized countries and developing countries, were taken from van der Hoek (1998). Animal manure is distributed over different management systems, including grazing, animal houses and storage, and other uses (mainly fuel); stored manure is used for spreading on cropland and grassland (Figure 5.4). Details on the approach to distributing the fertilizer and manure N are provided by Bouwman et al (2006b) and Beusen et al (2008).

Direct N_2O emission from fertilizer-N application and spreading of animal manure is calculated according to the residual empirical maximum likelihood (REML) model of Bouwman et al (2002b). Ammonia volatilization rates for animal barns and grazing systems are taken from Bouwman et al (1997), and volatilization from spreading of animal manure and N fertilizers is calculated using the REML model of Bouwman et al (2002a).

For indirect emissions we use the EFs of IPCC (2006). Indirect emissions include emissions associated with NO_3^- leaching calculated according to Van Drecht et al (2003), and emissions associated with NH_3 emissions from croplands, from applied fertilizer-N and manure-N.

Approximately 65 per cent of 112.4Tg yr^{-1} N in manure is generated from mixed and landless systems, whereas 26 per cent comes from pastoral systems (Table 5.2). The remaining 10 per cent ends outside the agricultural system (fuel, building material, etc.). Within the mixed and landless systems, 65 per cent of the manure is collected in animal barns and storage systems, and 35 per cent is excreted in pastures. A large amount of manure-N (33.3Tg yr^{-1}) is collected in animal barns and storage systems in mixed and landless systems and applied mainly to croplands, with a smaller amount to pastoral systems (Table 5.2).

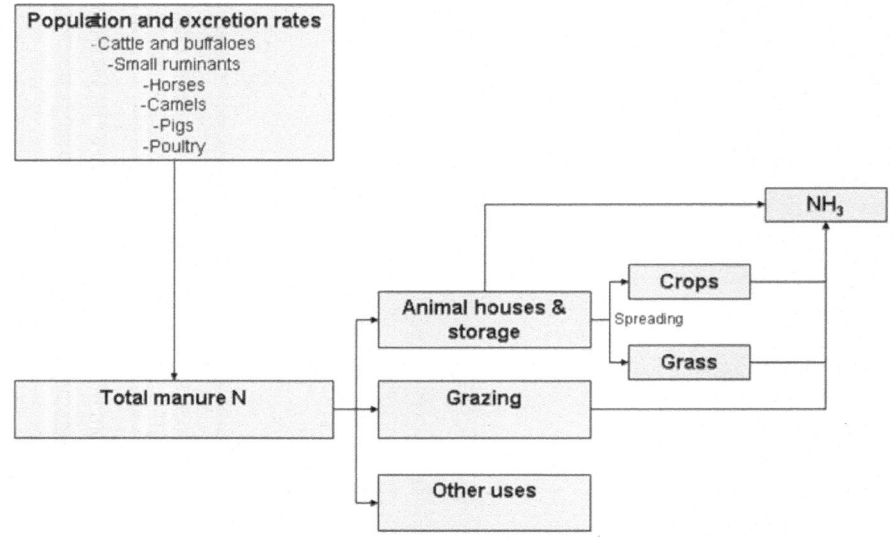

Figure 5.4 Distribution of animal manure over different systems
(pastoral and mixed systems) using a flow-path model approach

Note: This approach allows for assessing the impact of measures in one part of the N cascade on the next. For example, lower N excretion rates by improved feeding have an impact on the amount of manure-N in arable fields. Likewise, reducing NH₃ emission from animal houses and storage systems causes an increase in the availability of manure-N.
Source: Beusen et al (2008)

Total direct and indirect N_2O emissions from the annual 79.1Tg yr^{-1} of N fertilizer use and 33.3Tg yr^{-1} of manure-N spreading in global arable systems amount to 4Tg N_2O-N yr^{-1} for the standard inventory (Figure 5.5; Table 5.3). Using this model we calculated the effect of various mitigation options for reducing N_2O emissions discussed in the next section.

Mitigation options for nitrous oxide from arable systems

Optimizing fertilizer application with respect to nitrous oxide emission

Emissions of N_2O from agricultural systems will not decrease to zero when no N fertilizer is applied, but will remain at some background level, as in any non-agricultural system. For example, when 383 non-manured, non-N-fertilized cropping systems were evaluated for N_2O emissions, the average background emission was estimated at 0.55kg N_2O ha^{-1} (Helgason et al, 2005). Natural N deposition and biological fixation, and N mineralization from soil organic

Table 5.2 Global estimates of manure-N excretion in animal houses and storage systems and excretion during grazing, spreading of stored manure in cropland and grassland, and N fertilizer use in cropland and grassland for the year 2000

Agricultural system	Tg yr^{-1}	% of total
Animal manure management		
Mixed and landless systems	72.6	65
Housing and storage	47.1	42
Grazing	25.5	23
Pastoral systems	28.8	26
Housing and storage	1.2	1
Grazing	27.6	25
Outside agricultural system	11.0	10
Total	112.4	100
Spreading of manure (excluding NH$_3$ loss from animal houses and storage)		
Cropland in mixed systems	32.4	84
Cropland in pastoral systems	0.9	2
Grassland in mixed systems	5.3	14
Grassland in pastoral systems	0.0	0
Total	38.7	100
N fertilizer use		
Cropland	79.1	96
Grassland	3.4	4
Total	82.5	100
Biological N$_2$ fixation		
Cropland	21.9	

Source: Data for year 2000 using the Integrated Model for the Assessment of Global Environment (IMAGE) model framework, data and settings (Bouwman et al, 2006a).

matter, continue to drive the production of N$_2$O via nitrification or denitrification. In natural ecosystems, N$_2$O emissions are therefore commonly expressed on an area basis, not as a function of inputs.

Agricultural systems serve for the production of food, fibre, feed and biofuel crops. Production input factors, such as N fertilizers, are used to estimate N$_2$O emissions which can then be converted into GWP (IPCC, 2006; Stehfest and Bouwman, 2006). The conclusion could be drawn that reducing fertilizer input would be an adequate means of reducing N$_2$O emissions from agriculture. However, as agricultural activity will always remain essential for mankind to survive, a more revealing parameter is the N$_2$O emission per output factor such as grain yield, total N production or protein production. Evaluating conventional and no-tillage systems, Mosier et al (2006) calculated net global warming potential and greenhouse gas intensities in irrigated maize

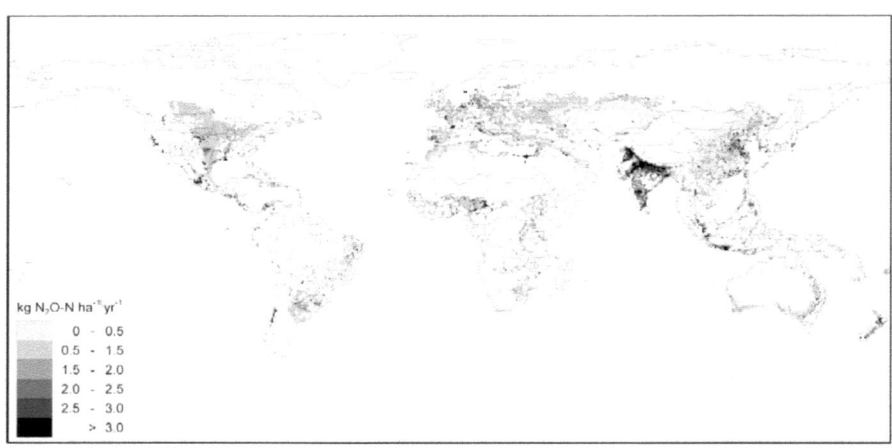

Figure 5.5 N$_2$O emission from global arable land

Source: Computed with the model of Bouwman et al (2002b) for the year 2000

Table 5.3 Direct and indirect N$_2$O emission from global arable systems for the standard 2000 situation, and for changing management in livestock and crop production systems

Case	N$_2$O-N emission (% change relative to standard)		
	Direct	Indirect Gg yr^{-1}	Total
Standard[1]	3582	442	4023
Reduce NH$_3$ loss from stored manure[2]	3599 (0%)	452 (+2%)	4051 (+1%)
Reduce grazing[3]	3706 (+3%)	483 (+9%)	4189 (+4%)
Substitute fertilizer by manure[4]	3191 (−11%)	301 (−32%)	3492 (−13%)
Improved animal diets[5]	3426 (−4%)	398 (−10%)	3825 (−5%)
Nitrification inhibitors[6]	−50%		

Note: [1] Standard is data for year 2000 using the model of Bouwman et al (2002b) and the Integrated Model for the Assessment of Global Environment (IMAGE) model framework, data and settings (Bouwman et al, 2006a). [2] As standard, with reduction of NH$_3$ loss from animal houses and storage by 20 per cent (from 20 to 16 per cent) leading to an increase of manure-N available for cropland of 33.3 to 34.9Tg yr^{-1}. [3] As standard, with reduction by 20 per cent of the time spent by ruminants in the meadow. This leads to an increase in the amount of manure collected in animal houses and storage systems, and an increase of manure-N available for cropland from 33.3 to 40Tg yr^{-1}. [4] As standard, with substitution of fertilizer-N by manure assuming effectiveness of manure of 60 per cent resulting in a reduction of N fertilizer use by 33 per cent (from 82 to 56Tg yr^{-1}). [5] As standard, with improved animal diets by 20 per cent (20 per cent less excretion), leading to a reduction of the total animal manure excretion from 112.4 to 89.9Tg yr^{-1}, and manure-N application to arable land from 33.3 to 26.2Tg yr^{-1}. [6] The impact of nitrification inhibitors is assumed to be a reduction of 50 per cent. However, there may also be an effect of nitrification inhibitors on NH$_3$ emission, crop growth and the surface N balance through N uptake, and leaching. We therefore do not present an estimate for the N$_2$O emission.

cropping systems and introduced the concept of greenhouse gas intensity (GHGI). The GHGI is obtained by dividing GWP by grain yield. A positive GHGI value indicates a net source of CO_2 equivalents per kg grain yield; a negative GHGI value indicates a net sink of greenhouse gas. Once emissions per unit of yield are known, adjustments to management practices can be made to optimize the balance between economic viability and environmental conservation, through selection of appropriate levels of input such as fertilizer-N.

The concept of linking GHG emissions to productivity can be expanded and link the amount of N_2O emitted to a unit of production. In this context, the following three relationships are determining factors controlling N_2O emission per unit yield for any given agricultural production system:

1 the amount of N fertilizer applied versus the amount of N_2O emitted;
2 the amount of N fertilizer applied versus N accumulation in the crop which affects fertilizer use efficiency; and
3 the amount of N in the crop versus yield.

Using a meta-analysis approach, Van Groenigen et al (2010) related the amount of N_2O emitted to the amount of N accumulated in the crop. The meta-analysis included 22 field experiments encompassing 188 individual observations that reported N_2O emissions and N accumulation in annual crops and grassland. To allow for a comparison between the different crops in the various studies, N_2O emissions were related to total N accumulation in the crop rather than crop yield. As there is a strong linear relationship between total crop N and grain yield (Cassman et al, 2003), N_2O emissions expressed per kg crop N would strongly correlate to N_2O emitted per unit of grain yield. Van Groenigen et al (2010) estimate that on average 16.7g N_2O kg^{-1} crop N is emitted when no or low rates of fertilizer-N were applied. The lowest relative N_2O emissions occurred, i.e. 6.1g N_2O kg^{-1} crop N, when on average 233kg N ha^{-1} of fertilizer-N was applied. Emissions increased again to 21.4g N_2O kg^{-1} crop N when the rate of fertilizer-N input increased to >300kg N ha^{-1}. The higher emissions of N_2O at zero N fertilizer input is caused by the combination of background N_2O emissions (i.e. without any N fertilizer input) and reduced yield/total crop N when a deficiency of available N is limiting crop growth. Higher N_2O emissions per kg crop N at high rates of N fertilizer input are expected because increased N fertilizer input leads to a diminishing return in yield and crop N per unit of N input, i.e. to a low fertilizer use efficiency (see below). Furthermore, rates of N fertilizer input exceeding crop N demand lead to a non-linear increase in the rates of N_2O emissions (Bouwman et al, 2002b).

The decrease in N_2O emissions per kg crop N when fertilizer-N input increased from zero or low levels to median levels has agronomic significance. Management practices could be developed toward maximizing yield produced with the lowest amount of N_2O emitted per unit of yield. Based on the result of this meta-analysis, both low and high fertilizer-N input would not lead to low N_2O emissions per unit of crop N produced. Median and near-optimum levels of fertilizer-N input in association with management practices that optimize the yield potential and increase fertilizer-N use efficiency would lead to a reduction in N_2O emissions as expressed per unit of crop N accumulated or grain yield.

Increasing fertilizer use

Currently, fertilizer-N use efficiency by agricultural crops is estimated to be approximately 50 per cent (Smil, 1999; Howarth et al, 2002; Ladha et al, 2005). In other words, on average half of the amount of the fertilizer-N applied cannot be recovered in the crop or in the soil and has to be considered to have been lost from the cropping system. Denitrification (N_2O and N_2), volatilization (NH_3) and leaching (NO_3) are the major pathways of N losses (Schlesinger, 1997; Mosier et al, 2001), causing a cascade of environmental and human health problems (Galloway et al, 2003). Therefore there is the potential for major improvements in N use efficiency through adopting fertilizer, soil water and crop management practices that focus on maximizing crop N uptake, with minimum N fertilizer losses and the optimum use of indigenous soil N (Ladha et al, 2005). Through optimizing N use efficiency, a total reduction in N_2O emissions will be achieved as well as a further reduction in N_2O emissions per kg yield or per kg crop N.

Improving fertilizer-N efficiency can be achieved through various management practices, which are often part of so-called best management practices (BMPs). BMPs try to reduce N input without a reduction in yield (McSwiney and Robertson, 2005; Adviento-Borbe et al, 2007). In intensively managed maize (*Zea mays*) cropping systems, using BMPs such as increasing planting densities, splitting N fertilizer applications and using higher N fertilizer rates, grain yields increased whereas emissions of N_2O did not increase even though fertilizer-N input had increased by 40 to 92 per cent compared to conventional cropping systems (Adviento-Borbe et al, 2007).

Improved fertilizer-N use efficiency can also be obtained by using site-specific N management practices. Here the application of synthetic fertilizer-N is matched with the natural variability of available soil N and the specific demand of N by the crop at that particular site that is required to achieve maximum yield potential (Raun and Schepers, 2008). The amount of fertilizer-N to apply during the growing season, adjusted for each m^2, can be based on optical sensors linked to an in-season estimate of yield. Using this technology, an improvement of 15 per cent in fertilizer-N use efficiency and grain yield of wheat was obtained compared to conventional management practices (Raun et al, 2002). Slow-release fertilizers allow for better matching of the supply with the crop demand during the growing season (Rao, 1987).

Management practices aimed at reducing NH_3 volatilization and leaching can also help to increase fertilizer use efficiency. Fertilizer application timing and mode influence the NH_3 volatilization and the efficiency of plant uptake, hence the availability of N for nitrification and denitrification. Timing and matching the N application with plant needs is important, because any prolongation of the period in which NH_4^+-based fertilizers can leach or undergo nitrification or NO_3^--based fertilizers can be denitrified, without competition from plant uptake, is likely to increase emissions of NO and N_2O (Ortiz-Monasterio et al, 1996; Smith et al, 1997; Chantigny et al, 1998).

However, some of these options may have an impact on processes further downstream along the N cascade. For example, subsurface application or injection of N fertilizers, to reduce NH_3 volatilization, leads to higher N_2O emission and leaching than broadcasting synthetic fertilizers and animal manure (Ellis et al, 1998; Kessavalou et al, 1998; Smith et al, 1998; Flessa and Beese, 2000).

Incorporation of biological nitrogen fixation in agricultural systems

An argument might be made that replacing synthetic fertilizer-N with biologically fixed N_2 will lead to lower synthetic N input and possibly lower N_2O emissions. Increases in biological fixed N in agro-ecosystems can be implemented by including cover or green manure crops in the rotation or by replacing non-N_2-fixing crops such as cereals with legumes. However, at the global scale, an overall increase in biologically fixed N_2 will also lead to an increase in N_2O emissions (Mosier et al, 1998; Sorai et al, 2007). For example, by including legumes in a pasture, N_2O emissions were estimated to increase by a factor of two to three (Duxbury et al, 1982). Like inorganic N fertilizer, N fixed by leguminous crops is converted by the plant into organic N that can be mineralized and nitrified following decomposition of legume residues, and subsequently denitrified. Biologically and synthetically fixed N are indistinguishable once in organic form and equally become potential sources for N_2O emissions. Although the actual process of biological N_2 fixation probably does not lead to a significant increase in N_2O emission, the release of N from root exudates during the growing season and the decomposition of legume residues will lead to an increase in N_2O emissions (Rochette and Janzen, 2005).

Better integration of animal manure or human nitrogen and phosphorus in the agricultural system

Better integration of animal manure in crop production systems, and recycling of N in human excreta and compost are further ways to reduce N fertilizer use. In practice only a fraction of this potential reduction in N fertilizer use is realized, in part because of the segregation of crop and livestock production systems and the lack of economic incentives for recycling. The concept of ecological sanitation aiming at closure of local material flow cycles (Langergraber and Muelleggera, 2005) could be a way to recycle human excreta, but is only applied in a few countries. Janssen and Oenema (2008) suggested that globally a large part (34–68 per cent) of synthetic N fertilizer could be substituted with manure-N. Replacing fertilizer-N with manure is not a 1:1 exchange, since they differ in plant N availability. Also, the P-to-N ratio in manure is higher than the optimum ratio for crops, and may therefore result in the accumulation of P in soil. Compound animal feed often contains additions of copper and zinc; as a result, accumulation of these metals has occurred in some countries (see references in Janssen and Oenema, 2008).

Here, we have calculated the global potential for recycling of animal manure, by replacing N fertilizer with stored manure in the inventory discussed above. Since the N in animal manure is partly present in organic forms, we assume that 60 per cent is effectively available for plant uptake. The remainder is lost through NH_3 volatilization or becomes part of the soil organic matter pool. Soil organic matter decomposes gradually and the N mineralized is lost through leaching and denitrification (Janssen and Oenema, 2008). Replacing synthetic N fertilizer with manure-N results in a reduction of global N fertilizer use by 33 per cent (from 82 to 56Tg yr^{-1}), and a reduction of N_2O emissions by 11 per cent (Table 5.3). Although N fertilizer use is reduced and replaced by animal manure-N, the emissions are not reduced proportionally. This is related to the background emissions, which are not reduced.

Improving animal diets

Changing animal diets can also lead to a reduction in N input, leading to lower rates of denitrification. Zebarth et al (1999) constructed a large-scale N budget for an area of approximately 70,000ha heavily used for animal husbandry including the production of feed. They found that through improved manure storage and fertilizer management practices, in combination with an improved animal diet by removing surplus dietary crude protein and balancing protein, carbohydrates and amino acids in the diet leading to a reduction in N in the excreta of between 20 and 25 per cent, the total N input across the region could be reduced by 17 per cent and the N surplus by 24 per cent. Moreover, it was estimated that the improved animal diet would reduce denitrification by 22 per cent compared to the reference scenario. If we apply improved animal diets by assuming a 20 per cent reduction of the N in animal excreta, manure N application to arable land falls from 33.3 to 26.2Tg yr^{-1} (Table 5.3). This causes a reduction of the global N_2O emission from arable land of 4 per cent.

Use of nitrification inhibitors

Nitrification inhibitors are often used to increase N fertilizer use efficiency and decrease NO_3^- losses by leaching and denitrification. There is increasing interest in the use of nitrification inhibitors that effectively block nitrification when applied to soils in conjunction with N fertilizers. In recent years, substantial progress has been made towards understanding the biochemistry of autotrophic NH_3 oxidation and the possible modes of action observed with the various inhibitors.

Nitrification inhibitors can effectively reduce N_2O emission, and yield is generally higher when inhibitors are applied with N fertilizer. For example, a meta-evaluation of the effectiveness of nitrapyrin (NP) in the mid-West of the US showed that, on average, NP combined with N fertilizer led to higher crop yield (7 per cent), increased soil N retention (28 per cent), less nitrate leaching (-16 per cent) and 51 per cent lower N_2O emissions compared to fields receiving N fertilizer without NP (Wolt, 2004). Alternative nitrification inhibitors such as DCD (dicyandiamide), ATS (ammonium thiosulphate) and DMPP (3,4-dimethyl pyrazole phosphate) can be expected to have a similar effect on N_2O emission. Research conducted recently in New Zealand showed that strategic application of a nitrification inhibitor to pasture, following simulated urine excretion by grazing dairy cattle, significantly reduced direct (-67 per cent) and indirect (-74 per cent) N_2O emissions over extensive periods (Kelliher et al, 2007). However, the performance of any of these nitrification inhibitors depends on physico-chemical properties, efficacy and persistence in various environments and management practices. Their use and performance in livestock systems are discussed in more detail in Chapter 6.

Summary and conclusions

Approaches estimating direct N_2O emissions range from using simple emission factors to statistical and mechanistic models. EFs are based on the relationship between N inputs and N_2O emission. Statistical models use different factors that influence N_2O emissions and are appropriate to depict spatial variability and differences due to climate, soil and management conditions. Finally, mechanistic models are designed to operate at the field to regional scales, and in general only use parameters that are also available at these larger scales. Improvement in the performance of N_2O simulation models at larger scales is not only hampered by insufficient understanding of the various processes, but also by a lack of available data on controlling variables such as hydraulic conductivity and porosity, and their inherent spatial heterogeneity.

A simple decrease in fertilizer-N input would lead to a concurrent reduction of N_2O emissions. Depending on the size of the decrease in fertilizer-N input, a dilemma will manifest itself as lower fertilizer-N input will have a negative effect on grain yield. A positive relationship between the amount of N fertilizer input and yield is well established (Cassman et al, 2003). Such a positive relationship between N fertilizer input and yield is also reflected at the global level as worldwide increases in the use of synthetic fertilizer-N have led to a significant increase in total yield and food production (FAO, 2008). Therefore, a significant reduction in N_2O emissions through a reduction in fertilizer-N input will most likely lead to lower food production. However, with an expected increase in world population to 8.9 billion in 2075 (Lutz et al, 2008), sustaining or increasing food production will become a necessity.

Simply reducing fertilizer use is therefore not an option, except in regions where there is over-fertilization, i.e. where reducing N inputs has no effect on crop yields. Generally, optimum and near optimum levels of fertilizer-N input in association with best management practices to optimize the yield potential and increase fertilizer-N use efficiency would lead to a reduction in N_2O emissions as expressed per unit of grain or food yield. Management practices can be used that increase the crop uptake of N directly, for example plant breeding, matching (and timing of) N supply with crop demand, etc. Other practices aim at reducing NH_3 volatilization, denitrification or leaching losses.

Furthermore, fertilizer-N can be replaced by N inputs in the crop rotation from biological N_2 fixation by leguminous crops such as soya beans, pulses, alfalfa, etc. However, generally N_2O emissions from agricultural systems including leguminous crops are comparable to those systems that rely on synthetic N fertilizer as their source of N. The option of including legumes in the system will reduce the need for synthetic N fertilizer but will not automatically lead to a reduction in N_2O emissions.

N fertilizer can also be replaced by better integrating animal manure in the arable production system, or by recycling human N and P. Our simple calculation shows that a considerable 13 per cent reduction of N_2O emission from the global arable land is possible by replacing fertilizer by animal manure (Table 5.3). The feasibility of this option depends on the spatial separation of livestock and crop production activities. Finally, measures in the production process in the livestock sector may also have an impact on N_2O emissions from cropland. Improving animal diets aimed at improving the N use efficiency by animals and thus reducing the N in animal manure (by 20 per cent) may lead to important reductions (5 per cent) in N_2O from arable land. Finally, the use of nitrification inhibitors is an effective way to mitigate perhaps 50 per cent of N_2O emissions from arable fields. However, these inhibitors are not used on a large scale yet.

In summary, there is no simple and generic approach for reducing N_2O emissions from arable land. Therefore, mitigation strategies need to be adapted to the local socio-economic situation, climate and soil conditions, the level of production and fertilizer input, and integration of livestock and crop production systems. Mitigation strategies should also account for the N-cascade effect, where a reduction in emissions for one part of the system may actually lead to an increase in emissions in another part of the N cascade.

References

Adviento-Borbe, M. A. A., Haddix, M. L., Binder, D. L., Walters, D. T. and Dobermann, A. (2007) 'Soil greenhouse gas fluxes and global warming potential in four high-yielding maize systems', *Global Change Biology*, vol 13, pp1972–1988

Beusen, A. H. W., Bouwman, A. F., Heuberger, P. S. C., Van Drecht, G. and Van Der Hoek, K. W. (2008) 'Bottom-up uncertainty estimates of global ammonia emissions from global agricultural production systems', *Atmospheric Environment*, vol 42, pp6067–6077

Bouwman, A. F. (1996) 'Direct emission of nitrous oxide from agricultural soils', *Nutrient Cycling in Agroecosystems*, vol 46, pp53–70

Bouwman, A. F., Lee, D. S., Asman, W. A. H., Dentener, F. J., Van Der Hoek, K. W. and Olivier, J. G. J. (1997) 'A global high-resolution emission inventory for ammonia', *Global Biogeochemical Cycles*, vol 11, pp561–587

Bouwman, A. F., Boumans, L. J. M. and Batjes, N. H. (2002a) 'Estimation of global NH_3 volatilization loss from synthetic fertilizers and animal manure applied to arable lands and grasslands', *Global Biogeochemical Cycles*, vol 16, 1024, doi:10.1029/2000GB001389

Bouwman, A. F., Boumans, L. J. M. and Batjes, N. H. (2002b) 'Modeling global annual N_2O and NO emissions from fertilized fields', *Global Biogeochemical Cycles*, vol 16, 1080 doi:10.1029/2001GB001812

Bouwman, A. F., Van Der Hoek, K. W., Eickhout, B. and Soenario, I. (2005) 'Exploring changes in world ruminant production systems', *Agricultural Systems*, vol 84, pp121–153

Bouwman, A. F., Kram, T. and Klein Goldewijk, K. (eds) (2006a) *Integrated Modelling of Global Environmental Change: An Overview of IMAGE 2.4*, Bilthoven, Publication 500110002/2006, Netherlands Environmental Assessment Agency, Bilthoven

Bouwman, A. F., Van Der Hoek, K. W. and Van Drecht, G. (2006b) 'Modelling livestock-crop-land use interactions in global agricultural production systems', in A. F. Bouwman, T. Kram, and K. Klein Goldewijk (eds) *Integrated Modelling of Global Environmental Change*, Netherlands Environmental Assessment Agency, Bilthoven, pp77–92

Bremner, J. M. (1997) 'Sources of nitrous oxide in soils', *Nutrient Cycling in Agroecosystems*, vol 49, pp7–16

Cassman, K., Dobermann, A., Walters, D. T. and Yang, H. (2003) 'Meeting cereal demand while protecting natural resources and improving environmental quality', *Annual Review of Environmental Resources*, vol 28, pp315–358

Chantigny, M. H., Prevost, D., Angers, D. A., Simard, R. R. and Chalifour, F. P. (1998) 'Nitrous oxide production in soils cropped to corn with varying N fertilization', *Canadian Journal of Soil Science*, vol 78, pp589–596

Chatskikh, D., Olesen, J. E., Berntsen, J., Regina, K. and Yamulki, S. (2005) 'Simulation of effects of soils, climate and management on N_2O emission from grasslands', *Biogeochemistry*, vol 76, pp395–419

Chen, D., Li, Y., Grace, P. and Mosier, A. R. (2008) 'N_2O emissions from agricultural lands: A synthesis of simulation approaches', *Plant and Soil*, vol 309, pp169–189

Crutzen, P. J., Mosier, A. R., Smith, K. A. and Winiwarter, W. (2008) 'N₂O release from agro-biofuel production negates global warming reduction by replacing fossil fuels', *Atmospheric Chemistry and Physics*, vol 8, pp389–395

Del Grosso, S. J., Parton, W. J., Mosier, A. R., Ojima, D. S., Kulmala, A. E. and Phongpan, S. (2000) 'General model for N₂O and N₂ gas emissions from soils due to denitrification', *Global Biogeochemical Cycles*, vol 14, pp1045–1060

Del Grosso, S. J., Ojima, D. S., Parton, W. J., Stehfest, E., Heistemann, M., Deangelo, B. and Rose, S. (2009) 'Global scale DAYCENT model analysis of greenhouse gas emissions and mitigation strategies for cropped soils', *Global and Planetary Change*, vol 67, pp44–50

Duxbury, J. M., Bouldin, D. R., Terry, R. E. and Tate, R. L. (1982) 'Emissions of nitrous oxide from soils', *Nature*, vol 298, pp462–464

Ellis, S., Yamulki, S., Dixon, E., Harrison, R. and Jarvis, S. C. (1998) 'Denitrification and N₂O emissions from a UK pasture soil following early spring application of cattle slurry and mineral fertiliser', *Plant and Soil*, vol 202, pp15–25

FAO (Food and Agriculture Organization of the United Nations) (2008) 'FAOSTAT database collections', Food and Agriculture Organization of the United Nations, Rome, http://faostat.fao.org/default.aspx

Farquharson, R. and Baldock, J. (2008) 'Concepts in modelling N₂O emissions from land use', *Plant and Soil*, vol 309, pp147–67

Firestone, M. K. and Davidson, E. A. (1989) 'Microbiological basis for NO and N₂O production and consumption in soils', in M. O. Andreae and D. S. Schimel (eds) *Exchange of Trace Gases between Terrestrial Ecosystems and the Atmosphere*, Wiley, Chichester, pp7–21

Flessa, H. and Beese, F. (2000) 'Laboratory estimates of trace gas emissions following surface application and injection of cattle slurry', *Journal of Environmental Quality*, vol 29, pp262–268

Forster, P., Ramaswamy, V., Artaxo, P., Berntsen, T., Betts, R., Fahey, D. W., Haywood, J., Lean, J., Lowe, D. C., Myhre, G., Nganga, J., Prinn, R., Raga, G., Schulz, M. and Van Dorland, R. (2007) 'Changes in atmospheric constituents and in radiative forcing', in S. Solomon, D. Qin, M. Manning, Z. Chen, M. Marquis, K. B. Averyt, M. Tignor, M. and H. L. Miller (eds) *Climate Change 2007: The Physical Science Basis. Contribution of Working Group I to the Fourth Assessment Report of the Intergovernmental Panel on Climate Change*, Cambridge University Press, Cambridge, pp129–234

Freibauer, A. and Kaltschmitt, M. (2003) 'Controls and models for estimating direct nitrous oxide emissions from temperate and sub-boreal agricultural mineral soils in Europe', *Biogeochemistry*, vol 63, pp93–115

Gabrielle, B., Laville, P., Hénault, C., Nicoullaud, B. and Germon, J. C. (2006) 'Simulation of nitrous oxide emissions from wheat-cropped soils using CERES', *Nutrient Cycling in Agroecosystems*, vol 74, pp133–146

Galloway J. N., Aber, J. D., Erisman, J. W., Seitzinger, S., Howarth, R. W., Cowling, E. B. and Cosby, B. J. (2003) 'The nitrogen cascade', *BioScience*, vol 53, pp341–356

Helgason B., Janzen, H., Chantigny, M., Drury, C., Ellert, B., Gregorich, E., Lemke, R., Pattey, E., Rochette, P. and Wagner-Riddle, C. (2005) 'Toward improved coefficients for predicting direct N₂O emissions from soil in Canadian agroecosystems', *Nutrient Cycling in Agroecosystems*, vol 72, pp87–99

Howarth, R. W., Boyer, E. W., Pabich, W. J. and Galloway, J. N. (2002) 'Nitrogen use in the United States from 1961–2000 and potential future trends', *Ambio*, vol 31, pp88–96

IFA/IFDC/FAO (2003) *Fertilizer Use by Crop*, 5th Edition, Food and Agriculture Organization of the United Nations, Rome

IPCC (Intergovernmental Panel on Climate Change) (2001) *Third Assessment Report. Working Group I*, Cambridge University Press, Cambridge

IPCC (2006) *2006 IPCC Guidelines for National Greenhouse Gas Inventories*, Institute for Global Environmental Strategies (IGES), Hayama, Japan

Janssen, B. H. and Oenema, O. (2008) 'Global economics of nutrient cycling', *Turkish Journal of Agriculture and Forestry*, vol 32, pp165–176

Kaiser, E. A. and Ruser, R. (2000) 'Nitrous oxide emissions from arable soils in Germany: An evaluation of six long-term field experiments', *Journal of Plant Nutrition and Soil Science (Zeitschrift für Planzenernährung und Bodenkunde)*, vol 163, pp249–259

Kelliher, F. M., Clough, T. J. and Clark, H. (2007) 'Developing revised emission factors for nitrous oxide emissions from agricultural pasture treated with nitrification inhibitors', report prepared by Landcare Research New Zealand Limited, Lincoln University and AgResearch for the New Zealand Ministry of Agriculture and Forestry, Wellington

Kessavalou, A., Doran, J. W., Mosier, A. R. and Drijber, R. A. (1998) 'Greenhouse gas fluxes following tillage and wetting in a wheat-fallow cropping system', *Journal of Environmental Quality*, vol 27, pp1105–1116

Klein Goldewijk, K., Van Drecht, G. and Bouwman, A. F. (2006) 'Contemporary global cropland and grassland distributions on a 5 by 5 minute resolution', *Journal of Land Use Science*, vol 2, pp167–190

Ladha, J. K., Pathal, H., Krupnik, T. J., Six, J. and Van Kessel, C. (2005) 'Efficiency of fertilizer nitrogen in cereal production: Retrospects and prospects', *Advances in Agronomy*, vol 87, pp85–156

Langergraber, G. and Muelleggera, E. (2005) 'Ecological sanitation – a way to solve global sanitation problems?', *Environment International*, vol 31, pp433–444

Li, C. and Aber, J. (2000) 'A process-oriented model of N_2O and NO emissions from forest soils: I. Model development', *Journal of Geophysical Research*, vol 105, pp4369–4384

Li, Y., Chen, D., Zhang, Y., Edis, R. and Ding, H. (2005) 'Comparison of three modeling approaches for simulating denitrification and nitrous oxide emissions from loam-textured arable soils', *Global Biogeochemical Cycles*, vol 19, GB3002

Li, Y., White, R., Chen, D., Zhang, J., Li, B., Zhang, Y., Huang, Y. and Edis, R. (2007) 'A spatially referenced water and nitrogen management model (WNMM) for (irrigated) intensive cropping systems in the North China Plain', *Ecological Modeling*, vol 203, pp395–423

Lutz, W., Sanderson, W. and Scherbov, S. (2008) 'The coming acceleration of global population ageing', *Nature*, vol 451, pp716–719

McKenney, D. J. and Drury, C. F. (1997) 'Nitric oxide production in agricultural soils', *Global Change Biology*, vol 3, pp317–326

McSwiney, C. P. and Robertson, G. P. (2005) 'Nonlinear response of N_2O flux to incremental fertilizer addition in a continuous maize (*Zea mays* L.) cropping system', *Global Change Biology*, vol 11, pp1712–1719

Mosier, A. R., Kroeze, C., Nevison, C., Oenema, O., Seitzinger, S. and Cleemput, O. V. (1998) 'Closing the global atmospheric N_2O budget: Nitrous oxide emissions through the agricultural nitrogen cycle', *Nutrient Cycling in Agroecosystems*, vol 52, pp225–248

Mosier, A., Bleken, M. A., Chaiwanakupt, P., Ellis, E., Freney, J. R., Howarth, R., Matson, P., Minami, K., Naylor, R., Weeks, K. and Zhu, Z.-L. (2001) 'Policy implications of human-accelerated nitrogen cycling', *Biogeochemistry*, vol 52, pp281–320

Mosier, A. R., Halvorson, A. D., Reule, C. A. and Liu, X. J. (2006) 'Net global warming potential and greenhouse gas intensity in irrigated cropping systems in Northeastern Colorado', *Journal of Environmental Quality*, vol 35, pp1584–1598

Neff, J. C., Keller, M., Holland, E. A., Weitz, A. W. and Veldkamp, E. (1995) 'Fluxes of nitric oxide from soils following the clearing and burning of a secondary tropical rainforest', *Journal of Geophysical Research*, vol 100, pp25913–25922

Ortiz-Monasterio, J. I., Matson, P. A., Panek, J. and Naylor, R. L. (1996) 'Nitrogen fertilizer management: Consequences for N_2O and NO emissions in Mexican irrigated wheat', *Proceedings of 9th Nitrogen Workshop*, Technische Universität Braunschweig, Braunschweig, Germany

Parton, W. J., Mosier, A. R., Ojima, D. S., Valente, D. W., Weier, K. and Kulmala, A. E. (1996) 'Generalized model for N_2 and N_2O production from nitrification and denitrification', *Global Biogeochemical Cycles*, vol 10, pp401–412

Parton, W. J., Holland, E. A., Del Grosso, S. J., Hartman, M. D., Martin, R. E., Mosier, A. R., Ojima, D. S. and Schimel, D. S. (2001) 'Generalized model for NO_x and N_2O emissions from soils', *Journal of Geophysical Research*, vol 106, pp17403–17420

Rao, D. L. N. (1987) 'Slow-release urea fertilizers – effect on floodwater chemistry, ammonia volatilization and rice growth in an alkali soil', *Fertilizer Research*, vol 13, pp209–221

Raun, W. R. and Schepers, J. S. (2008) 'Nitrogen management for improved efficiency', in J. S. Schepers and W. R. Raun (eds) *Nitrogen in Agricultural Systems*, American Society of Agronomy, Madison, WI, pp675–694

Raun, W. R., Solie, J. B., Johnson, G. V., Stone, M. L., Mullen, R. W., Freeman, K. W., Thomason, W. E. and Lukina, E. V. (2002) 'Improving nitrogen use efficiency in cereal grain production with optical sensing and variable rate application', *Agronomy Journal*, vol 94, pp815–820

Riley, W. J. and Matson, P. A. (2000) 'NLOSS: A mechanistic model of denitrified N_2O and N evolution from soil', *Soil Science*, vol 165, pp237–249

Rochette, P. and Janzen, H. (2005) 'Towards a revised coefficient for estimating N_2O emissions from legumes', *Nutrient Cycling in Agroecosystems*, vol 73, pp171–179

Schlesinger, W. H. (1997) *Biogeochemistry: An Analysis of Global Change*, 2nd edition, Academic Press, San Diego, CA

Seré, C. and Steinfeld, H. (1996) *World Livestock Production Systems: Current Status, Issues and Trends*, Food and Agriculture Organization of the United Nations, Rome

Smil, V. (1999) 'Nitrogen in crop production: An account of global flows', *Global Biogeochemical Cycles*, vol 13, pp647–662

Smith, K. A., McTaggart, I. P. and Tsuruta, H. (1997) 'Emissions of N_2O and NO associated with nitrogen fertilization in intensive agriculture, and the potential for mitigation', *Soil Use and Management*, vol 13, pp296–304

Smith, K. A., McTaggart, I. P., Dobbie, K. E. and Conen, F. (1998) 'Emissions of N_2O from Scottish agricultural soils, as a function of fertilizer N', *Nutrient Cycling in Agroecosystems*, vol 52, pp123–130

Sorai, M., Yoshida, N. and Ishikawa, M. (2007) 'Biogeochemical simulation of nitrous oxide cycle based on the major nitrogen processes', *Journal of Geophysical Research*, vol 112, G01006, doi:10.1029/2005JG000109

Stehfest, E. (2005) 'Modelling of global crop production and resulting N_2O emissions', PhD thesis, Kassel, Germany, International Max Planck Research School on Earth System Modelling, Center for Environmental Systems Research (www.earthsystemschool.de/fileadmin/user_upload/Documents/Theses/Thesis_Stehfest.pdf).

Stehfest, E. and Bouwman, A. F. (2006) 'N_2O and NO emission from agricultural fields and soils under natural vegetation: Summarizing available measurement data and modeling of global annual emissions', *Nutrient Cycling in Agroecosystems*, vol 74, pp207–228

Tiedje, J. M. (1988) 'Ecology of denitrification and dissimilatory nitrate reduction to ammonium', in A. J. B. Zehnder (ed) *Biology of Anaerobic Microorganisms*, Wiley and Sons, New York, pp179–244

Van Der Hoek, K. W. (1998) 'Nitrogen efficiency in global animal production', in K. W. van Der Hoek, J. W. Erisman, S. Smeulders, J. R. Wisniewski and J. Wisniewski (eds) *Nitrogen, the Confer-N-s*, Elsevier, Amsterdam, pp127–132

Van Drecht, G., Bouwman, A. F., Knoop, J. M., Beusen, A. H. W. and Meinardi, C. R. (2003) 'Global modeling of the fate of nitrogen from point and nonpoint sources in soils, groundwater and surface water', *Global Biogeochemical Cycles*, vol 17, art. no 1115, doi:10.129/2003GB002060

Van Groenigen, J. W., Velthof, G., Oenema, O., Van Groenigen, K. J. and Van Kessel, C. (2010, in press) 'Toward an agronomic assessment of N_2O emissions: A case study for arable crops', *European Journal of Soil Science*

Veldkamp, E. and Keller, M. (1997) 'Nitrogen oxide emissions from a banana plantation in the humid tropics', *Journal of Geophysical Research*, vol 102, pp15,889–15,898

Weitz, A. M., Linder, E., Frolking, S., Crill, P. M. and Keller, M. (2001) 'N_2O emissions from humid tropical agricultural soils: Effects of soil moisture, texture and nitrogen availability', *Soil Biology and Biochemistry*, vol 33, pp1077–1093

Wolt, J. D. (2004) 'A meta-evaluation of nitrapyrin agronomic and environmental effectiveness with emphasis on corn production in the Midwestern USA', *Nutrient Cycling in Agroecosystems*, vol 69, pp23–41

Xu-Ri and Prentice, I. C. (2008) 'Terrestrial nitrogen cycle simulation with a dynamic global vegetation model', *Global Change Biology*, vol 14, pp1745–1764

Youssef, M. A., Skaggs, R. W., Chescheir, G. M. and Gilliam, J. W. (2005) 'The nitrogen simulation model, DRAINMOD-N II', *Transactions of the American Society of Agricultural Engineers*, vol 48, pp611–626

Zebarth, B. J., Paul, J. W. and Van Kleeck, R. (1999) 'The effect of nitrogen management in agricultural production on water and air quality: Evaluation on a regional scale', *Agriculture, Ecosystems and Environment*, vol 72, pp35–52

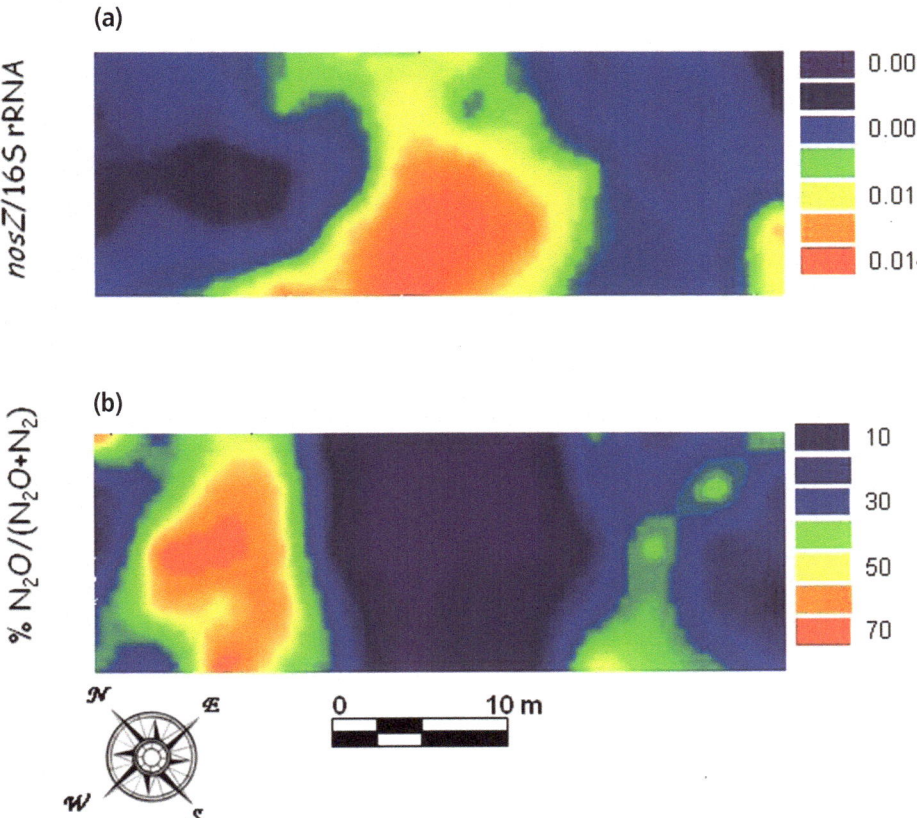

Plate 2.1 Spatial distribution in a pasture of (a) the proportion of bacteria within the total bacterial community (nosZ/16S rRNA ratio) capable of reducing N_2O to N_2, and (b) the percentage of the terminal product of denitrification being N_2O rather than N_2. Adapted from Philippot et al (2009a).

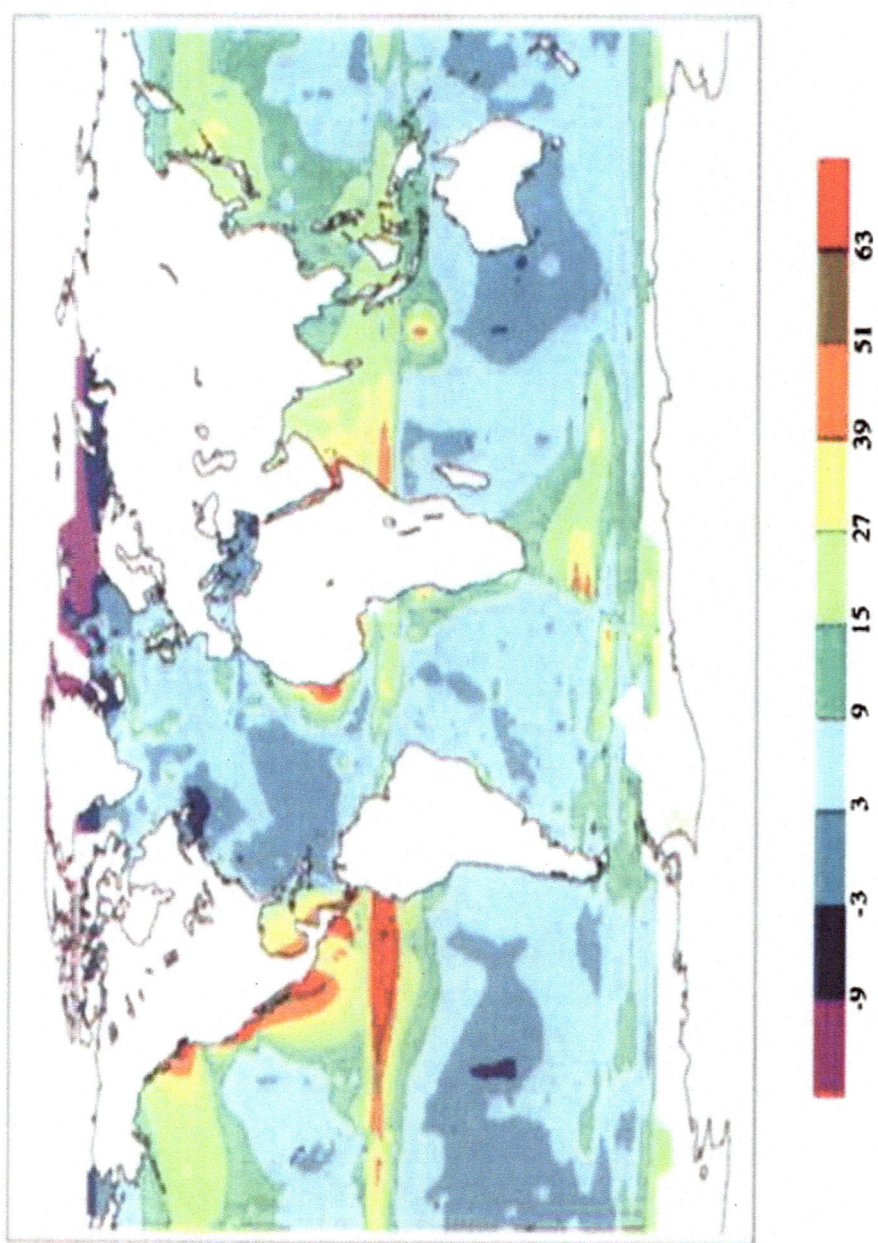

(a)

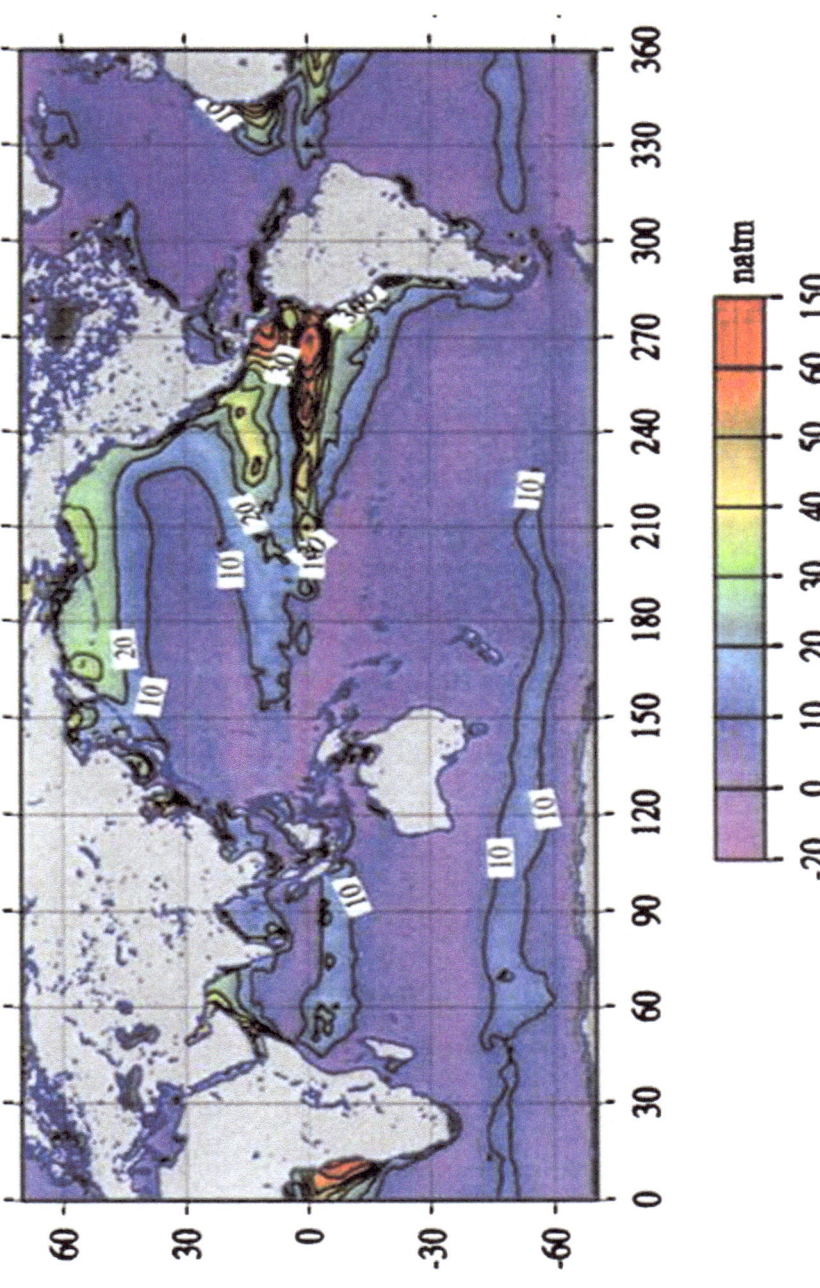

Plate 3.1 Maps of ΔpN_2O (in natm (c. Pa $\times 10^{-4}$)) in the surface layer of the world's oceans: (a) map by Nevison et al (1995) and (b) map by Suntharalingam and Sarmiento (2000). Note that the colour coding is non-linear and different for both maps.

(a)

60N
30N
EQ
30S
60S
180W
120W
60W
0
60E
120E
180E
5000
2000
1000
750
500
250
100
50
25
5

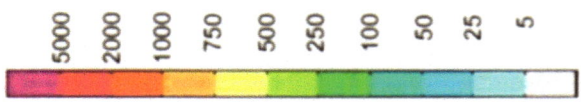

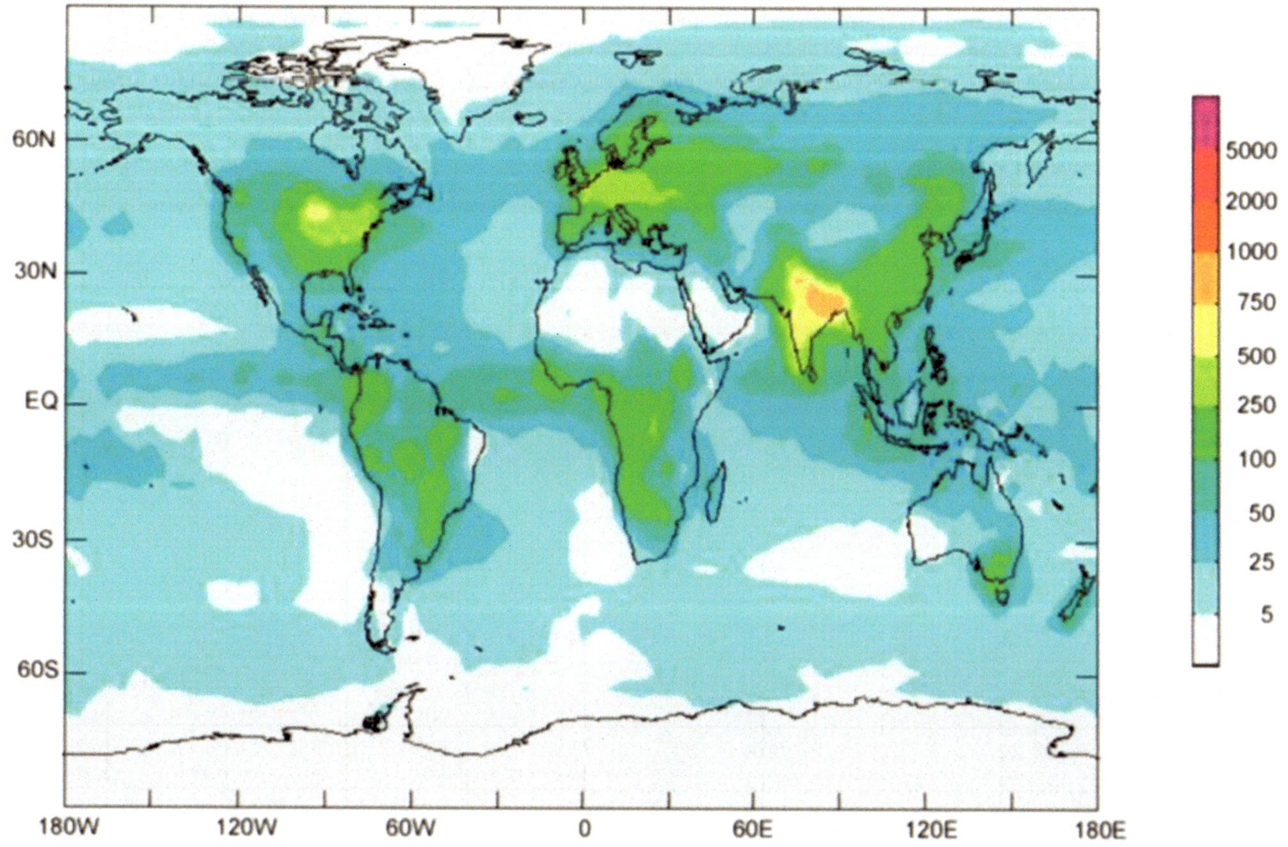

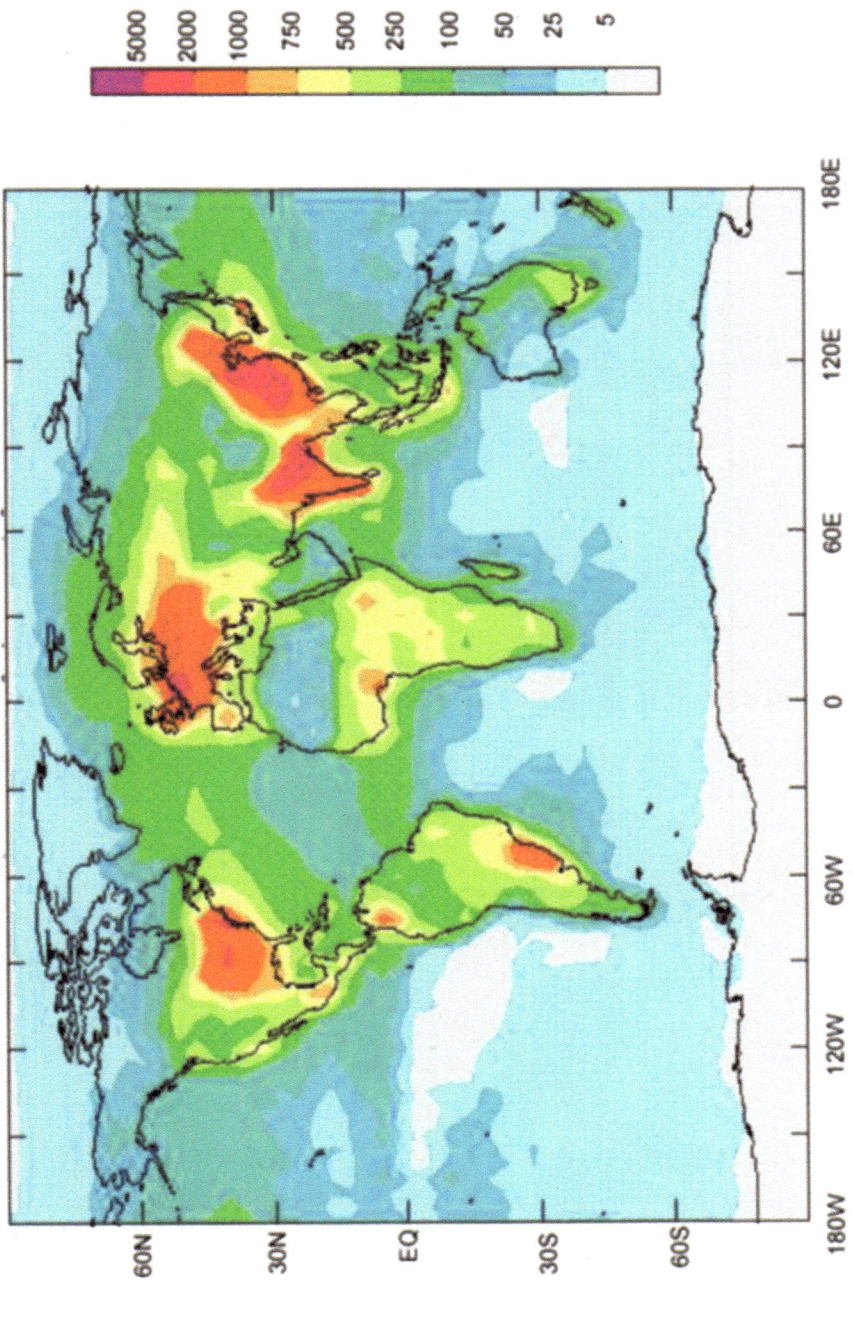

Plate 4.1 Spatial patterns of total inorganic nitrogen deposition in (a) 1860 and (b)early 1990s, in units of mg N m^{-2} yr^{-1} (= kg N km^{-2} yr^{-1}). Note the large increases in many parts of the world, but note also that in some regions deposition of nitrogen pollution from the atmosphere remains low. Reproduced with kind permission from Springer Science + Business Media, from Figure 2 of J. N. Galloway et al, *Biogeochemistry*, vol 70, pp153–226, 2004, copyright 2004, Kluwer Academic Publishers.

Plate 4.2 Getting ice cores from the Greenland ice cap. Gas chromatographic analysis of air entrapped n the ice makes it possible to determine the composition of the atmosphere in past ages, including the concentrations of N_2O and other greenhouse gases. Courtesy of NASA GSFC and copyright holder Reto Stöckli.

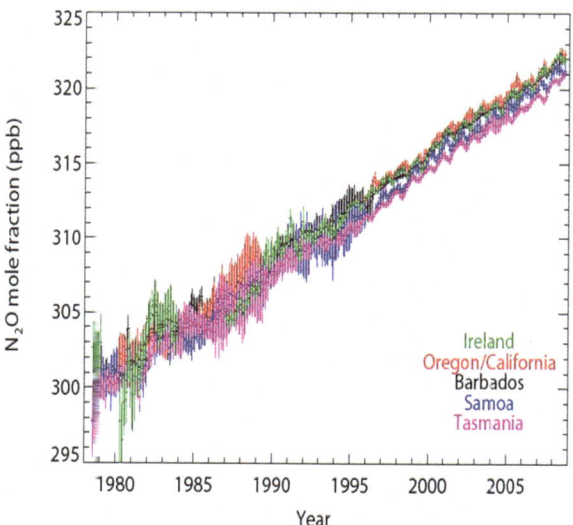

Plate 4.3 Increase in N_2O concentrations in the atmosphere over the last three decades. Data from the ALE/GAGE/AGAGE database, http://agage.eas. gatech.edu/data.htm.

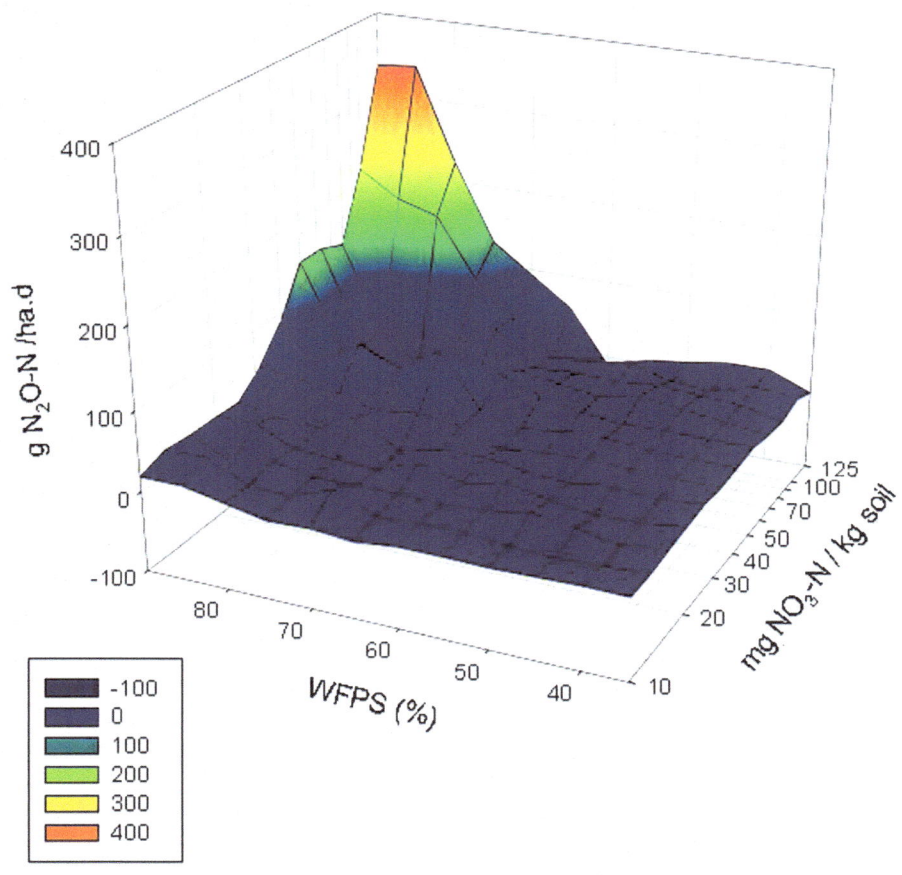

Plate 6.1 The effect of water-filled pore space and soil NO_3^- content on N_2O emissions from a pastoral soil (Letica et al, 2010).

Plate 7.1 Tropical land-use change: conversion of forest to agricultural use, near Puerto Maldonado, Peru. Photograph by Michael Zimmermann, University of Edinburgh.

Plate 8.1 Automated chamber system for measuring N_2O and NO emissions, Höglwald forest research site, SW Germany.

6

Nitrous Oxide Emissions from the Nitrogen Cycle in Livestock Agriculture: Estimation and Mitigation

Cecile de Klein, Richard Eckard and Tony van der Weerden

Introduction

Livestock agriculture accounts for 65 per cent of global N_2O emissions (Steinfeld et al, 2006). This N_2O is largely produced by the soil microbial processes of nitrification and denitrification, which convert ammonium and nitrate nitrogen cycling through the animal/plant/soil system into N_2O (Figure 6.1). In livestock agriculture, N enters this system mainly through N fertilizer applications and biological N fixation by legumes. However, a relatively large proportion of these N inputs is subsequently recycled within the system via grazing ruminants that generally utilize very little of the N in feed and excrete between 75 to 90 per cent of ingested N (Whitehead, 1995). This N is excreted either directly on to pasture as urine and dung, or in animal housing systems and then applied to land as effluent or manure.

Consumption of livestock products has increased substantially in the past decade, with exponential increases in areas such as Latin America and East Asia (Steinfeld et al, 2006). This increase in demand for animal-based protein has been driven by increasing affluence and urbanization of a growing and ageing population, which in turn has driven the intensification of the livestock sector. Current projections are that, by 2050, the world population will reach 9 billion and that both milk and meat production will grow exponentially to meet the demand. The growth and intensification of livestock systems will have a major environmental impact, including increased N_2O emissions, and will require sustainable mitigation options for minimizing emissions whilst maintaining the required livestock production levels.

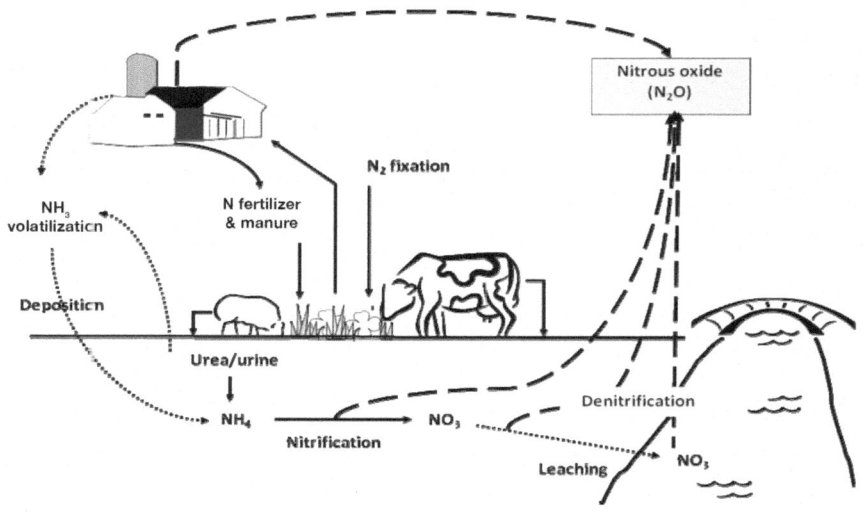

Figure 6.1 Schematic overview of the N_2O emissions from the N cycle
in livestock agriculture

Source: de Klein et al (2008)

Livestock systems can broadly be categorized into pastoral, confinement and hybrid systems (Kleinman and Soder, 2008). Pastoral systems are most common in the Southern Hemisphere: in Australia, New Zealand, Africa and Latin America. These systems are land-based and livestock graze pasture or fodder crops generally all year round. The use of imported feed or supplements is limited. The pastures are often legume-based and both biological N fixation and N fertilizer are important N inputs to the system (Wassermann, 1979; Ledgard and Steele, 1992; Ledgard, 2001).

Confinement systems are 'landless' (Oenema, 2006) and primarily import feed and supplements for their livestock that is (predominantly) housed or confined in barns or feedlots. These systems are most common in North America and parts of Europe where animals are typically kept at high densities and largely grain-fed. The N inputs into these systems are mainly through imported feed and supplements.

Hybrid systems, as referred to here, are those that combine both pastoral and confinement elements, with animals spending roughly equal amounts of time either grazing pastoral land or being housed indoors and fed on imported supplements. These systems are typical in European countries such as The Netherlands, the UK and France. The main N inputs into these hybrid systems are typically N fertilizer and N in imported feed and supplements.

The three livestock systems each have different impacts on the N cycle and on the key sources of N_2O. For example, in pastoral systems, N excreta deposited to pasture by grazing animals can contribute up to 85 per cent of the total N_2O emissions, whereas in confined or hybrid systems the main sources are N fertilizer and animal manure (Figure 6.2).

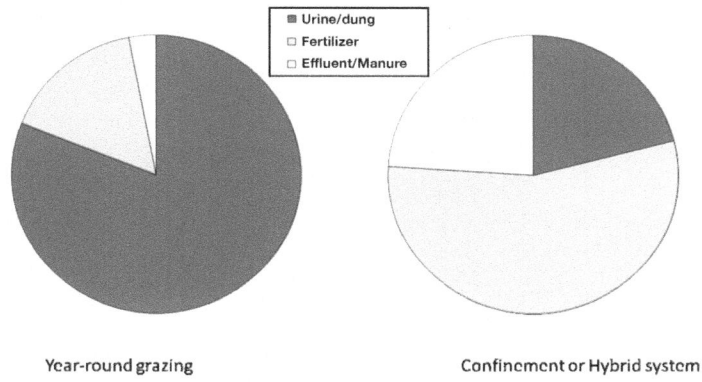

Figure 6.2 Relative contribution of N sources to N_2O emissions from pastoral and confinement or hybrid livestock systems

Source: Based on data from the national inventory reports of New Zealand (Ministry for the Environment, 2008) and the US (Environmental Protection Agency, 2008)

Globally, N_2O emissions from grazing are estimated to contribute 45 per cent or more of total livestock emissions, while emissions from manure collected in confinement systems and following application to land are estimated at around 20 and 10 per cent, respectively (Oenema et al, 2005; Steinfeld et al, 2006). Emissions from N fertilizer use on pastures are around 10 per cent of global livestock N_2O emissions, largely from high-rainfall dairy production systems. The most effective options for reducing N_2O emissions from livestock systems are those that target the main source of N_2O for a given system.

This chapter describes:

- current approaches for estimating N_2O emissions from livestock systems.
- N_2O and total greenhouse gas profiles from different systems.
- a summary of current N_2O mitigation strategies for different livestock systems; and
- estimates of the impact of a range strategies on total greenhouse gas emissions.

The chapter concludes with a discussion of the future trends in livestock agriculture covering issues such as greenhouse gas intensity versus total greenhouse gas emissions, impact of climate change, and accounting for mitigation strategies.

Estimating nitrous oxide emissions from livestock agriculture

Nitrous oxide measurement techniques and estimation procedures

There are broadly two methods for measuring N_2O emissions: flux chamber and micrometeorological techniques. In chamber techniques, N_2O emission rates are commonly determined by enclosing the atmosphere above the source (for example soil, manure or water body) and measuring the increase in headspace N_2O concentration over time. These flux chamber techniques have been widely used to measure N_2O emissions from soil because they are relatively cheap and simple, are useful for relative comparisons between adjacent treatments and they allow process-based studies of N_2O emission from soils. Their main disadvantage is that they measure N_2O emission over a relatively small area and that a large number of measurements and chambers are required to deal with the large spatial and temporal variability in emissions. Recent advances in flux chamber techniques include (1) continuous N_2O analysis using gas chromatography (GC) (Kiese et al, 2003), tunable-diode laser, Fourier transform infrared (FTIR) (Kelly et al, 2008) or photo-acoustic infrared spectroscopy (Dinuccio et al, 2008) and (2) the use of chambers that open and close automatically (for example Breuer et al, 2000; Kiese et al, 2003; Kelly et al, 2008).

Micrometeorological techniques involve measurements of N_2O in the atmosphere at two or more points above the soil surface, in combination with meteorological measurements (for example wind speed, wind direction and air temperature) (Denmead et al, 2000). These techniques measure N_2O emissions on a field scale, and thus spatially integrate N_2O flux measurements. Their disadvantages are, however, that they require large homogeneous field sites, are less reliable with low wind speed and high atmospheric stability, and require expensive N_2O analysis equipment (Fowler et al, 1997; Phillips et al, 2007). Phillips et al (2007) measured N_2O fluxes from a flood-irrigated dairy pasture, using a micrometeorological method coupled to a tunable-diode laser, in order to detect rapid changes in N_2O fluxes in response to irrigation, grazing and N fertilizer application (Figure 6.3). N_2O fluxes remained low following each irrigation event while WFPS was close to saturation (>95 per cent). However, two to three days after irrigation, as soil moisture decreased below ~95 per cent WFPS, N_2O emissions increased rapidly and remained high for one to two days. As WFPS approached ~75 per cent, there was an initially rapid decrease in fluxes, followed by a gradual decrease to background levels at a WFPS of about 65 per cent. The magnitude of each N_2O flux response to irrigation was greatly influenced by the N input, via either grazing or N fertilizer. This pattern was similar for each irrigation or significant rainfall

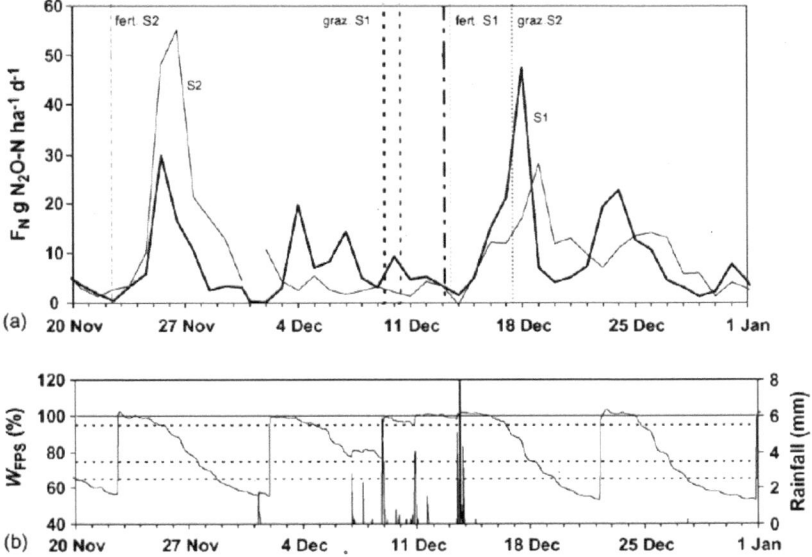

Figure 6.3 (a) Daily averaged N₂O fluxes from a pasture for two adjacent flood-irrigation bays (S1 and S2); and (b) half-hourly WFPS (8cm depth) in response to irrigation and rainfall

Note: Dotted and dashed vertical lines in (a) represent grazing and fertilization events at each site. Dotted horizontal lines in (b) indicate 95 per cent, 75 per cent and 65 per cent WFPS
Source: Phillips et al (2007)

event. Micrometeorological methods are well suited to measuring real-time fluxes over a large area in response to management intervention. However, since enclosures and chamber techniques are by far the most commonly deployed method for measuring soil N₂O fluxes, some key aspects for their deployment are discussed below.

Flux chamber deployment

Fluxes are calculated by determining the increase in headspace N_2O concentration over a given period (typically about one hour), and hourly N_2O emissions (mg N m^{-2} h^{-1}) corrected for temperature and the ratio of cover volume to surface area (de Klein et al, 2003). These hourly emission rates are converted to daily fluxes for each enclosure, and then integrated over time using a trapezoidal calculation to estimate total emissions from the enclosed area over a sampling period.

An increase in the headspace N_2O concentration may decrease the concentration gradient within the soil and atmosphere, resulting in a declining flux. This could lead to an underestimation of the calculated flux when relying solely on linear models (Conen and Smith, 2000; Davidson et al, 2002). Non-linear models often result in less biased estimates of the rate of change in headspace N_2O concentrations compared to linear models (Healy et al, 1996).

It is important to recognize that deploying enclosures onto the soil surface often modifies the flux that is to be measured (Rochette and Eriksen-Hamel, 2008). These workers assessed 356 studies employing enclosures by using 16 criteria to assess chamber design and technique. They conclude that confidence in absolute flux values was considered 'very low' or 'low' in about 60 per cent of the studies. To ensure N_2O flux data is of high quality, Rochette and Eriksen-Hamel (2008) recommend the following six criteria when using enclosures:

1   Use an insulated and vented base-and-chamber design.
2   Avoid chamber heights lower than 10cm.
3   Have a minimum insertion depth of 5cm for the chamber base.
4   If samples are stored for analysis off-site, then use pressurized fixed-volume containers of known efficiency for air sample storage.
5   Include a minimum of three discrete air samples during deployment, including one at time zero.
6   Test non-linearity of changes in headspace concentration with time for estimating dC/dt at time zero.

Similarly Meyer et al (2001) suggested the following six requirements for accuracy in designing chambers:

1   Ensure that there is sufficient air speed over the soil surface to overcome boundary layer effects resulting from the chamber physical design.
2   Ensure adequate turbulent mixing of the free volume within large (for example around 100-litre) chambers to avoid concentration gradients forming.
3   Ensure that chamber materials are inert.
4   Ensure equality of ambient pressure within and outside the chamber, which is also achieved by venting a chamber as suggested by Rochette and Eriksen-Hamel (2008).
5   Minimize interference of the chamber with key soil and plant environmental variables.
6   Quantify the rate of flow of air into and out of the chamber when measuring headspace concentrations using continuous or near-continuous N_2O analysis.

Sampling time and sampling frequency

The high spatial and temporal variability of N_2O emissions (Figure 6.3) hampers the estimation of N_2O emissions from chamber-based flux measurements. Significant emission events may be underestimated, or indeed missed, when sampling at intervals of several days compared with frequent sampling at regular short time intervals (Smith and Dobbie, 2001), unless an 'event-related' sampling regime (for example more intensive sampling following N inputs or significant rainfall) is adopted. For example, Parkin (2008) evaluated the impact of sampling frequency on actual cumulative fluxes, and found that a three-day sampling regime resulted in deviations in cumulative emission estimates being within ±10 per cent. This deviation increased with the increase in time interval between sampling, with 21-day sampling intervals resulting in deviations of between +60 per cent and −40 per cent of the actual cumulative flux. Smith and Dobbie (2001) observed that cumulative N_2O emission values from sampling at eight-hour intervals were on average 14 per cent higher than values based on samples collected once every three to seven days. This difference was not significant at the 95 per cent confidence interval level, as both overestimations and underestimations occurred. It is noteworthy that their temporal variation was smaller than the corresponding spatial variation. Less frequent sampling may be useful for relative comparisons from adjacent treatments, subjected to similar conditions. However, using these data to calculate annual fluxes or emission factors should only be applied with caution, as significant peaks of N_2O emission may have been missed. Adopting 'event-related' sampling regimes reduces the error associated with temporal variability, and thus improves the accuracy of determining actual fluxes or emission factors.

Consideration of diurnal variation in soil temperatures and fluxes is required when selecting an appropriate time of day for collecting chamber headspace samples. Fluxes calculated from samples collected between 10am and 12 noon were not significantly different from those based on sampling at eight-hour intervals during a period when diurnal variation in temperatures was small (Smith and Dobbie, 2001). Diurnal temperature and flux data collected from a grassland soil in northern Germany (Dittert et al, 2005) would also suggest that midday sampling is representative of the 24-hour period. In situations where diurnal fluctuations in temperature and flux are likely to be much larger, auto-chambers programmed to sample at a high frequency could be deployed (Smith and Dobbie, 2001).

Modelling

Nitrous oxide emissions are highly variable, both in space and time, and the estimation of N_2O emissions from individual livestock farms based on actual measurements is very costly and impractical. Models have therefore become an important means for improving our understanding of the complex interactions between drivers of N_2O emissions, for estimating N_2O emissions from livestock systems and for evaluating practices that can reduce emissions. These models range from relatively simple national inventory or annual accounting models to detailed process-based models, and use factors and constants that are derived from measurements under experimental and controlled conditions.

The main example of the inventory approach is the IPCC methodology for estimating national greenhouse gas inventories (IPCC, 2006), while the DNDC model (Li et al, 1992a, 1992b) is one of the best known biophysical models for estimating N_2O emissions. These inventory and modelling approaches are discussed below.

Inventory/accounting models

The IPCC inventory methodology has been developed for estimating the greenhouse gas emissions of a nation. In its simplest form, the IPCC inventory methodology estimates N_2O emissions from the size of a given source of emissions (for example amount of N fertilizer applied) multiplied by N_2O emission factors associated with that source. The inventory methodology includes both direct and indirect N_2O emissions. Direct emissions are those that occur from N sources within the farm system. Indirect emissions are those from nitrogen that is lost through nitrate leaching or ammonia volatilization and subsequently emitted as N_2O from surface waters or following re-deposition of NH_3 to land (Figure 6.1). The IPCC *Guidelines for National Greenhouse Gas Inventories* (IPCC, 1997, 2006) provide the methodology as well as default values for the N_2O emission factors and N loss fractions, but recommend that countries use 'country-specific' factors for key sources of N_2O.

A detailed description of the revised 1996 IPCC methodology for N_2O is provided by Mosier et al (1998). These 1996 guidelines were again revised in 2006 (IPCC, 2006) based on more recent research findings, and new knowledge and insights. The main changes between the 1996 and 2006 N_2O methodologies include updated default values for some of the emission factors, removal of one source of N_2O and inclusion of various new sources. In addition, adjustments to the methodology were made to remove some inconsistencies with respect to N inputs from crop residue and N leaching estimates. Table 6.1 summarizes the 2006 changes to the IPCC inventory methodology for N_2O (see also Chapter 4).

Background emissions from improved pastures

Neither the 1996 nor the 2006 IPCC guidelines include estimates of 'background' N_2O emissions from improved pastures as an anthropogenic source of N_2O. Although the 1996 guidelines make reference to the fact that background emissions from agricultural soils may be higher than historic natural emissions as a result of enhanced mineralization of soil organic matter, only N_2O emissions from added sources (for example N fertilizer, urine and manure applications) are included in the methodology. Yet evidence exists that N_2O emissions from improved pastures without additional N inputs (often referred to as 'control sites' in experimental trials to determine N_2O emission factors) are significantly higher than emissions from unimproved sites and should therefore be considered as anthropogenic (Bouwman, 1996; Stehfest and Bouwman, 2006). For example, Bouwman et al (2002) provided a summary of available measurement data of N_2O emissions from fertilized fields (>900 observations) and suggested that annual N_2O emissions from unfertilized pastures were between 0 and 1.5kg N ha^{-1}. This is similar to 'background' emissions of 1kg N ha^{-1} suggested by the equation Bouwman (1996) presented:

$$N_2O\text{-}N \text{ (kg ha}^{-1}) = 1 + (1.25 \times N \text{ input in kg ha}^{-1})/100 \qquad (6.1)$$

which was the origin of the default N_2O emission factor for N fertilizer in the

Table 6.1 Overview of the main changes between the 1996 and 2006 IPCC default inventory methodologies

IPCC factor	Change	Source/references
The N_2O emission factor for animal excreta deposited during grazing ($EF_{3\,PRP}$)	Default value for sheep and 'other animals' reduced from 0.02 to 0.01kg N_2O-N (kg N)$^{-1}$ Remains at 0.02kg N_2O-N (kg N)$^{-1}$ for cattle (dairy, non-dairy and buffalo), poultry and pigs	IPCC (2006); C. A. M. de Klein (2004, unpublished)
The N_2O emission factor for leached N (EF_5)	Default value reduced from 0.025 to 0.0075kg N_2O-N/per kg N lost in leachate or runoff water	Hiscock et al (2003); Dong et al (2004); Reay et al (2004); Sawamoto et al (2005); Clough et al (2006)
Biological N fixation as a source of N_2O (F_{BN})	Removed from inventory as a direct source of N_2O. N_2O emissions induced by the growth of legume crops/forages may be estimated solely as a function of the above- and below-ground N inputs from crop/forage residues	Rochette and Janzen (2005)
The annual amount of N input from crop residue (F_{CR})	Now includes the contribution of below-ground N to the total input of N from crop residues	IPCC (2006)
	Now also includes the N released following pasture renewal	van der Weerden et al (1999); Davies et al (2001)
N mineralization associated with C loss due to land-use change and management practices (F_{SOM})	Now included as a new source of N_2O	Smith and Conen (2004)
Amounts of applied mineral N fertilizers (F_{SN}) and organic N fertilizers (F_{ON})	The default methodology no longer adjusts F_{SN} and F_{ON} for the amounts of NH_3 and NO_x volatilized after application to soil before applying the N_2O EF	IPCC (2006)
N sources from which the amount of leached N and subsequent indirect N_2O emissions is estimated	Now includes N inputs from crop residue (F_{CR}) and from N mineralization associated with C loss due to land-use change (F_{SOM})	IPCC (2006)

1996 IPCC guidelines. Recently, Rochette et al (2008) used the same regression approach as Bouwman (1996) and Bouwman et al (2002) to estimate the N_2O emission factor for N fertilizer from Canadian field studies. Their results suggested average background emissions of 0.1, 0.6 and 0.8kg N_2O-N ha^{-1} yr^{-1} for Canadian agricultural soils in three different regions.

Some countries have adapted the IPCC default methodology to include 'background' emissions. For example, in Sweden a 'background' EF of 0.5kg N_2O-N/ha/year for mineral soils (Kasimir-Klemedtsson, 2001) is used. This value was determined on the basis of the country-specific EF for inorganic N fertilizer inputs of 0.8 per cent, which is lower than the IPCC default value of 1.25 per cent. The reason for including these 'background' emissions in the Swedish inventory is to account for the effect of N mineralization following decomposition of soil organic matter after cultivation events (Kasimir-Klemedtsson, 2001). This thus recognizes N fertilizer inputs and soil cultivation as two separate sources of N_2O. As mentioned above, this was also recognized in the 2006 revision of the IPCC guidelines and cultivation effects are now included as a separate source in the 2006 IPCC guidelines. However, in addition to the cultivation effect, 'background' emissions from uncultivated improved land are higher than those from natural land, and these anthropogenic emissions are currently not accounted for in the default IPCC methodology. As a result, the New Zealand and Australian inventories do not include a correction for background N_2O emissions. The Australian National Inventory Report further argues that, in contrast to European and North American agriculture, there has been little accumulation of soil nitrogen from previous cropping that might otherwise predispose these soils to substantial background N_2O emission rates (DCC, 2008).

Farm-scale accounting

The IPCC methodology has been incorporated into on-farm inventory models to estimate N_2O emissions at a farm scale. Examples of such models are the nutrient budgeting model OVERSEER® (Wheeler et al, 2003, 2008); the Australian Farm Greenhouse Gas Accounting Tools (Eckard, 2008) and various whole-farm models such as DairyWise or Farmgreenhouse gas (Olesen et al, 2006; Schils et al, 2007; van Groenigen et al, 2008). These models generally operate at an annual time step.

The nutrient budgeting model OVERSEER® (Wheeler et al, 2003) is extensively used throughout New Zealand as a tool for optimizing on-farm nutrient management as well as implementing resource requirements from local authorities to minimize N and P losses to waterways. The model includes a greenhouse gas module for estimating on-farm emissions of methane, nitrous oxide and carbon dioxide (Wheeler et al, 2008), and is based on the New Zealand IPCC (NZ-IPCC) inventory methodology (Ministry for the Environment, 2008). It thus estimates N_2O emissions from the size of N inputs in the system multiplied by an EF for each input. The model can operate in two modes. One mode uses the NZ-IPCC values for the EFs and fractions for N leaching and ammonia volatilization. The other mode is more 'site-specific' and uses disaggregated EFs for animal excreta based on animal type and soil drainage class. This mode also uses the nutrient budgeting model to estimate N leaching and volatilization losses, instead of using the NZ-IPCC fractions. Both modes use farm-specific input variables on animal production, fertilizer use and pasture quality, and can account for N mitigation strategies such as optimized N fertilizer form and timing, nitrification inhibitor use, low N feed, and destocking or housing options (Wheeler et al, 2008).

Schils et al (2007) reviewed a range of whole-farm models for estimating greenhouse gas emissions from dairy systems, including DairyWise, Farm-greenhouse gas, $SIMS_{DAIRY}$ and FarmSim. These models often use IPCC inventory methodologies for estimating CH_4 and N_2O emissions (and to some extent CO_2), but the authors concluded that they should not be seen as a replacement for the IPCC methodology, which is clearly aimed at providing transparent and consistent reports on national greenhouse gas emissions. The whole-farm models provide powerful tools for evaluating the impact of mitigation strategies at an individual farm level (Schils et al, 2007). In addition, these models are useful tools for examining the changes in IPCC EFs and fractions that would result from adoption of mitigation strategies, and can thus help to adjust the IPCC inventory methodology to account for these mitigations.

Biophysical models

Modelling provides a valuable complement to measurement, extending limited temporal and spatial measurements to other climatic and edaphic conditions, regions and scales. In recent years, the development of biophysical N_2O emission models has received much attention and a large number of models have been developed to simulate N_2O fluxes from natural and managed ecosystems (Dalal et al, 2003; Chen et al, 2008).

In contrast to the empirical inventory models, detailed biophysical, process-based, daily time-step models tend to simulate changes in environmental factors affecting the emissions (for example N transformations, moisture, temperature and available carbon) from key input variables (for example climate, soil type, system and management practices), thereby predicting the soil N pools available for nitrification and denitrification to N_2O. These models apply largely mechanistic algorithms to estimate the effect of the proximal factors on N_2O emissions and reflect our understanding of drivers of N_2O emissions at a process level. Whole-farm systems models tend to use a combination of accounting, empirical and mechanistic modelling approaches based on daily, monthly or annual time steps.

Some highly mechanistic models, such as DNDC (Li et al, 1992a) and several others (Smith, 1980; Grant and Pattey, 1999; Riley and Matson, 2000) simulate microbial growth rates and solute and gas transport through the soil profile and aggregates. Other models simulate nitrification and denitrification as a function of frequently measured and modelled variables such as soil water, temperature, NO_3^-, NH_4^+ and soil respiration, for example DAYCENT (Del Grosso et al, 2006), EcoMod and DairyMod (Johnson et al, 2008) and WNMM (Li et al, 2007). More generalized models have been developed to simulate N_2O fluxes at regional and global scales (Parton et al, 1996; Potter et al, 1996). The annual time-step models are more generally applicable but are surrounded by great uncertainties, while the more complex daily time-step models are highly site-, climate- and system-specific, often requiring significant parameterization.

The different N_2O models described above have different purposes, from annual accounting to evaluating potential mitigation strategies and/or improving understanding of the drivers of N_2O emissions at a process level. The level of complexity and the detail of the required input data generally increase with the extent to which a model accounts for soil, climate and management variability. Depending on the purpose of the model, a balance is often required between available input data and the reliability of the estimates. Although N_2O emission models have been further developed and improved in recent years, the spatial and temporal variability in emissions and complex interactions between the driving variables challenge our ability to predict emissions (Calanca et al, 2007). For example, in a recent comparison between the process-based models DAYCENT and DNDC for estimating N_2O emissions from cropping systems, Smith W. N. et al (2008) concluded that the DNDC model accurately predicted average seasonal N_2O emissions, whereas DAYCENT underestimated them by up to 58 per cent. However, as neither model accurately simulated the timing of individual emission events, the authors concluded that improvements to the soil moisture and nitrogen transformation modules of the models are required before enhancements are made to the N_2O emission routines. Another contributing factor affecting the simulation of the timing of N_2O emissions is that most mechanistic models include algorithms describing N_2O *production* in the soil, but do not include processes and time steps for the actual N_2O *emission* from the soil surface.

Uncertainties

Regardless of the approach or model used, uncertainties associated with N_2O estimates for livestock systems are very large due to the complexity of the N_2O emissions and their drivers in different systems. These uncertainties are associated with both the actual emission estimates, and the assessment of the effectiveness of mitigation strategies. When discussing uncertainties in N_2O emission estimates, Oenema et al (2005) distinguished between fundamental and operational uncertainties. These authors described fundamental uncertainties as being associated with the methodology or structure (and upscaling) of the estimation procedure, whereas operational uncertainties relate to those associated with EFs and/or the input variables that underpin the estimates (for example fertilizer use, manure production, N excretion etc.). The uncertainties associated with EFs are partly due to uncertainties arising from the measurement methodology. Careful deployment of the measurement methodology and procedures, as discussed above for soil chamber measurements, will help to minimize some of the uncertainties associated with EFs. However, accounting for the large spatial and temporal variability of N_2O emissions will remain a key challenge when seeking to minimize uncertainties associated with upscaling small-scale measurements to regional or national inventory estimates. These uncertainties can only be addressed by applying appropriate measurement technologies and methods, including online measurement from automated chambers or micrometeorological systems, with sufficient spatial replication, fetch size and measurement frequency to ensure that spatial and temporal variability issues are addressed.

Mitigating nitrous oxide emissions from livestock agriculture

As discussed in Chapter 2, N_2O emissions are highly variable in space and time, due to the numerous controlling factors of nitrification and denitrification. However, the two key variables that regulate N_2O emissions from soils of livestock systems are generally soil N availability and soil aeration (de Klein and Eckard, 2008). The combined effect of these two factors is illustrated by measurements of N_2O emissions from a sheep-grazed pastoral soil in New Zealand (Plate 6.1) and denitrification losses from a grazed ryegrass/clover pasture in Australia (Figure 6.4). These results clearly indicate that the coincidence of high soil nitrate N and high soil moisture content (low aeration) greatly enhanced N_2O emissions. Most options for mitigating N_2O emissions from livestock systems therefore focus on reducing the availability or input of soil N, particularly under wet conditions (Figure 6.5).

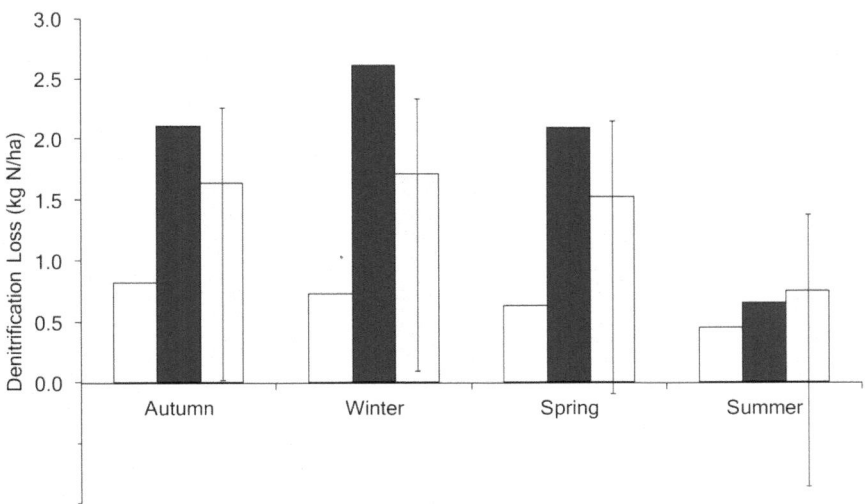

Figure 6.4 Total N lost by denitrification from a grazed perennial ryegrass and white clover pasture following zero (white bars) or 50kg N/ha in each of four seasons applied as urea (dark grey bars) or ammonium nitrate (light grey bars)

Source: Eckard et al (2003)

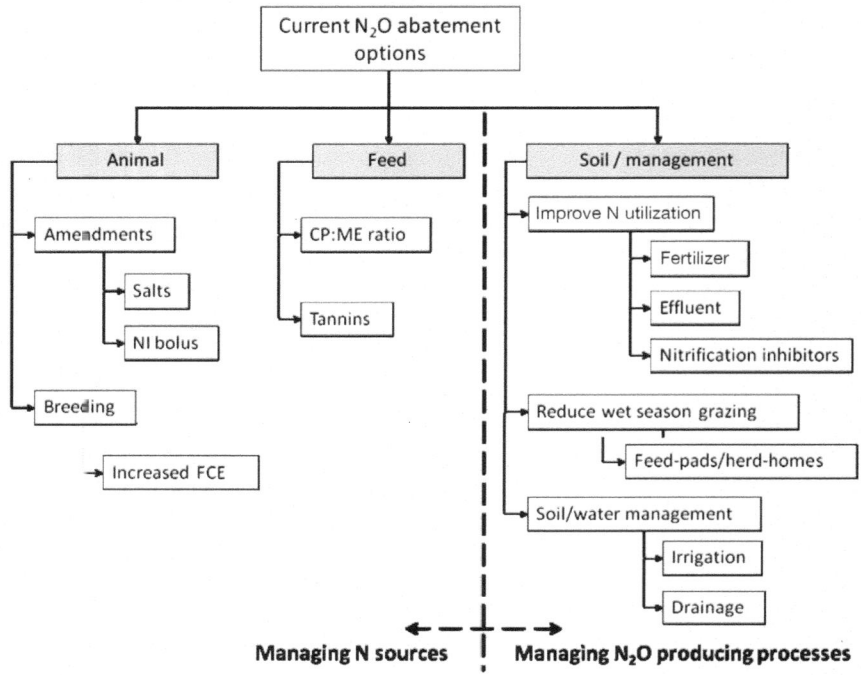

Figure 6.5 Schematic overview of current N₂O abatement options in livestock agriculture

Temperature can also be important, with higher N_2O emissions occurring at higher temperatures. However, its effect is generally more pronounced at a regional or national scale, i.e. temperature-induced differences in N_2O emissions are generally observed between regions rather than at a farm scale, and temperature is therefore not often considered as a factor to target in relation to N_2O mitigation at a farm level.

Mitigation options

Animal- and feed-based interventions

Ruminants excrete between 75 per cent and 95 per cent of the N they ingest; excess dietary N is mainly excreted in the urine, whereas dung N excretion remains relatively constant (Castillo et al, 2000; Eckard et al, 2007). Of the dietary N consumed by ruminants, <30 per cent is utilized for production, with >60 per cent lost from the grazing system (Whitehead, 1995). The effective N application rate within a urine patch from a dairy cow is commonly between 800 and 1300kg N/ha (Eckard and Chapman, 2006), and N is deposited at concentrations that are orders of magnitude greater than what soil–plant systems can efficiently utilize. Therefore, strategies for reducing N_2O emissions should also focus on improving the efficiency of N cycling through the soil–plant–animal system.

Conceptually, if animal urine in grazing systems was spread more evenly across the paddock, the effective N application rate would be reduced, potentially reducing N_2O emissions. Although no specific animal technologies have currently been developed, this may require some physical intervention with the animal, yet be practically and ethically conceived, that will cause the urine to be distributed more evenly (de Klein and Eckard, 2008).

Breeding and diet

Genetic manipulation or animal breeding may improve the N conversion efficiency within the rumen, or produce animals that urinate more frequently or walk while urinating, all leading to lower N concentrations or greater spread of urine (de Klein and Eckard, 2008). Coffey (1996) reported that an improvement in the feed conversion efficiency ratio of 0.01kg dry matter (DM) intake per kg live weight gain could result in a 3.3 per cent reduction in nutrient excretion, assuming a similar growth rate and nutrient retention. Therefore, breeding for more efficient feed conversion should produce animals that partition more of their intake into production and less into N excretion, thereby reducing potential N_2O losses.

Ruminants on lush spring pasture commonly ingest protein in excess of their requirements, but are usually energy-limited. The excess protein is used as an energy source and a relatively larger proportion of the ammonia produced in the rumen is excreted as urea (Whitehead, 1995). Therefore, balancing the protein-to-energy ratios in the diets of ruminants is important for minimizing the N_2O emissions resulting from excess urinary N excretion. Misselbrook et al (2005) showed that dairy cows fed a 14 per cent crude protein (CP) diet excreted 45 per cent less urinary N than did dairy cows fed a 19 per cent CP diet. Similarly, van Vuuren et al (1993) showed that supplementing cows on a perennial ryegrass diet with low protein/high sugar supplements reduced the amount of total N and urinary N excreted by 6–9 per cent and 10–20 per cent, respectively, compared with those of cows fed an all-grass diet. More recently, Miller et al (2001) reported that dairy cows on a novel 'high-sugar' variety of perennial ryegrass excreted 18 per cent less N in total and 29 per cent less urinary N.

Improving N efficiency and reducing excess urinary N can be achieved through either breeding animals for more efficient feed conversion, breeding forage crops that use N more efficiently and have a higher energy-to-protein ratio, or balancing high-protein forages with high-energy supplements.

Tannins

Condensed tannins (CT) form complexes with proteins in the rumen, protecting them from microbial digestion, resulting in either the more efficient digestion of amino acids in the abomasum and lower intestine, or the excretion of the CT-protein complex in the dung (Min et al, 2003; de Klein and Eckard, 2008). Carulla et al (2005) showed that sheep fed a CT extract from *Acacia mearnsii* (black wattle) increased their partitioning of N from the urine to the faeces, reducing urinary N by 9.3 per cent as a proportion of the total N excreted Grainger et al (2009) added a CT extract from black wattle to the diet of lactating dairy cattle and observed a 45–59 per cent reduction in urinary N excretion, with 18–21 per cent more N in the faeces. Similarly, Misselbrook et al (2005) showed that dairy cows on a 3.5 per cent CT diet excreted 25 per cent less urinary N, 60 per cent more dung N, and 8 per cent more N overall compared with cows on a 1 per cent CT diet.

Faecal N is mainly present in a complexed organic form and is thus less volatile than urinary N, which is largely urea that is mineralized to NH_4^+ and to NO_3^-. As well as being vulnerable to leaching, this urinary N accounts for about 60 per cent of N_2O emissions from pasture (de Klein and Ledgard, 2005). The CT-protein complex in dung is also more recalcitrant to breakdown in the soil, because mineralization of the complex is inhibited and the faeces decompose more slowly than faeces that do not contain CT (Fox et al, 1990; Palm and Sanchez, 1991; Somda and Powell, 1998; Niezen et al, 2002). By reducing N excretion in the urine, the risk of subsequent N_2O emission from this highly concentrated N source is reduced (de Klein and Eckard, 2008). Currently, CT extracts are expensive because there is no large commercial demand for their production in agriculture. Because many forage plants contain CT, plant breeding may be a way to introduce CT into the diet of animals when daily supplementation is not practical or economic. However, further research is required to identify suitable and cost-effective high-tannin forages and tannin extracts with which to supplement the diets of ruminants.

Salt supplementation increases water intake in ruminants, both reducing their urinary N concentration and inducing more frequent urination events, thus spreading urinary N more evenly across grazed pasture (Ledgard et al, 2007b). In a laboratory study, van Groenigen et al (2005) found that reducing the N concentration of urine tended to reduce N_2O emissions from incubated soil cores by 5–10 per cent. However, no field measurements of actual N_2O emissions have been reported in the context of breeding or salt supplementation, and this area requires further research.

Soil/management interventions

Fertilizer and effluent inputs

The rate, source and timing of applications of N fertilizer are important management factors affecting the efficiency of pasture growth responses, and thus potential N_2O losses. When conditions are suitable for denitrification, N_2O emissions increase exponentially with the rate of N applied in any single application (Mosier et al, 1983; Whitehead, 1995; Eckard and Chapman 2006). In a modelling study, Eckard and Chapman (2006) predicted that annual N_2O emissions will increase exponentially as the annual rate of N fertilizer is increased, with the rate of increase being faster for nitrate than for urea fertilizer. Galbally (2005) reported N_2O emissions of 1.0–1.2kg N_2O–N ha^{-1} yr^{-1} from unfertilized irrigated dairy pastures in temperate Australia, increasing to 2.4kg N_2O–N ha^{-1} yr^{-1} after three applications of 50kg N/ha per year.

Nitrate-based N fertilizer generally produces high N_2O emissions relative to ammoniated N sources when applied to actively growing pasture. In a review, de Klein et al (2001) cited N_2O emission factors for N fertilizers of <0.1–1.9 per cent (median 0.5 per cent) when N was applied as urea fertilizer, and <0.1–12 per cent (median 3.2 per cent) when N was applied as calcium nitrate. Similarly, Eckard et al (2003) reported that average N_2O losses were higher with ammonium nitrate (12.9 per cent) than with urea. In South Africa, Australia and New Zealand, urea or diammonium phosphate is the main source of N applied to intensive pastures, with recommended rates not exceeding 50–60kg of fertilizer-N per hectare in any single application per grazing rotation (Ledgard, 1986; Eckard et al, 1995; Eckard and Franks, 1998). Although the total amount of N fertilizer used can be reduced and the timing of its application can be optimized to the soil moisture conditions (Luo et al, 2007), any further N_2O abatement potential in the rate of application and the source of N fertilizer may be limited in pasture-based grazing systems.

The rate, timing and placement of animal effluent applied to soils all affect potential N_2O emissions (for example Luo et al, 2008c). The N_2O emissions from effluent applied to soils are generally lower than those from urine patches, if the effluent is applied at the recommended rates and at the appropriate times of the year (Chadwick, 1997; de Klein et al, 2001; Saggar et al, 2004). Saggar et al (2004) demonstrated that N_2O emissions from effluent are higher when applied to wet soil than when applied to drier soil, and that emission peaks generally occur within 24 hours of application. The timing of the application of effluent relative to the application of N fertilizer can also affect N_2O emissions; N_2O emissions are lower when the N fertilizer is applied at least three days after the effluent, rather than together with the effluent (Stevens and Laughlin, 2002).

Effluent application techniques can also affect N_2O emissions. For example, the injection or incorporation of effluent into the soil can increase direct N_2O emissions but reduce ammonia (NH_3) volatilization (Chadwick, 1997; Saggar et al, 2004), resulting in lower indirect N_2O emissions. Effluent injection is also likely to increase the overall efficiency of the use of effluent N and could thus reduce the N fertilizer requirement and the associated N_2O emissions (Chadwick, 1997).

Nitrification inhibitors

Nitrification inhibitors are chemical compounds that inhibit the oxidation of NH_4^+ to NO_3^- in soil and thus reduce N_2O emissions from NH_4^+-based fertilizers and from urine (Di and Cameron, 2002). The most widely used are nitrapyrin and DCD (de Klein and Eckard, 2008). Fertilizers coated with nitrification inhibitors have been shown to be effective in reducing nitrification and N_2O emissions by up to 80 per cent, as reviewed by de Klein et al (2001). Applied as a spray or as a granular formulation, nitrification inhibitors can also effectively reduce N_2O emissions from animal urine by 61–91 per cent, with pasture yield increases of 0–36 per cent (Di et al, 2007; Kelly et al, 2008; Smith L. C. et al, 2008; Monaghan et al, 2009). However, some of these studies have been conducted under optimal conditions for N_2O production and over short periods, so the potential on-farm abatement is likely to be more conservative in a grazing system than the published data suggest (de Klein and Eckard, 2008). Nitrification inhibitor can also reduce nitrate leaching emissions from urine patches by 17–68 per cent (for example Di and Cameron, 2002, 2005) or from grazed pasture by 21–56 per cent (Monaghan et al, 2009), thus reducing indirect N_2O emissions.

Novel approaches to placing nitrification inhibitors where they are most effective could include either feeding the inhibitors to animals, with the inhibitor excreted in the urinary stream, or breeding plants that exude natural inhibitors from their roots. Ledgard et al (2007a) demonstrated that ruminants supplemented with a nitrification inhibitor (DCD) excreted the inhibitor unaltered in the urine. Further research is required to quantify the N_2O abatement potential of this approach, including a slow-release delivery mechanism, because this has great potential for the abatement of N_2O from urine in grazing systems. Recently, Subbarao et al (2006) reported the release of a natural nitrification inhibitor from the roots of *Brachiaria humidicola*, raising the possibility of breeding plants that synthesize their own inhibitors.

Apart from directly reducing N input into grazing systems, nitrification inhibitors are currently the only well-published technology available for reducing the loss of N from soils. Although their use has historically been limited, predominantly by cost, with future constraints on emissions envisaged in many countries they are likely to form a significant part of any comprehensive abatement strategy to reduce N_2O emissions from both urinary and N fertilizer inputs to pasture systems.

Grazing management

Restricting grazing on seasonally wet soils not only reduces urinary N returns, but also reduces soil anaerobicity caused by hoof compaction. De Klein et al (2006) and Luo et al (2008b) reported a total reduction in direct and indirect N_2O emissions of 7–11 per cent from farm systems under restricted grazing regimes during relatively wet months, and following the application of effluent from feed pods or stand-off pads. Schils et al (2006) reported that the combined effect of reduced N fertilizer use and reduced grazing time on case study farms in The Netherlands was a reduction in N_2O emissions of 10 per cent for the total farm system. These studies show that increased N utilization increases N efficiency and reduces N losses while increasing production.

Irrigation and drainage

Nitrous oxide fluxes from flood-irrigated dairy pasture rose rapidly two to three days after irrigation, when the soil WFPS was >95 per cent (Phillips et al,

2007). The emissions remained high for a further one to two days before gradually subsiding to background levels as the soil moisture decreased. The N_2O emissions remained low immediately following irrigation, which was most likely the result of complete denitrification, producing mainly N_2 emissions (Phillips et al, 2007). Irrigation through extended dry seasons may in fact reduce N_2O emissions in the later wetter season by reducing the build-up of unutilized soil NO_3^- through increased plant uptake (Jordan and Smith, 1985).

As denitrification is enhanced under conditions of low soil aeration, reducing the waterlogging of pastures will reduce potential N_2O emissions. A common practice in the management of seasonally wet soils has been to introduce surface or subsurface drains, but the impact of this management practice on N_2O emissions is not straightforward. Waterlogged soils will denitrify more efficiently than well-drained soils but improved drainage will increase N loss through the drains, only to denitrify in a wetland or elsewhere in the landscape (de Klein and Eckard, 2008). However, if the improved drainage merely reduces the WFPS of the soil to below saturation (about 80 per cent), but it remains above wilting point (40 per cent), this may actually increase N_2O emissions (Granli and Bockman, 1994) and could lead to increased nitrate leaching. In some cases, stimulating denitrification has been recommended as a way of reducing nitrate leaching in nitrate-sensitive areas (Russelle et al, 2005). These data highlight the need for further research into the compromise between managing irrigation for efficient plant growth and N_2O emissions, and the compromise between improved drainage and enhanced NO_{3-} leaching, for a range of soils, systems and environmental objectives.

Matching interventions to the farm system

Strategies for mitigating N_2O emissions can broadly be categorized into those that reduce the total amount of N returned to or cycling through the soil, and those that utilize the available N more efficiently. The first group of mitigation options will always reduce N_2O emissions, regardless of the source of the N_2O within a system. Examples of such mitigation options are

- breeding to produce animals that have a higher nitrogen use efficiency,
- reducing the amount of N fertilizer inputs,
- reducing the amount of N in the diet, and
- feeding animal amendments, such as tannins, that can reduce the amount of urine N excreted.

Mitigation options that utilize available N more efficiently include fertilizer and effluent management, balancing energy and protein intakes, wet-season grazing management and nitrification inhibitors. The most effective options for reducing N_2O emissions are those that target the key source(s) of N_2O and these thus vary with the livestock system in which they are used. The main characteristics of livestock systems determining the relative importance of N_2O sources are (1) the number of grazing days, with effluent management being a more dominant N_2O source with fewer grazing days; and (2) N fertilizer use, which becomes a less dominant source in legume-based pasture systems. Table 6.2 summarizes published results on N_2O emissions and their relative importance within a system. It should be noted that comparisons between countries need to be treated with caution as each publication uses a different methodology for estimating the N_2O emissions. For example, Lovett et al (2008) use EFs for animal urine and dung deposited during grazing in Ireland (0.56 per cent and 0.19 per cent, respectively) that are much lower than values used by others (generally between 1 and 2 per cent, with no distinction between urine or dung). As a result, the relative contribution of the urine and dung deposited during grazing is much lower and fertilizer use and indirect emissions are the main sources. However, if these reduced EFs are acceptable country-specific values for Ireland, mitigation options targeting the urine and dung deposition will have a lower impact in Irish dairy systems than options that target N fertilizer use.

Table 6.2 Relative contribution (%) of different sources to total N_2O emissions from different livestock systems in New Zealand (NZ), The Netherlands (NL), Ireland and Germany

	NZ		NL dairy		Ireland dairy		Germany	
	dairy[a,b]	sheep/ beef[a,c]	grass N[d]	grass/ clover[e]			beef[h]	organic cattle[i]
Grazing days	365	365	195	203	250[f]	149[g]	0	182
N fertilizer use	23	13	31	5	40	36	54	0
Urine/dung deposited during grazing	52	71	45	63	11	7	0	24
Effluent applied to land	8	0	11	14	9	14	46	76
Indirect emissions	17	16	13	18	40	42	ng	ng
Total N_2O	100	100	100	100	100	100	100	100

Note: ng: not given. [a] estimated using OVERSEER® (Wheeler et al, 2003); [b] NZ dairy farm: year-round grazing grass/clover pasture; 100kg fertilizer N/ha/yr; 2.8 cows/ha; 350kg milk solids/cow i.e. around 4060 litres milk/cow; [c] NZ sheep/beef farm: year-round grazing on grass/clover pasture; 24kg fertilizer N/ha/yr; 16 stock units/ha; [d] From Schils et al (2005). NL dairy farm grass N: part housing, part grazing grass-only pasture; 275kg fertilizer N/ha/yr; 2.2 cows/ha; 8095kg fat and protein corrected milk production (FPCM)/cow; i.e. around 7600 litres milk/cow. [e] From Schils et al (2005). NL dairy farm grass/clover: part housing, part grazing grass/clover pasture; 69kg fertilizer N/ha/yr; 1.9 cows/ha; 8294kg FPCM/cow; i.e. around 7800 litres milk/cow; [f] From Lovett et al (2008). Ireland dairy system: 250 days of 24-hour grazing grass-only pasture; 330kg fertilizer N/ha/yr; 2.1 cows/ha; 6237kg milk/cow; [g] From Lovett et al (2008). Ireland dairy system: 149 days of 24-hour grazing grass-only pasture; 238kg fertilizer N/ha/yr; 1.7 cows/ha; 5753kg milk/cow; [h] From Flessa et al (2002). German cattle farm: year-round housing, 188kg fertilizer N/ha/yr; [i] From Flessa et al (2002). German organic cattle farm: six-month grazing, no synthetic N fertilizer.

The effect of fertilizer management to reduce total N_2O emissions is most effective in non-legume-based systems, for example the hybrid, grass-only and confinement systems commonly employed in the Northern Hemisphere.

Similarly, effluent management options will have a bigger impact on reducing N_2O emissions in hybrid and confinement systems, while improved wet-season soil and grazing management is a key option in pastoral systems. Nitrification inhibitors can be an effective mitigation option in all systems, targeting urine and dung in grazed pastures and effluent or fertilizer emissions in hybrid/confined systems. Although nitrification inhibitor use is being promoted in New Zealand and Australia to reduce N_2O emissions from urine patches, this technology has not been as readily adopted in the Northern Hemisphere as an effective N_2O mitigation option targeting N fertilizer or effluent.

Table 6.3 Greenhouse gas emissions (t CO_2-eq per farm system per year) from case study dairy farms (base farm) in four catchments in New Zealand and under two mitigation strategies (use of wintering pad or a nitrification inhibitor)

Catchment	Greenhouse gas	Base farm	Base farm with mitigation strategy	
			Wintering pad	Nitrification inhibitor
Toenepi	N_2O	175	162 (−7)	92 (−47)
	CH_4	393	407	418
	CO_2	205	236	221
	Total GHG	774	804 (+4) (−5)*	731 −6)(−15)*
Waiokura	N_2O	247	244 (−1)	134 (−46)
	CH_4	476	532	507
	CO_2	231	275	252
	Total GHG	955	1051 (+10)(−2)*	893 (−6)(−17)*
Waikakahi	N_2O	766	751 (−2)	463 (−40)
	CH_4	1305	1301	1567
	CO_2	872	916	1001
	Total GHG	2943	2969 (+1) (+1)*	3031 (+3)(−15)*
Bog Burn	N_2O	647	638 (−1)	309 (−52)
	CH_4	1277	1273	1410
	CO_2	682	718	742
	Total GHG	2606	2629 (+1) (+1)*	2460 (+6)(−14)*

Note: GHG = greenhouse gas; * relative change in greenhouse gas emission intensity (greenhouse gas emissions/unit of product); values in brackets represent the relative change (per cent) in emissions compared to the case study farm
Source: de Klein and Monaghan (2005)

In the pastoral systems employed in New Zealand and southern Australia, improved wet season grazing management (for example the use of wintering pads) and the use of nitrification inhibitors are key N_2O mitigation strategies (Di et al, 2007; Luo et al, 2008a). An assessment of their impact on whole-farm N_2O emissions from four key New Zealand dairying regions showed that these strategies could reduce N_2O emissions by 1–7 per cent (wintering pads) and 40–50 per cent (nitrification inhibitors) (Table 6.3). In hybrid systems in

Europe, key N_2O mitigation strategies include reducing N fertilizer use, reduced grazing, diet manipulation and improved effluent management (for example Schils et al, 2005; Lovett et al, 2008). Schils et al. (2005) estimated that reduced N fertilizer use and reduced grazing times (from 20 to 16 hours/day) each could reduce N_2O emissions from Dutch dairy farms by about 5 per cent. Mitigation options for confinement systems obviously need to focus on effluent or manure management systems, and examples include improved utilization of animal manure as fertilizer and (an)aerobic digestion of manure. Amon et al (2001) showed that manure from housed dairy cows emitted 35–40 per cent less N_2O when aerobically composted instead of being stacked.

Mitigation options that increase the efficiency of N within the soil/plant system are likely to increase pasture and/or animal productivity, which in turn is likely to increase methane emissions (for example increased stocking rates). Therefore, to fully assess the impact of N_2O mitigation options on total greenhouse gas emissions requires a whole-farm system analysis that accounts for all greenhouse gas emissions. The whole-farm models discussed above are useful tools for assessing the farm level impact of N_2O mitigation options. For example, Schils et al (2005) showed that although reduced N fertilizer use could reduce N_2O emissions by about 5 per cent, total greenhouse gas emissions were reduced by 3 per cent. Reduced grazing times did not reduce total greenhouse gas emissions, even though N_2O was reduced by 5 per cent. Similarly, de Klein and Monaghan (2005) estimated that the wintering pad N_2O mitigation options could increase total greenhouse gas emissions by 1–10 per cent (Table 6.3) due to an increase in CO_2 emissions associated with fuel use, supplementary feed production and fertilizer manufacturing and use. The effect of nitrification inhibitors on total on-farm) greenhouse gas emissions ranged from a 6 per cent reduction to a 6 per cent increase (Table 6.3) due to an increase in methane emissions associated with the utilization of the increased pasture production. However, expressed per unit of product, total greenhouse gas emissions from the systems using nitrification inhibitors were reduced by around 15 per cent, indicating that nitrification inhibitors increased the efficiency of the farming systems. This is an important consideration for identifying management strategies that have the largest reduction in environmental emissions for a given production level. However, as production levels continue to increase, the reduction in greenhouse gas per unit of product needs to be greater than the increase in production (products per ha) to ensure that net greenhouse gases are reduced.

The effectiveness of a mitigation strategy also depends on how it is adopted within a system and the choices a farmer makes. For example, studies have shown that the use of nitrification inhibitors can potentially increase pasture growth by between 0 per cent and 36 per cent (Di et al, 2007, Kelly et al, 2008; Smith L. C. et al, 2008). Using standard inventory methods Eckard (2008) estimated that 72 per cent of total greenhouse gas emissions from an average dairy farm in Southern Australia was due to enteric CH_4, while total N_2O equated to 28 per cent of total CO_2-equivalent emissions. Half of these N_2O emissions derived from urine deposition on pasture. A nitrification inhibitor applied to this pasture may reduce N_2O emissions from urine by 61 to 91 per cent (Di et al, 2007; Kelly et al, 2008; Smith L. C. et al, 2008), which equates to a whole-farm abatement of between 8.5 to 12.5 per cent CO_2 equivalent (CO_2-eq). If the nitrification inhibitor increased pasture yield by, say, 25 per cent, the farmer could either increase stocking rate by 25 per cent to utilize this extra DM or reduce N fertilizer inputs while keeping the stocking rate constant. With the former choice, whole-farm greenhouse gas emissions would increase by between 11.5 and 7.5 per cent as a result of increased enteric CH_4 emissions. However, whole-farm emissions would reduce by between 12.0 and 16.0 per cent CO_2-eq if the farmer chose to reduce N fertilizer inputs. Therefore, in the above example, recommendations for the use of nitrification inhibitors as an abatement strategy need to emphasize the potential to reduce N inputs, and/or to harvest a greater grass surplus for dry-season feeding, rather than importing additional feed or fertilizer onto the farm.

Discussion

The spatial and temporal variability of N_2O emissions in livestock agriculture continues to challenge researchers tasked with the development of accurate measurement techniques and realistic models for estimating N_2O emissions, and with evaluating the impact of mitigation strategies. Furthermore, emerging conflicts in meeting greenhouse gas reduction targets, as well as future trends in livestock agriculture and the impact of climate change on N_2O emissions, will no doubt provide additional challenges for researchers and policy makers in our quest to meet the global N_2O and greenhouse gas emission targets. Here we discuss some of these issues.

Emerging conflicts

Greenhouse gas intensity versus total emissions

Recent FAO reports show that world food demand is increasing, with a lack of basic supply resulting in hunger in some regions and changing food demand due to increasing affluence in other regions. The rate of productivity growth in agriculture is predicted to decrease from 2.3 per cent between 1961 and 2000, to 1.5 per cent between 2000 and 2030, and further to 0.9 per cent between 2030 and 2050 (FAO, 2009). In contrast, the number of people dependent on each hectare of arable land is predicted to increase from 2.4 to 4.5 and 6.4 over the same periods (FAO, 2009). In short, the world needs more protein and this requires a given amount of N input to produce. Whilst we can improve the efficiency of N conversion and potentially reduce N_2O emissions per unit of production, total N use in agriculture will need to increase to produce the protein needed to feed a growing world population. This inevitably means an increase in total global N_2O emissions. The primary food producers only have control over the efficiency per unit of production and we may need to design policies that reward increased food production with an acceptable greenhouse gas emission intensity. A decrease in greenhouse gas intensity can either be achieved by improving the efficiency of a livestock sector, or by putting more emphasis on livestock sectors with inherently low greenhouse gas emissions per unit of product. For example, Verge et al (2008, 2009) estimated that total greenhouse gas emissions for beef production in Canada was ~10kg CO_2-eq per kg of live weight, while those for pork production were ~ 2kg CO_2-eq per kg of live weight.

These greenhouse gas emissions intensity arguments are likely to be tabled in future World Trade talks, for example looking at where to produce food with the lowest carbon footprint. In parallel, research should continue to focus on improving the efficiency of conversion of nutrient N into products, whilst minimizing N_2O losses, as efficiency gains obtained with current N_2O mitigation strategies are not likely to be able to match the rate of productivity increase expected.

Mitigation and climate change adaptation

In many parts of the world, future climate change projections show warmer winters with increased growth potential, but drier summers with more variable rainfall and extreme heat (for example Cullen et al, 2009). A logical adaptation to this changing environment would be to increase N fertilizer applications in the warmer winter period, thus compensating for reduced summer production. However, it is obvious that this strategy could well increase total N_2O emissions. Thus there are a number of possible emerging conflicts between strategies for adapting to a changed climate as opposed to strategies to reduce total N_2O emissions. To avoid these conflicting messages to producers requires both mitigation and adaptation outcomes to be considered when developing policy and farmer advice.

Recognition of on-farm mitigation in national inventories

To credit countries for adoption of mitigation strategies, their effect on the N_2O inventory needs to be accounted for. Analogous to the adoption of country-specific EFs, the incorporation of mitigation options into the IPCC inventory methodology is subject to scrutiny by international expert review teams (ERTs). The ERTs will judge the soundness of the assumptions and assessment based on scientific evidence. However, the current structure of the IPCC methodology, i.e. estimating N_2O emissions from the size of N inputs in the system multiplied by an EF for each input, requires these EFs to be determined under relatively controlled conditions, so that the effect of a mitigation strategy on N_2O emissions from each source can be separated. However, these controlled conditions are not likely to represent the full farm system and care should be taken when translating the results of these controlled experiments into a whole system's response to a mitigation strategy. To fully understand the impact of a mitigation strategy and to appropriately account for this impact on inventories requires a combination of experiments under controlled conditions, determining the effect of EFs individually, coupled with field measurements and/or modelling at a paddock or farm level to fully understand the whole system's response to these strategies.

This highlights the importance of farm-scale inventory methodologies to underpin an agricultural emission trading scheme (ETS). Both Australia and New Zealand intend to impose a form of emissions constraint on their agricultural sectors, to assist with meeting the current and future emissions reduction targets. Currently the imposition of an ETS on agriculture is limited by lack of appropriate and agreed measurement and estimation methods. Australia's National Carbon Accounting System (NCAS) accounts for greenhouse gas emissions from land-based sectors. The NCAS includes an IPCC Tier 3 modelling framework that combines remote sensing of land cover change, land use and management data, climate and soil data, greenhouse gas accounting tools, and spatial and temporal ecosystem modelling (FullCAM), which also includes algorithms for accounting for N_2O spatially. However, it is critical that these spatial inventory methods

- align with actual on-farm mitigation action,
- are sufficiently flexible to incorporate new mitigation practices,
- use methods that are agreed by both policy and farmers alike,
- are easy to use and do not incur significant costs of compliance to farmers. Failing to achieve this is likely to result in farmers deliberately working to undermine the system, misguided mitigation efforts and significant increases in transaction costs associated with the implementation of ETS policy.

Future trends in livestock agriculture

Changing consumer demands and preferences

Whilst a clear trend is evident for increased animal-based protein in more affluent societies (Figure 6.6), red-meat consumption is likely to decline in most Western countries due to a combination of health and animal welfare concerns and increasing environmental awareness. It is also now clear that the total CO_2-eq from crop production is lower than that of dairy or red-meat production. Coupled with this, an increasing public awareness of animal welfare, an increase in vegetarianism and a recognition of the relatively high greenhouse gas footprint of ruminant production are likely to bring about a change in protein consumption away from those based on ruminant livestock production to ones based on cropping and mono-gastric production systems, in areas where these alternative systems are viable. In a recent study, Stehfest et al (2009) explored the potential impact of dietary changes on greenhouse gas emissions. These authors concluded that a global transition to a low-meat diet would not only benefit human health, but would also halve the estimated mitigation costs required to stabilize the greenhouse gas concentration of the atmosphere to 450ppm CO_2-eq. This stabilization target is expected to limit

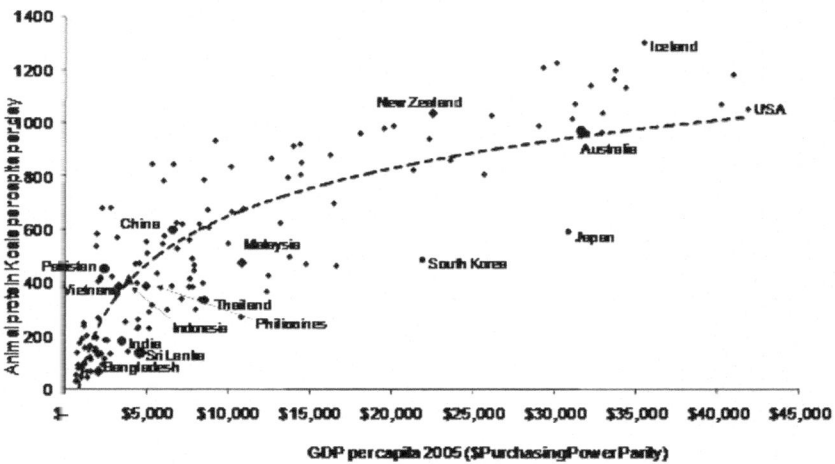

Figure 6.6 The effect of GDP per capita on animal-based protein consumption (including fish and dairy)

Note: Purchasing power parity ($ppp) is used by the United Nations and the FAO to convert international incomes to a common standard US$
Source: FAO (2009)

temperature increases to less than 2°C compared to pre-industrial levels (IPCC, 2007). The study did not account for any possible socio-economic changes as a result of a low-meat diet (for example effect of health changes on gross domestic product (GDP) and population numbers) or any agro-economic consequences, such as land prices, which could offset some of the estimated benefits. However, the results clearly highlight the potential greenhouse gas benefits of a low-meat diet, which, coupled with the health benefits, should be enough to give this objective serious consideration.

Impacts of climate change

Impacts on productivity

Although there is some level of certainty associated with increasing temperature and climate change, predictions of rainfall change are far less clear. It is evident that there will be some regions where rainfall will increase and seasonal patterns will change, and other regions where rainfall will decrease. These changing patterns of rainfall will affect N_2O emissions, particularly where associated with extreme events such as protracted droughts (for example in Australia) and extreme flooding. For example, N fertilizer use in southern and eastern Australia has decreased since 1997 with decreasing rainfall. This has inevitably resulted in less total N_2O emission, but also impacted on potential production from both pasture and crop production.

Plant responses to elevated carbon dioxide

Assuming no nutrient and water limitations, plants have been shown to increase their growth rates in the presence of elevated atmospheric CO_2 concentrations. The three principal physiological plant responses to elevated CO_2 concentration are:

1 an increase in leaf photosynthetic potential (especially in C3 pastures);
2 a decrease in stomatal, and therefore canopy, conductance;
3 a decrease in plant nitrogen content (Long et al, 2004).

Elevated CO_2 levels can induce a biogeochemical feedback mechanism, resulting in a reduction in nitrogen and/or phosphorus availability. This mechanism, referred to as progressive nutrient limitation (PNL) has been examined in recent studies (Finzi et al, 2006; Gill et al, 2006; Hungate et al, 2006) that suggest that the CO_2 fertilization effect on plant growth could be restricted due to the reduced availability of N and/or P. In legume-based pasture, where external N inputs occur via biological N fixation, phosphorus is likely to be the main limiting factor on plant growth. Although the PNL effect can be alleviated by external inputs of these nutrients, a better understanding of the full impacts of elevated CO_2 and/or increased temperatures on pastoral grazing systems is required to allow the development of management practices that are adapted to global changes (Soussana and Lüscher, 2007).

The net effect of warmer winter temperatures, higher plant growth rates under elevated CO_2, and PNL may mean that more N fertilizer is applied to crops and pastures in the future. This increased N input may well result in greater N_2O emissions in areas with increased rainfall, but may have limited impact in regions where water increasingly becomes a limiting factor.

Conclusions

Estimating and mitigating N_2O emissions from livestock agriculture continues to challenge researchers. Recent advances in measurement techniques and procedures have made significant improvements in the estimation of N_2O emissions. This has enhanced both the refinement of accounting and modelling approaches and the development of effective mitigation technologies, which need to continue to go hand in hand to ensure that the effect of new mitigation technologies can be accounted for in national inventories. Mitigation research on the N_2O reduction potential of soil/management interventions is generally further advanced than that of animal and feed-based intervention technologies. However, the potential of these latter interventions to reduce urinary N excretion provides a good indication of their efficacy as an N_2O abatement strategy. Of the currently available technologies, nitrification inhibitors, managing animal diets and fertilizer management show the best potential for reducing emissions in the short term. However, the growing world population will require a continued growth of livestock production and reducing global N_2O emissions will not only depend on the development of new N_2O abatement technologies, but also on policy decisions and consumer dietary changes. A move to policies that reward low-greenhouse-gas-intensity food production and a global transition to a low-meat diet could have a pronounced impact on agricultural greenhouse gas emissions.

References

Amon, B. Amon, T. T., Boxberger, J. and Alt, C. (2001) 'Emissions of NH_3, N_2O and CH_4 from dairy cows housed in a farmyard manure tying stall (housing, manure storage manure spreading)', *Nutrient Cycling in Agroecosystems*, vol 60, pp103–113

Bouwman, A. F. (1996) 'Direct emission of nitrous oxide from agricultural soils', *Nutrient Cycling in Agroecosystems*, vol 46, pp53–70

Bouwman, A. F., Boumans, L. J. M. and Batjes, N. H. (2002) 'Emissions of N_2O and NO from fertilized fields: Summary of available measurement data', *Global Biogeochemical Cycles*, vol 16, pp6-1–6-13

Breuer, L., Papen, H. and Butterbach-Bahl, K. (2000) 'N_2O emission from tropical forest soils of Australia', *Journal of Geophysical Research*, vol 105, pp26353–26367

Calanca, P., Vuichard, N., Campbell, C., Viovy, N., Cozic, A., Fuhrer, J. and Soussana, J. F. (2007) 'Simulating the fluxes of CO_2 and N_2O in European grasslands with the Pasture Simulation Model (PaSim)', *Agriculture, Ecosystems and Environment*, vol 121, pp164–174

Carulla, J. E., Kreuzer, M., Machmüller, A. and Hess, H. D. (2005) 'Supplementation of *Acacia mearnsii* tannins decreases methanogenesis and urinary nitrogen in forage-fed sheep', *Australian Journal of Agricultural Research*, vol 56, pp961–970

Castillo, A. R., Kebreab, E., Beever, D. E. and France, J. (2000) 'A review of efficiency of nitrogen utilisation in lactating dairy cows and its relationship with environmental pollution', *Journal of Animal and Feed Sciences*, vol 9, pp1–32

Chadwick, D. (1997) 'Nitrous oxide and ammonia emissions from grassland following applications of slurry: Potential abatement practices', in S. C. Jarvis. and B. F. Pain (eds) *Gaseous Nitrogen Emissions from Grasslands*, CAB International, Wallingford, pp257–264

Chen, D., Li, Y., Grace, P. and Mosier, A. R. (2008) 'N_2O emissions from agricultural lands: A synthesis of simulation approaches', *Plant and Soil*, vol 309, pp169–189

Clough, T. J., Bertram, J. E., Sherlock, R. R., Leonard, R. L. and Nowicki, B. L. (2006) 'Comparison of measured and EF5-r-derived N_2O fluxes from a spring-fed river', *Global Change Biology*, vol 12, pp352–363

Coffey, M. T. (1996) 'Environmental challenges as related to animal agriculture – swine', in E. T. Kornegay (ed) *Nutrient Management of Food Animals to Enhance and Protect the Environment*, Lewis Publishers, CRC, Boca Raton, FL, pp29–39

Conen, F. and Smith, K. A. (2000) 'An explanation of linear increases in gas concentration under closed chambers used to measure gas exchange between soil and the atmosphere', *European Journal of Soil Science*, vol 51, pp111–117

Cullen, B. R., Johnson, I. R., Eckard, R. J., Lodge, G. M., Walker, R. G., Rawnsley, R. P. and McCaskill, M. R. (2009) 'Climate change impacts on Australian pasture systems', *Crop and Pasture Science*, vol 60, pp933–942

Dalal, R. C., Wang, W., Robertson, G. P. and Parton, W. J. (2003) 'Nitrous oxide emission from Australian agricultural lands and mitigation options: A review', *Australian Journal of Soil Research*, vol 41, pp165–195

Davidson, E. A., Savage, K., Verchot, L. V. and Navarro, R. (2002) 'Minimizing artifacts and biases in chamber-based measurements of soil respiration', *Agricultural and Forest Meteorology*, vol 113, pp21–37

Davies, M. G., Smith, K. A. and Vinten, A. J. A. (2001) 'The mineralisation and fate of nitrogen following ploughing of grass and grass-clover swards', *Biology and Fertility of Soils*, vol 33, pp423–434

DCC (Department of Climate Change) (2008) 'National Inventory Report 2006 – Volume 1', The Australian Government Submission to the UN Framework Convention on Climate Change, June 2008, Department of Climate Change, Canberra

de Klein, C. A. M. and Eckard, R. J. (2008) 'Targeted technologies for nitrous oxide abatement from animal agriculture', *Australian Journal of Experimental Agriculture*, vol 48, pp14–20

de Klein, C. A. M. and Ledgard, S. F. (2005) 'Nitrous oxide emissions from New Zealand agriculture – key sources and mitigation strategies', *Nutrient Cycling in Agroecosystems*, vol 72, pp77–85

de Klein, C. A. M. and Monaghan, R. M. (2005) 'The impact of potential nitrous oxide mitigation strategies on the environmental and economic performance of dairy systems in 4 New Zealand catchments', *Environmental Sciences – Journal of Integrative Environmental Research*, vol 2, pp351–360

de Klein, C. A. M., Sherlock, R. R., Cameron, K. C. and van der Weerden, T. J. (2001) 'Nitrous oxide emissions from agricultural soils in New Zealand – a review of current knowledge and directions for future research', *Journal of The Royal Society of New Zealand*, vol 31, pp543–574

de Klein, C. A. M., Barton, L., Sherlock, R. R., Li, Z. and Littlejohn, R. P. (2003) 'Estimating a nitrous oxide emission factor for animal urine from some New Zealand pastoral soils', *Australian Journal of Soil Research*, vol 41, pp381–399

de Klein, C. A. M., Smith, L. C. and Monaghan, R. M. (2006) 'Restricted autumn grazing to reduce nitrous oxide emissions from dairy pastures in Southland, New Zealand', *Agriculture, Ecosystems and Environment*, vol 112, pp192–199

Del Grosso, S. J., Parton, W. J., Mosier, A. R., Walsh, M. K., Ojima, D. S. and Thornton, P. E. (2006) 'DAYCENT national-scale simulations of nitrous oxide emissions from cropped soils in the United States', *Journal of Environmental Quality*, vol 35, pp1451–1460

Denmead, O. T., Leuning, R., Jamie, I. and Griffith, D. W. T. (2000) 'Nitrous oxide emissions from grazed pastures: Measurements at different scales', *Chemosphere-Global Change Science*, vol 2, pp301–312

Di, H. J. and Cameron, K. C. (2002) 'The use of a nitrification inhibitor, dicyandiamide (DCD), to decrease nitrate leaching and nitrous oxide emissions in a simulated grazed and irrigated grassland', *Soil Use and Management*, vol 18, pp395–403

Di, H. J. and Cameron, K. C. (2005) 'Reducing environmental impacts of agriculture by using a fine particle suspension nitrification inhibitor to decrease nitrate leaching from grazed pastures', *Agriculture, Ecosystems and Environment*, vol 109, pp202–212

Di, H. J., Cameron, K. C. and Sherlock, R. R. (2007) 'Comparison of the effectiveness of a nitrification inhibitor, dicyandiamide, in reducing nitrous oxide emissions in four different soils under different climatic and management conditions', *Soil Use and Management*, vol 23, pp1–9

Dinuccio, E., Berg, W. and Balsari, P. (2008) 'Gaseous emissions from the storage of untreated slurries and the fractions obtained after mechanical separation', *Atmospheric Environment*, vol 42, pp2448–2459

Dittert, K., Lampe, C., Gasche, R., Butterbach-Bahl, K., Wachendorf, M., Papen, H., Sattelmacher, B. and Taube, F. (2005) 'Short-term effects of single or combined application of mineral N fertilizer and cattle slurry on the fluxes of radiatively active trace gases from grassland soil', *Soil Biology and Biochemistry*, vol 37, pp1665–1674

Dong, L. F., Nedwell, D. B., Colbeck, I. and Finch, J. (2004) 'Nitrous oxide emission from some English and Welsh rivers and estuaries', *Water, Air, and Soil Pollution: Focus*, vol 4, pp127–134

Eckard, R. J. (2008) 'The dairy, beef, sheep and grains farm greenhouse accounting tools', www.greenhouse.unimelb.edu.au/Tools.htm

Eckard, R. J. and Chapman, D. (2006) 'Modelling nitrous oxide abatement strategies in intensive pasture systems', *International Congress Series*, vol 1293, pp76–85

Eckard, R. J. and Franks, D. R. (1998) 'Strategic nitrogen fertilizer use on perennial ryegrass and white clover pasture in north-western Tasmania', *Australian Journal of Experimental Agriculture*, vol 38, pp155–160

Eckard, R. J., Bartholomew, P. E. B. and Tainton, N. M. (1995) 'The yield response of annual ryegrass *Lolium multiflorum* to varying nitrogen fertilizer application strategies', *South African Journal of Plant and Soil*, vol 123, pp112–116

Eckard, R. J., Chen, D., White, R. E. and Chapman, D. F. (2003) 'Gaseous nitrogen loss from temperate perennial grass and clover dairy pastures in south-eastern Australia', *Australian Journal of Agricultural Research*, vol 54, pp561–570

Eckard, R. J., Chapman, D. F. and White, R. E. (2007) 'Nitrogen balances in temperate perennial grass and clover dairy pastures in south-eastern Australia', *Australian Journal of Agricultural Research*, vol 58, pp1167–1173

Environmental Protection Agency (2008) *Inventory of US Greenhouse Gas Emissions and Sinks: 1990–2006*, US Environmental Protection Agency, Washington, DC

FAO (2009) 'FAO Corporate Statistical Database', http://faostat.fao.org/

Finzi, A. C., Moore, D. J. P., DeLucia, E. H., Lichter, J., Hofmockel, K. S., Jackson, R. B., Kim, H. S., Matamala, R., McCarthy, H. R., Oren, R., Pippen, J. S. and Schlesinger, W. H. (2006) 'Progressive nitrogen limitation of ecosystem processes under elevated CO_2 in a warm-temperate forest', *Ecology*, vol 87, pp15–25

Flessa, H., Ruser, R., Dörsch, P., Kamp, T., Jimenez, M. A., Munch, J. C. and Beese, F. (2002) 'Integrated evaluation of greenhouse gas emissions (CO_2, CH_4, N_2O) from two farming systems in southern Germany', *Agriculture, Ecosystems and Environment*, vol 91, pp175–189

Fowler, D., Skiba, U. and Hargreaves, K. J. (1997) 'Emissions of nitrous oxide from grasslands', in S. C. Jarvis and B. F. Pain (eds) *Gaseous Nitrogen Emissions from Grasslands*, CAB International, Wallingford, pp147–164

Fox, R. H., Myers, R. J. K. and Vallis, I. (1990) 'The nitrogen mineralization rate of legume residues as influenced by their polyphenol, lignin and nitrogen contents', *Plant and Soil*, vol 129, pp251–259

Galbally, I. (2005) 'A study of environmental and management drivers of non-CO_2 greenhouse gas emissions in Australian agro-ecosystems', *Environmental Sciences*, vol 22, pp133–142

Gill, R. A., Anderson, L. J., Polley, H. W., Johnson, H. B. and Jackson, R. B. (2006) 'Potential nitrogen constraints on soil carbon sequestration under low and elevated atmospheric CO_2', *Ecology*, vol 87, pp41–52

Grainger, C., Clarke, T., Auldist, M. J., Beauchemin, K. A., McGinn, S. M., Waghorn, G. C. and Eckard, R. J. (2009) 'Potential use of *Acacia mearnsii* condensed tannins to reduce methane emissions and nitrogen excretion from grazing dairy cows', *Canadian Journal of Animal Science*, vol 89, pp241–251

Granli, T. and Bockman, O. C. (1994) 'Nitrous oxide from agriculture', *Norwegian Journal of Agricultural Science,* Supplement no. *12*, pp1–128

Grant, R. F. and Pattey, E. (1999) 'Mathematical modeling of nitrous oxide emissions from an agricultural field during spring thaw', *Global Biogeochemical Cycles*, vol 13, pp679–694

Healy, R. W., Striegl, R. G., Russell, T. F., Hutchinson, G. L. and Livingston, G. P. (1996) 'Numerical evaluation of static-chamber measurements of soil-atmosphere gas exchange: Identification of physical processes', *Soil Science Society of America Journal*, vol 60, pp740–747

Hiscock, K. M., Bateman, A. S., Mühlherr, I. H., Fukada, T. and Dennis, P. F. (2003) 'Indirect emissions of nitrous oxide from regional aquifers in the United Kingdom', *Environmental Science and Technology*, vol 37, pp3507–3512

Hungate, B. A., Johnson, D. W., Dijkstra, P., Hymus, G., Stiling, P., Megonigal, J. P., Pagel, A. L., Moan, J. L., Day, F., Li, J., Hinkle, C. R. and Drake, B. G. (2006) 'Nitrogen cycling during seven years of atmospheric CO_2 enrichment in a scrub oak woodland', *Ecology*, vol 87, pp26–40

IPCC (Intergovernmental Panel on Climate Change) (1997) *Revised 1996 IPCC Guidelines for National Greenhouse Gas Inventories: Volume 3 Reference Manual*, IPCC/OECD/IEA, Paris

IPCC (2006) *2006 Guidelines for National Greenhouse Gas Inventories*, National Greenhouse Gas Inventories Programme, IGES, Hayama, Japan

IPCC (2007) *Climate Change 2007: Mitigation of Climate Change Contribution of Working Group III to IPCC Fourth Assessment Report*, Cambridge University Press, Cambridge

Johnson, I. R., Chapman, D. F., Snow, V. O., Eckard, R. J., Parsons, A. J., Lambert, M. G. and Cullen, B. R. (2008) 'DairyMod and EcoMod: Biophysical pastoral simulation models for Australia and New Zealand', *Australian Journal of Experimental Agriculture*, vol 48, pp621–631

Jordan, C. and Smith, R. V. (1985) 'Factors affecting leachate of nutrients from an intensively managed grassland in County Antrim, Northern Ireland', *Journal of Environmental Management*, vol 20, pp1–15

Kasimir-Klemedtsson, Å. (2001) 'Metodik för skattning av jordbrukets emissioner av lustgas' (Methodology for estimating the emissions of nitrous oxide from agriculture), Report 5170, Swedish Environmental Protection Agency, Stockholm

Kelly, K. B., Phillips, F. A. and Baigent, R. (2008) 'Impact of dicyandiamide application on nitrous oxide emissions from urine patches in northern Victoria, Australia', *Australian Journal of Experimental Agriculture*, vol 48, pp156–159

Kiese, R., Hewett, B., Graham, A. and Butterbach-Bahl, K. (2003) 'Seasonal variability of N₂O emissions and CH₄ uptake by tropical rainforest soils of Queensland, Australia', *Global Biogeochemical Cycles*, vol 17, doi:10.1029/2002GBO02014

Kleinman, P. J. A. and Soder, K. (2008) 'The impact of hybrid dairy systems on air, soil and water quality: Focus on nitrogen and phosphorus cycling', in R. W. McDowell (ed) *Environmental Impacts of Pasture-Based Farming*, CAB International, Wallingford, pp249–276

Ledgard, S. F. (1986) 'Nitrogen fertiliser use on pastures and crops', Ministry of Agriculture and Fisheries, Ruakura, New Zealand

Ledgard, S. F. (2001) 'Nitrogen cycling in low input legume-based agriculture, with emphasis on legume/grass pastures', *Plant and Soil*, vol 228, pp43–59

Ledgard, S. F. and Steele, K. W. (1992) 'Biological nitrogen fixation in mixed legume/grass pastures', *Plant and Soil*, vol 141, pp137–153

Ledgard, S. F., Menneer, J. C., Dexter, M. M., Kear, M. J., Lindsey, S., Peters, J. S. and Pacheco, D. (2007a) 'A novel concept to reduce nitrogen losses from grazed pastures by administering soil nitrogen process inhibitors to animals: A study with sheep', *Agriculture, Ecosystems and Environment*, vol 124, pp148–158

Ledgard, S. F., Welten, B., Menneer, J. C., Betteridge, K., Crush, J. R. and Barton, M. D. (2007b) 'New nitrogen mitigation technologies for evaluation in the Lake Taupo catchment', *Proceedings of the New Zealand Grasslands Association*, vol 69, pp 117–121

Letica, S. A., de Klein, C. A. M., Hoogendoorn, C. J., Tillman, R. W., Littlejohn, R. P. and Rutherford, A. J. (2010) 'Short term measurement of N₂O emissions from sheep-grazed pasture receiving increasing rates of fertiliser N in Otago, New Zealand', *Animal Production Science*, vol 50, pp17–24

Li, C., Frolking, S. and Frolking, T. A. (1992a) 'A model of nitrous oxide evolution from soil driven by rainfall events: 1. Model structure and sensitivity', *Journal of Geophysical Research*, vol 97, pp9759–9776

Li, C., Frolking, S. and Frolking, T. A. (1992b) 'A model of nitrous oxide evolution from soil driven by rainfall events: 2. Model applications', *Journal of Geophysical Research*, vol 97, pp9777–9783

Li, Y., Chen, D. L., White, R. E., Zhang, J. B., Li, B. G., Zhang, Y. M., Huang, Y. F. and Edis, R. (2007) 'A spatially referenced water and nitrogen management model (WNMM) for (irrigated) intensive cropping systems in the North China Plain', *Ecological Modelling*, vol 203, pp395–423

Long, S., Ainsworth, E., Rogers, A. and Ort, D. (2004) 'Rising atmospheric carbon dioxide: Plants FACE the future', *Annual Review of Plant Biology*, vol 55, pp591–628

Lovett, D. K., Shalloo, L., Dillon, P. and O'Mara, F. P. (2008) 'Greenhouse gas emissions from pastoral based dairying systems: The effect of uncertainty and management change under two contrasting production systems', *Livestock Science*, vol 116, pp260–274

Luo, J., Ledgard, S. F. and Lindsey, S. B. (2007) 'Nitrous oxide emission from urea application', *New Zealand Journal of Agricultural Research*, vol 50, pp1–11

Luo, J., Ledgard, S. F., de Klein, C. A. M., Lindsey, S. B. and Kear, M. (2008a) 'Effects of dairy farming intensification on nitrous oxide emissions', *Plant and Soil*, vol 309, pp227–237

Luo, J., Ledgard, S. F. and Lindsey, S. B. (2008b) 'A test of a winter farm management option for mitigating nitrous oxide emissions from a dairy farm', *Soil Use and Management*, vol 24, pp121–130

Luo, J., Saggar, S., Bhandral, R., Bolan, N., Ledgard, S. F., Lindsey, S. B. and Sun, W. (2008c) 'Effects of irrigating dairy-grazed grassland with farm dairy effluent on nitrous oxide emissions', *Plant and Soil*, vol 309, pp119–130

Meyer, C. P., Galbally, I. E., Wang, Y. P., Weeks, I. A., Jamie, I. and Griffith, D. W. T. (2001) 'Two automatic chamber techniques for measuring soil-atmosphere exchanges of trace gases and results of their use in the OASIS Field Experiment', CSIRO Atmospheric Research Technical Paper No. 51, CSIRO Atmospheric Research, Aspendale, Australia

Miller, L. A., Moorby, J. M., Davies, D. R., Humphreys, M. O., Scollan, N. D., MacRae, J. C. and Theodorou, M. K. (2001) 'Increased concentration of water-soluble carbohydrate in perennial ryegrass (*Lolium perenne L.*): Milk production from late-lactation dairy cows', *Grass and Forage Science*, vol 56, pp383–394

Min, B. R., Barry, T. N., Attwood, G. T. and McNabb, W. C. (2003) 'The effect of condensed tannins on the nutrition and health of ruminants fed fresh temperate forages: A review', *Animal Feed Science and Technology*, vol 106, pp3–19

Ministry for the Environment (2008) *New Zealand's Greenhouse Gas Inventory 1990–2006*, Ministry for the Environment, Wellington, New Zealand

Misselbrook, T. H., Powell, J. M., Broderick, G. A. and Grabber, J. H. (2005) 'Dietary manipulation in dairy cattle: Laboratory experiments to assess the influence on ammonia emissions', *Journal of Dairy Science*, vol 88, pp1765–1777

Monaghan, R. M., Smith, L. C. and Ledgard, S. F. (2009) 'The effectiveness of a granular formulation of dicyandiamide (DCD) in limiting nitrate leaching from a grazed pasture', *New Zealand Journal of Agricultural Research*, vol 52, pp145–159

Mosier, A. R., Parton, W. J. and Hutchinson, G. L. (1983) 'Modelling nitrous oxide evolution from cropped and native soils', in R. Hallberg (ed) *Environmental Biogeochemistry*, Publishing House/FRN, Elsevier B. V., Amsterdam, The Netherlands, pp229–241

Mosier, A., Kroeze, C., Nevison, C., Oenema, O., Seitzinger, S. and van Cleemput, O. (1998) 'Closing the global N_2O budget: Nitrous oxide emissions through the agricultural nitrogen cycle: OECD/IPCC/IEA phase II development of IPCC guidelines for national greenhouse gas inventory methodology', *Nutrient Cycling in Agroecosystems*, vol 52, pp225–248

Niezen, J. H., Waghorn, G. C., Graham, T., Carter, J. L. and Leathwick, D. M. (2002) 'The effect of diet fed to lambs on subsequent development of *Trichostrongylus colubriformis* larvae in vitro and on pasture', *Veterinary Parasitology*, vol 105, pp269–283

Oenema, O. (2006) 'Nitrogen budgets and losses in livestock systems', *International Congress Series*, vol 1293, pp262–271

Oenema, O., Wrage, N., Velthof, G. L., van Groenigen, J. W., Dolfing, J. and Kuikman, P. J. (2005) 'Trends in global nitrous oxide emissions from animal production systems', *Nutrient Cycling in Agroecosystems*, vol 72, pp54–65

Olesen, J. E., Schelde, K., Weiske, A., Weisbjerg, M. R., Asman, W. A. H. and Djurhuus, J. (2006) 'Modelling greenhouse gas emissions from European conventional and organic dairy farms', *Agriculture, Ecosystems and Environment*, vol 112, pp207–220

Palm, C. A. and Sanchez, P. A. (1991) 'Nitrogen release from the leaves of some tropical legumes as affected by their lignin and polyphenolic contents', *Soil Biology and Biochemistry*, vol 23, pp83–88

Parkin, T. B. (2008) 'Effect of sampling frequency on estimates of cumulative nitrous oxide emissions', *Journal of Environmental Quality*, vol 37, pp1390–1395

Parton, W. J., Mosier, A. R., Ojima, D. S., Valentine, D. W., Schimel, D. S., Weier, K. and Kulmala, A. E. (1996) 'Generalized model for N_2 and N_2O production from nitrification and denitrification', *Global Biogeochemical Cycles*, vol 10, pp401–412

Phillips, F. A., Leuning, R., Baigent, R., Kelly, K. B. and Denmead, O. T. (2007) 'Nitrous oxide fluxes measured from an intensively managed irrigated pasture using micrometeorological techniques', *Agricultural and Forest Meteorology*, vol 143, pp92–105

Potter, C. S., Matson, P. A., Vitousek, P. M. and Davidson, E. A. (1996) 'Process modeling of controls on nitrogen trace gas emissions from soils worldwide', *Journal of Geophysical Research*, vol 101, pp1361–1377

Reay, D. S., Smith, K. A. and Edwards, A. C. (2004) 'Nitrous oxide in agricultural drainage waters following field fertilisation', *Water, Air, and Soil Pollution: Focus*, vol 4, pp437–451

Riley, W. J. and Matson, P. A. (2000) 'NLOSS: A mechanistic model of denitrified N_2O and N_2 evolution from soil', *Soil Science*, vol 165, pp237–249

Rochette, P. and Eriksen-Hamel, N. S. (2008) 'Chamber measurements of soil nitrous oxide flux: Are absolute values reliable?', *Soil Science Society of America Journal*, vol 72, pp331–342

Rochette, P. and Janzen, H. H. (2005) 'Towards a revised coefficient for estimating N_2O emissions from legumes', *Nutrient Cycling in Agroecosystems*, vol 73, pp171–179

Rochette, P., Worth, D. E., Lemke, R. L., McConkey, B. G., Pennock, D. J., Wagner-Riddle, C. and Desjardins, R. L. (2008) 'Estimation of N_2O emissions from agricultural soils in Canada: I. Development of a country-specific methodology', *Canadian Journal of Soil Science*, vol 88, pp641–654

Russelle, M. P., Browne, B. A., Turyk, N. B. and Pearson, B. (2005) 'Denitrification under pastures on permeable soils helps protect ground water quality [abstract]', in F. P. O'Mara, R. J. Wilkins, L. 't Mannetje, D. K. Lovett, P. A. M. Rogers and T. M. Boland (eds) *Proceedings of the XXth International Grassland Congress, 26 June–1 July, 2005, Dublin, Ireland*, Wageningen Academic Publishers, Wageningen, The Netherlands, p692

Saggar, S., Hedley, C. B., Bolan, N. S., Bhandral, R. and Luo, J. (2004) 'A review of emissions of methane, ammonia, and nitrous oxide from animal excreta deposition and farm effluent application in grazed pastures', *New Zealand Journal of Agricultural Research*, vol 47, pp513–544

Sawamoto, T., Nakajima, Y., Kasuya, M., Tsuruta, H. and Yagi, K. (2005) 'Evaluation of emission factors for indirect N_2O emission due to nitrogen leaching in agro-ecosystems', *Geophysical Research Letters*, vol 32, pp1–4

Schils, R. L. M., Verhagen, A., Aarts, H. F. M. and Sebek, L. B. J. (2005) 'A farm level approach to define successful mitigation strategies for greenhouse gas emissions from ruminant livestock systems', *Nutrient Cycling in Agroecosystems*, vol 71, pp163–175

Schils, R. L. M., Verhagen, A., Aarts, H. F. M., Kuikman, P. J. and Sebek, L. B. J. (2006) 'Effect of improved nitrogen management on greenhouse gas emissions from intensive dairy systems in the Netherlands', *Global Change Biology*, vol 12, pp382–391

Schils, R. L. M., Olesen, J. E., del Prado, A. and Soussana, J. F. (2007) 'A review of farm level modelling approaches for mitigating greenhouse gas emissions from ruminant livestock systems', *Livestock Science*, vol 112, pp240–251

Smith, K. A. (1980) 'A model of the extent of anaerobic zones in aggregated soils, and its potential application to estimates of denitrification', *Journal of Soil Science*, vol 31, pp263–277

Smith, K. A. and Conen, F. (2004) 'Impacts of land management on fluxes of trace greenhouse gases', *Soil Use and Management*, vol 20, pp255–263

Smith, K. A. and Dobbie, K. E. (2001) 'The impact of sampling frequency and sampling times on chamber-based measurement of N_2O emissions from fertilized soils', *Global Change Biology*, vol 7, pp933–945

Smith, L. C., de Klein, C. A. M. and Catto, W. D. (2008) 'Effect of dicyandiamide applied in a granular form on nitrous oxide emissions from a grazed dairy pasture in Southland, New Zealand', *New Zealand Journal of Agricultural Research*, vol 51, pp387–396

Smith, W. N., Grant, B. B., Desjardins, R. L., Rochette, P., Drury, C. F. and Li, C. (2008) 'Evaluation of two process-based models to estimate soil N_2O emissions in Eastern Canada', *Canadian Journal of Soil Science*, vol 88, pp251–260

Somda, Z. C. and Powell, J. M. (1998) 'Seasonal decomposition of sheep manure and forage leaves in soil', *Communications in Soil Science and Plant Analysis*, vol 29, pp2961–2979

Soussana, J. F. and Lüscher, A. (2007) 'Temperate grasslands and global atmospheric change: A review', *Grass and Forage Science*, vol 62, pp127–134

Stehfest, E. and Bouwman, L. (2006) 'N$_2$O and NO emission from agricultural fields and soils under natural vegetation: Summarizing available measurement data and modeling of global annual emissions', *Nutrient Cycling in Agroecosystems*, vol 74, pp207–228

Stehfest, E., Bouwman, L., van Vuuren, D. P., den Elzen, M. G. J., Eickhout, B. and Kabat, P. (2009) 'Climate benefits of changing diet', *Climatic Change*, vol 95, pp83–102

Steinfeld, H., Greber, P. T. W., Castel, V., Rosales, M. and de Haan, C. (2006) 'Livestock's long shadow – environmental issues and options', Food and Agricultural Organization of the United Nations, Rome

Stevens, R. J. and Laughlin, R. J. (2002) 'Cattle slurry applied before fertilizer nitrate lowers nitrous oxide and dinitrogen emissions', *Soil Science Society of America Journal*, vol 66, pp647–652

Subbarao G. V., Ishikawa, T., Ito, O., Nakahara, K., Wang, H. Y. and Berry, W. L. (2006) A bioluminescence assay to detect nitrification inhibitors released from plant roots: A case study with *Brachiaria humidicola*', *Plant and Soil*, vol 288, pp101–112

van der Weerden, T. J., Sherlock, R. R., Williams, P. H. and Cameron, K. C. (1999) 'Nitrous oxide emissions and methane oxidation by soil following cultivation of two different leguminous pastures', *Biology and Fertility of Soils*, vol 30, pp52–60

van Groenigen, J. W., Velthof, G. L., Van Der Bolt, F. J. E., Vos, A. and Kuikman, P. J. (2005) Seasonal variation in N$_2$O emissions from urine patches: Effects of urine concentration, soil compaction and dung', *Plant and Soil*, vol 273, pp15–27

van Groenigen, J. W., Schils, R. L. M., Velthof, G. L., Kuikman, P. J., Oudendag, D. A. and Oenema, O. (2008) 'Mitigation strategies for greenhouse gas emissions from animal production systems: synergy between measuring and modelling at different scales', *Australian Journal of Experimental Agriculture*, vol 48, pp46–53

van Vuuren, A. M., van der Koelen, C. J., Valk, H. and de Visser, H. (1993) 'Effects of partial replacement of ryegrass by low protein feeds on rumen fermentation and nitrogen loss by dairy cows', *Journal of Dairy Science*, vol 76, pp2982–2993

Verge, X. P. C., Dyer, J. A., Desjardins, R. L. and Worth, D. (2008) 'Greenhouse gas emissions from the Canadian beef industry', *Agricultural Systems*, vol 98, pp126–134

Verge, X. P. C., Dyer, J. A., Desjardins, R. L. and Worth, D. (2009) 'Greenhouse gas emissions from the Canadian pork industry', *Livestock Science*, vol 121, pp92–101

Wassermann, V. (1979) 'Some views on the potential for legume-based pastures in South Africa', *Proceedings of the Grassland Society of South Africa*, vol 14, pp19–22

Wheeler, D. M., Ledgard, S. F., de Klein, C. A. M., Monaghan, R. M., Carey, P. L., McDowell, R. W. and Johns, K. L. (2003) 'OVERSEER® nutrient budgets 2 – moving towards on-farm resource accounting', *Proceedings of the New Zealand Grasslands Association*, vol 65, pp191–194

Wheeler, D. M., Ledgard, S. F. and de Klein, C. A. M. (2008) 'Using the OVERSEER nutrient budget model to estimate on-farm greenhouse gas emissions', *Australian Journal of Experimental Agriculture*, vol 48, pp99–103

Whitehead, D. C. (1995) *Grassland Nitrogen*, CAB International, Wallingford

7

Nitrous Oxide Emissions from Land-Use and Land-Management Change

Franz Conen and Albrecht Neftel

Introduction

World population has quadrupled during the 20th century. This has been accompanied by a dramatic increase in agricultural productivity, based to a major extent on industrial nitrogen fixation by the Haber-Bosch process developed at the beginning of that century. From a global perspective, population growth was, and still is, larger than the growth in agricultural productivity per unit land area. Consequently, agriculture continues to expand into previously pristine ecosystems. Also, emerging and growing demands for new products, such as biofuels from sugar cane, maize or palm oil, result in land-use changes. Increasing affluence in many parts of the world boosts demands for meat and milk products, driving a conversion of large areas of tropical forests into grasslands and soya bean fields. Land-use change has also taken place in temperate and boreal regions, where pristine grasslands have been converted to cropland, and wetlands have been drained to establish commercial forestry, pastures or cropland. Such changes often involve the clearance of forest vegetation, drainage, tillage of soil, and the introduction of new crops and animals, often combined with the use of mineral fertilizers and pesticides. Each activity affects plant and animal compositions and density, microclimate, hydrology and soil properties. Consequently, cycles of water, C, N, P and other nutrients are disturbed. If sustainable, a new type of land use may lead to different equilibria of pools and fluxes of nutrients. Where land use is not sustainable, it will be transitory and often results in abandonment and secondary forest succession (Hirsch et al, 2004). During transition from one type of land use to another, factors affecting N_2O emissions may change. Emissions may also be different from those in the pristine situation, once a new equilibrium has been reached.

N_2O is the product of microbial transformations of a range of N compounds in soil. Related processes and pathways are discussed in Chapter 2. In the context of land-use change, it is important to remember that the major control of N_2O emissions is the amount of mineral N (ammonium [NH_4^+] and nitrate [NO_3^-]) turned over by the soil microbial population (described by the hole-in-the-pipe model – see Chapter 5). Microbial turnover of N is a function of total N in the system, microbial activity and, importantly, the competition between plants and microbes for N. Any N taken up by plants no longer has the potential to be turned into N_2O before it is returned again to the soil. Advances in understanding the competition between plants and the soil microbial population, and how it affects turnover in N-limited and more fertile soils, has led to a new paradigm of N mineralization (Schimel and Bennett, 2004), which may help us to understand the effect of land-use changes on N_2O emissions. Batjes (1996) estimated that between 133 and 140Pg of organic N are stored in the organic matter of soils globally. Organic N becomes available to plants and microbes in successive steps, whereby the first ones play a minor role in most agricultural systems. The first of these steps is the depolymerization of organic matter by extracellular enzymes (Chapin et al, 2002). No N_2O is produced in this process. The resulting monomers (amino acids, amino sugars, etc.) are suggested to be the main forms of N taken up by plants and microbes when N availability is low. Plants access these compounds with the help of mycorrhizal fungi (Hobbie and Hobbie, 2008). With increasing N availability, amino acids are mineralized to NH_4^+, which is again taken up quickly by plants and microorganisms. Again, no N_2O is produced in the process. Where heterotrophic nitrifiers oxidize organic N to NO_3^- (heterotrophic nitrification of organic N), N_2O may be produced (Bateman and Baggs, 2005). The presence of N-fixing plants or external inputs of organic and mineral fertilizers in an ecosystem can substantially increase N availability, lead to increasing concentrations of NH_4^+ in soil and reduce the competition between plants and microbes. Autotrophic nitrifiers will flourish, derive energy from oxidizing NH_4^+ to NO_3^- and generate some N_2O as a by-product. When N availability increases even further, the proportion of NH_4^+ transformed into NO_3^- increases, as does the concentration of NO_3^-, which may be taken up by plants or reduced stepwise by O_2-limited microorganisms to N_2O or N_2 – the denitrification process. There are more pathways of N_2O production, discussed elsewhere (for example Bothe et al, 2007, and in Chapter 2). In summary, the number of possible N transformations in the course of which N_2O may be produced, as well as the total amounts of N_2O generated, increase with increasing N availability. Understanding the influence of land-use changes on N availability enables us to infer consequent effects on N_2O emissions.

A major factor governing N availability in ecosystems is the C:N ratio of plant litter and soil organic matter, which depends on the vegetation and can change with land use. In a tropical forest, for example, the C:N ratio of litter can be 25, whereas it reaches values of 60 to 70 in pastures replacing it (Wick et al, 2005). Larger C:N ratios are associated with smaller rates of mineral N release, transformation and ultimately N_2O emission. This has to do with the C:N ratio of the microbial decomposer community and its C use efficiency (Manzoni et al, 2008). Soil microbial biomass has a relatively stable C:N ratio of around 8.6 across different ecosystems (Cleveland and Liptzin, 2007). Let us assume that from one unit of C in organic matter, decomposers assimilate 0.4 and respire 0.6 units (Chapin et al, 2002). Then, N is in excess of the microbial requirement for a C:N ratio of 8.6 and excreted as NH_4^+ when the C:N ratio of the decomposed organic matter is below 21.5 (8.6:0.4). The proportion of excreted N increases strongly with decreasing C:N ratio. Springob and Kirchmann (2003) determined an increase in N release by about a factor of five when C:N ratios in soil organic matter dropped from 15 to 10. Equally, large C:N ratios result in small rates of mineral N release during decomposition and, subsequently, in smaller rates of mineral N turnover and N_2O formation.

Forest conversion

The largest land-use changes are currently taking place in the tropics. Plate 7.1 shows a typical example of forest clearance in Peru. An estimated 13.7×10^6ha of tropical forests are expected to be cleared annually between 2008 and 2012, while only 2.6×10^6ha of cleared land in the tropics are annually afforested (IPCC, 2000). In the Brazilian Amazon region alone, deforestation during the past two decades has been estimated to be 1 to 4×10^6ha annually, with 70 per cent of this area being converted to pasture (Serrão, 1992; Melillo et al, 2001). Tropical forest soils are thought to be the largest single natural source of N_2O to the atmosphere (Stehfest and Bouwman, 2006). Therefore, it is important to know the effect of land-use change on these emissions. In this section we look at the N cycle and N_2O emissions in pristine tropical forests, at what happens during forest conversion and at the different outcomes depending on the type of land use succeeding the primary forest.

Tropical lowland soils are often millions of years old and highly weathered. Weathering of rock and soil minerals releases plant-available P, but reductions in weathering inputs with time, combined with continuing losses of dissolved forms of P, lead eventually to P limitation. In contrast, N inputs originate from atmospheric N_2. This pool of N is abundant but its transformation into a form available to organisms is limited by the energy demand of the transforming processes. Atmospheric N_2 is made available to the ecosystem by N-fixing microorganisms, often living in symbiosis with certain trees, and by atmospheric N_2 oxidation to NO_x, for example by lightning or volcanic activity. Within 20,000 years, the increasing abundance of N and the decrease in P availability results in P, instead of N, becoming the limiting factor, a reversal of the initial situation (Hedin et al, 2003). Changes in N pools and transformations along a chronosequence on Hawaii indicate an easing of N limitations relatively early in forest development (Hedin et al,

2003). Concentrations of NO_3^- were low relative to those of NH_4^+ in forest on 300- and 2100-year-old substrate, suggesting a strong competition between plants and microorganisms for NH_4^+ and little nitrification at these stages in a succession. Accordingly, N_2O fluxes were negligible (< 0.01kg N_2O-N ha^{-1} yr^{-1}). However, in forests on substrates 20,000 years old, exchangeable NO_3^- increased and remained high in all older forests. Also, net nitrification rates had increased substantially up to 20,000 years. At the older sites, nearly 100 per cent of mineralized N was converted to NO_3^-. Consequently, fluxes of N_2O increased up to 1.1kg N_2O-N ha^{-1} yr^{-1} (Hedin et al, 2003). This is close to the mean of 77 observations (0.85kg N_2O-N ha^{-1} yr^{-1}) summarized by Stehfest and Bouwman (2006). Much larger values have been reported for individual sites, ranging up to annual fluxes of 6–7kg N_2O-N ha^{-1} yr^{-1} in the undisturbed Tapajos National Forest, near Santarem, Brazil (Keller et al, 2005).

Vegetation in a tropical rainforest has been found to take up three times as much mineral N as is taken up by microbes. Even so, 0.7 per cent of mineral N derived from organic matter decomposition was emitted in the form of N_2O (Templer et al, 2008). Forest clearance initially reduces the uptake of mineral N by vegetation. Consequently, availability of mineral N to the microbial population may drastically increase and so may N_2O emissions. Indeed, Melillo et al (2001), in western Amazonia, found pasture emissions to be 2.5 times those from the forest soils for the first two years, but decreasing to below the forest rate in pastures three or more years old. A similar observation was made by Neill et al (2005) in Rhondonia, Brazil, where annual emissions from forests were 1.7–4.3kg N_2O-N ha^{-1} yr^{-1}, whereas young pastures (one to three years) emitted 3.1–5.1kg N_2O-N ha^{-1} yr^{-1} and older pastures (≥ six years) between 0.1 and 0.4kg N_2O-N ha^{-1} yr^{-1}. Earlier, Keller et al (1993), in a very fertile region of Costa Rica, had found N_2O emissions (driven by rapid N mineralization and denitrification in moist soils) very much greater than those from the Amazon (Figure 7.1), decreasing to below the forest emissions only after 18 years after conversion to pasture. Veldkamp et al (1999), in a follow-up study at the same sites, recorded fluxes only about one-third of those found earlier, but the work was done in relatively dry periods, thus possibly giving an underestimation of the long-term mean. Forest clearance and establishment of coffee gardens in Sumatra show a similar pattern (Verchot et al, 2006): initially increasing N_2O emissions followed by a decrease, although not below forest emissions within the first ten years (Figure 7.1). Also much shorter transient increases in N_2O lasting only a few months have been seen following the slash-and-burn of secondary forest (Weitz et al, 1998). Some studies even report no such effect at all. In eastern Amazonia, Verchot et al (1999) recorded emissions from pastures only one-eighth to one-third of those from primary forest, similar to later observations by Wick et al (2005) in central Amazonia.

The latter study identified an increase in litter C:N ratio from around 25 in the forest to between 40 and 70 in two-year-old pastures and a strong decline in microbial biomass following deforestation. Wick et al (2005) presume pro-

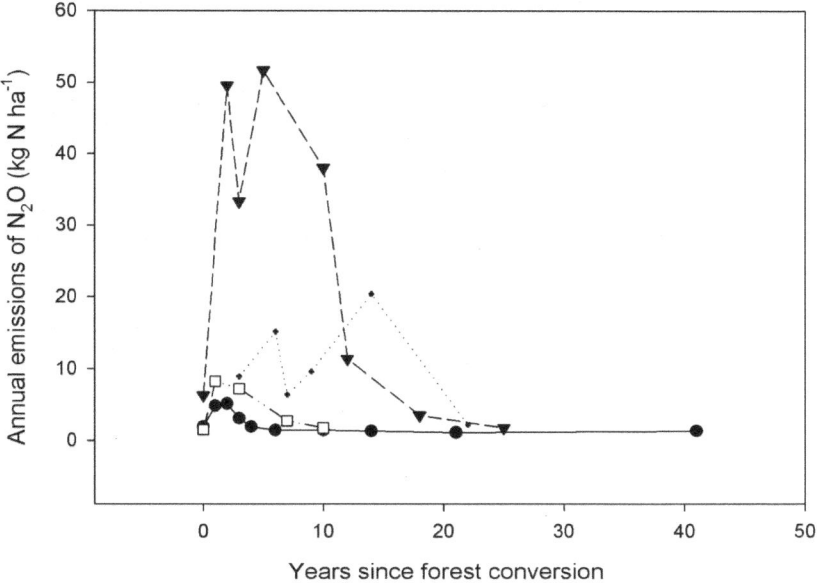

Figure 7.1 Emissions of N_2O from tropical forest (time 0) and during the first decades following clearance for pasture and coffee gardens

Source: ▼ Keller et al (1993); ◆ Veldkamp et al (1999); ● Melillo et al (2001); □ Verchot et al (2006)

gressive N limitation with pasture age to be probably caused by a combination of early loss of N through gaseous emission and leaching, reduced N fixation and the immobilization of N by grass litter and roots with a high C:N ratio. Consequently, rates of N mineralization and nitrification are smaller under pasture than under forest (Neill et al, 1999). Losses of N are further enhanced by the burning of biomass during forest clearance and on established pastures, which is a commonly used way to stimulate regrowth of fresh and palatable forage. Fire turns most of the organic N in the biomass into N_2 but some NO_x and N_2O is also produced pyrogenically. Detailed measurements from tropical dry deciduous forests in India revealed that 0.05 per cent to 0.07 per cent of N can be released in this way in the form of N_2O (Prasad et al, 2002).

The overall effect of land-use change on regional N_2O emissions is difficult to assess. A gradual decline in fertility as a consequence of reduced N availability leads to degradation of pastures and the described decrease in N_2O emissions. An estimated 12×10^6ha of pastures in the Amazon region show some degree of degradation (do Carmo and Cerri, 2007). Modelling by Melillo et al (2001) indicates that the total contribution of the Amazon Basin to global N_2O emissions has decreased slightly as a consequence of the land-use changes. Where grassland management is no longer economically viable, pastures are abandoned and secondary forest may establish. Such a development can reverse the situation regarding N limitation and N_2O emissions (Davidson et al, 2007). Concentrations of N in litter increase with secondary forest establishment. Forests older than 20 years can already show signs of N abundance, as indicated by ratios of soil extractable NO_3^- to NH_4^+ greater than 1. Almost in parallel, annual emissions of N_2O can grow from near 0.1kg N_2O-N ha^{-1} to 1–2kg N_2O-N ha^{-1} (Davidson et al, 2007).

There are different ways by which forests are converted into agricultural land. Instead of slashing and burning the initial vegetation, it can be chopped and mulched. Davidson et al (2008) followed both systems during the first two years of establishment. They found that the chop-and-mulch system conserves plant nutrients and offers farmers new economic opportunities. However, it enhances N_2O emissions by 50 per cent, although the overall greenhouse gas emissions in terms of CO_2 equivalent over a 100-year timeframe are six times smaller than under slash-and-burn management, mainly because of reduced CH_4 emissions from fire.

Growing N-fixing trees, such as acacia, is another way to increase N cycling and raise productivity. Arai et al (2008) showed in a study on Sumatra, Indonesia, that in acacia plantations the ratio of NO_3^- to NH_4^+ was larger than in secondary forest, suggesting an easing of N shortage by the presence of acacia. At the same time, N_2O emissions increased from 0.33kg N_2O-N ha^{-1} yr^{-1} in the secondary forest to 2.56kg N_2O-N ha^{-1} yr^{-1} in the acacia plantations. However, in terms of CO_2 equivalent, N_2O emissions from these plantations cancelled out less than 10 per cent of the CO_2 uptake during the growth of these trees.

Agricultural sustainability can also be increased by adding residues from N-fixing shrubs during the fallow period. A study in western Kenya by Millar (2002, cited in Mutuo et al, 2005) has shown that this raises N_2O emissions from 1kg N_2O-N ha^{-1} yr^{-1} to between 1.4 and 4.9kg N_2O-N ha^{-1} yr^{-1}. Emission factors for residues from N-fixing shrubs were estimated to be between 0.5 per cent and 1.9 per cent, compared to an emission factor of only 0.2 per cent from the natural-fallow residues with probably a much higher C:N ratio.

Other plantations commonly replacing tropical forest are rubber and oil palm, and these have been studied by Ishizuka et al (2005). They found that N_2O emissions from such plantations in Jambi province, Indonesia, were spatially very heterogeneous but were of a similar magnitude to those from forests (0.76kg N_2O-N ha^{-1} yr^{-1}) and a factor of five larger than from grassland. Comparable emissions from rubber plantation and primary forest were also determined in Southwest China (Werner et al, 2006). Melling et al (2007) followed N_2O emissions over one year on tropical peatland in Sarawak, Malaysia, where a mixed peat swamp forest emitted least N_2O (0.7kg N_2O-N ha^{-1} yr^{-1}), followed by a palm oil plantation (1.2kg N_2O-N ha^{-1} yr^{-1}) and a sago (*Metroxylon sagu*) ecosystem (3.3kg N_2O-N ha^{-1} yr^{-1}). In warm humid tropics, such as Central Sulawesi, Indonesia, cacao agroforestry is a sustainable form of agriculture. However, N_2O emissions (3.1kg N_2O-N ha^{-1} yr^{-1}) may be larger than from maize (0.8kg N_2O-N ha^{-1} yr^{-1}) or from secondary forest (2.2kg N_2O-N ha^{-1} yr^{-1}) (Veldkamp et al, 2008). Larger emissions of N_2O per unit land area in more sustainable forms of land use are usually accompanied by a larger productivity (for example income generated per unit land area). Consequently, the emission of N_2O per unit product may often be much lower than in less sustainable systems with smaller emissions but also a much lower productivity.

To summarize, N_2O emissions from tropical forests tend to be large compared to those from other natural ecosystems. Large NO_3^-/NH_4^+ ratios in soil indicate N abundance and high rates of nitrification and denitrification, both N_2O-producing processes. Where forests are converted, emissions may transiently increase for a number of years. However, burning of vegetation and replacement with grassland, where C:N ratios in litter are much greater than in forest, often degrades the fertility of the ecosystem because of N limitation and N_2O emissions decline below values previously observed under forest. Restoration, for example through N-fixing shrubs, restores previous rates of N cycling but also leads to larger N_2O emissions. Rubber, oil palm, coffee or cacao plantations may be more profitable and sustainable than grassland. Yet they tend to have similar, sometimes larger, emissions than the forests they succeed.

Conversion of grassland to cropland

Conversion of pristine grassland to cropland in temperate regions mainly occurred between the mid-19th and the mid-20th centuries. Major conversions include the ploughing of prairie by settlers in the US (Figure 7.2) and later the large-scale land reclamation in the steppes of the Soviet Union during the 1950s.

Cultivation of grassland accelerates mineralization of soil organic matter and the release of mineral N, part of which is turned into N_2O. Experimental ploughing up of a grass sward in southeast Scotland, for example, yielded 449kg N ha^{-1} by mineralization, and associated emissions of 2.0kg N_2O-N ha^{-1} over 18 months. Even larger N_2O emissions were observed after ploughing of a grass-clover sward, where 244kg N ha^{-1} were released by mineralization and 4.5kg as N_2O-N ha^{-1} (Davies et al, 2001). The N_2O EFs for these two sites were therefore 0.45 and 1.6 per cent, respectively. These values fall within the uncertainty range of the IPCC default value of 0.3 to 3 per cent (default value: 1 per cent) for the direct emission of N_2O from agricultural soils. There are data available on the rates of decline of soil C and N after bringing land into cultivation (for example Tiessen et al, 1982; Bowman et al, 1990; Lobe et al, 2001) (Figure 7.3), but no associated measurements of N_2O emissions.

Smith and Conen (2004) estimated the likely scale of emissions following conversion of grassland to cropland by analogy with the IPCC methodology at

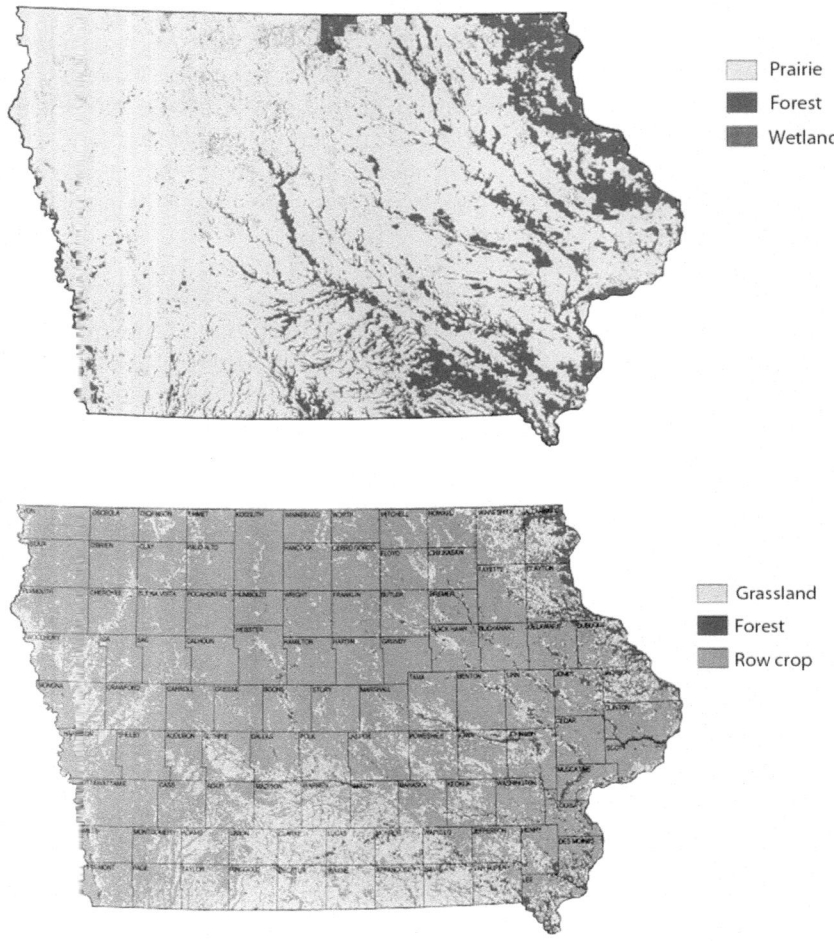

Figure 7 2 Dramatic, swift and almost complete change of prairie grassland to cropland in Iowa between the (upper panel) 1850s and (lower panel) 1990s

Source: Adapted from Iowa Department of Natural Resources (2000)

that time. They assumed that mineral N liberated by mineralization of soil organic matter and plant remains, following land-use change, can be regarded as a comparable potential source of N_2O to N being applied in the form of synthetic N fertilizer. The IPCC (2006) has subsequently adopted the same approach, in which the NH_4^+ and NO_3^- resulting from the mineralization of soil organic matter following a change in land use is deemed to be of the same value as a substrate for microorganisms producing N_2O by nitrification or denitrification as the NH_4^+ and NO_3^- in an application of synthetic N fertilizer. By way of illustration, Smith and Conen (2004) applied this approach

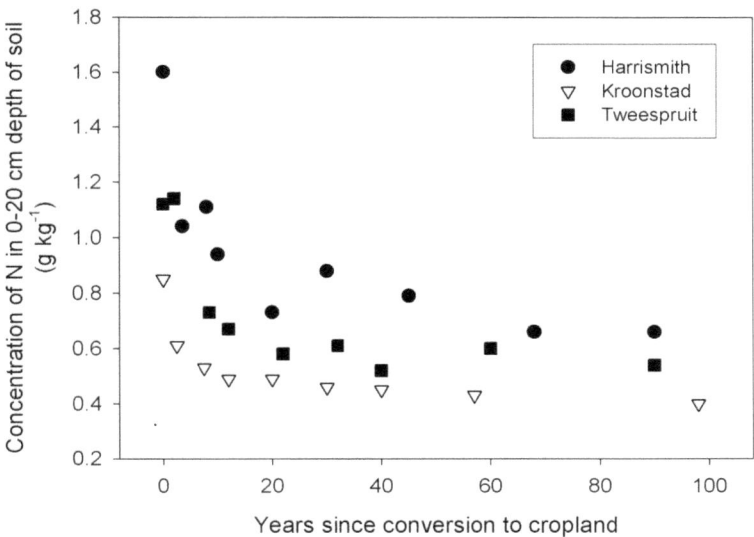

Figure 7.3 Loss of N from soil after the cultivation of grassland in three ecosystems of the South African Highveld

Note: It can be expected (on the basis of IPCC methodology) that 1 per cent of the total N lost has been emitted as N_2O.
Source: Data from Lobe et al (2001)

to the data from Lobe et al (2001) for the mineralization of N in savanna grassland soils of the Highveld of South Africa. The average rate of N loss from the soil organic matter in three soils was 7.8 per cent of the rate of C loss, corresponding to 90–170kg N ha^{-1} yr^{-1} over the first eight years after cultivation. The net change in emission from these soils over this period predicted by applying the then current IPCC default EF of 1.25 per cent was 1.1–2.1kg N_2O-N ha^{-1} yr^{-1}; applying the updated (IPCC, 2006) default factor of 1 per cent, the emission would be 0.9–1.7kg N_2O-N ha^{-1} yr^{-1}. This emission augments the global warming impact already caused by the loss of soil C in the form of CO_2 through the ploughing of these grasslands by another 10 per cent. The historical loss of soil C and its impact on atmospheric CO_2 enrichment may be mitigated by soil management conducive to enhancing soil organic matter contents, such as reversion to grassland or reduced tillage. However, once the N_2O is released during the conversion of grasslands to cropland it cannot be similarly withdrawn from the atmosphere.

Drainage and cultivation of organic soils

Moving even further away from the tropics, we find another form of land-use change that had a major impact on N_2O emissions. Boreal and subarctic peatlands have accumulated around 455Pg C during the postglacial period

(Gorham, 1991). A mean C:N ratio of 25.8 (Batjes, 1996) indicates a pool of organic N around 17.6Pg in these soils. While vegetation adapted to flooded conditions may thrive in a wetland, a lack of oxygen in flooded soil severely limits the mineralization of shed litter by microorganisms. Consequently, organic matter, and the N that is a part of it, has accumulated. Emissions of N_2O from undisturbed wetlands are small (Martikainen et al, 1993; Brumme et al, 1999) for at least four reasons. First, small rates of mineralization mean little availability of NH_4^+. Second, oxygen limitation restricts nitrification, one of the pathways by which N_2O can be produced. Consequently, third, very little NO_3^- is generated that may be turned into N_2O by denitrification. Fourth, under oxygen limitation, N_2O is itself a welcome electron acceptor to the microbial community and isturned efficiently into N_2 (Vieten et al, 2009). If one desires to grow agricultural crops, grass or commercial forest on organic soils, then these soils have to be drained. Once the water level has been lowered, the layers of organic matter above that level become aerated; supplied with atmospheric oxygen, microbial decomposers proliferate and rapid mineralization sets in. Rogiers et al (2008) have estimated drainage-induced mineralization equivalent to 5–9Mg C ha^{-1} yr^{-1} in a subalpine sedge peat in Switzerland. Similar rates have been observed in other parts of the world. Mineralization produces NH_4^+, and the availability of oxygen permits nitrification, which produces NO_3^- and some N_2O. Denitrification rates increase and so does total N_2O production. Because of a better aeration compared to the situation previously, denitrification is now less complete and a larger proportion of the reduced NO_3^- is released as N_2O rather than as N_2.

Drainage of organic soils and associated effects on N_2O emissions have been well documented in Scandinavia. Weslien et al (2009) describe the historical development in Sweden, where in the 1930s around 0.65–0.70 × 10^6ha of organic soil were farmed. Economic and political reasons led to the abandonment of 0.40–0.45 × 10^6ha by the 1990s. Nowadays, most of the abandoned area is probably under forest. Alm et al (2007) describe an even more dramatic situation in Finland. Of the original 10.4 × 10^6ha of pristine ecosystems, 5.4–5.7 × 10^6ha have been drained for forestry and 0.7–1.0 × 10^6ha for agriculture. Further major areas of drained organic soils are found in the Baltic states and in Russia (Rydin and Jeglum, 2006). Emissions of N_2O from drained organic soils are, per unit area, much larger than those discussed for tropical forests in the first section. The IPCC (2006) default value for temperate organic crop and grassland soils is 8kg ha^{-1} yr^{-1} with an uncertainty envelope of 2–24kg ha^{-1} yr^{-1}. Maljanen et al (2007) have summarized field measurements on cultivated organic soils in Finland. They range, in kg N_2O-N ha^{-1} yr^{-1}, for barley fields from 5.4 to 24.1, for grass from 1.7 to 11.0 and for fallow from 3.8 to 37.0. Although it is common practice on highly fertile drained organic soils to add no N fertilizer at all to cereal production (Kasimir Klemedtsson et al, 2009), part of the reported emissions may be associated with the use of mineral N fertilizer as part of the cultivation. However, such fertilization seems to make little difference to the overall emission total. Comparing cultivated with abandoned fields, Maljanen et al (2007) found emissions to be similar, even 30 years after abandonment. A lack of correlation between emissions and time since abandonment suggests that enhanced emissions will still continue for a much longer time.

Afforestation of abandoned organic cropland does not ameliorate the situation because availability of mineral N still remains high (Mäkiranta et al, 2007). Mineralization of organic matter and subsequent nitrification continues to provide a source of NO_3^- for denitrification (Weslien et al, 2009), probably the dominant process of N_2O production in cultivated organic soils (Maljanen et al, 2003). Nevertheless, greater assimilation and storage of CO_2 by forest, compared to crops, may improve the greenhouse gas balance, even if N_2O emissions remain unchanged (Weslien et al, 2009).

In the previous section we discussed evidence for about 1 per cent of mineralized N being emitted in the form of N_2O. This assumption is endorsed by a large number of studies (summarized in IPCC, 2006) on mineral soils. However, a larger proportion than 1 per cent of mineral N may be emitted as N_2O from organic soils. Because of generally more acidic conditions in organic soils, compared to most mineral soils, reduction of N_2O to N_2 may be severely inhibited (Stevens and Laughlin, 1998; Simek et al, 2002). Data in Maljanen et al (2007) and Weslien et al (2009) support this hypothesis.

Maljanen et al (2007) report CO_2 and N_2O emissions from different land uses on cultivated organic soil. One type of land use is bare soil without any plants. Here, CO_2 emissions are a good proxy for mineralization of soil organic matter. Where plants are present, part of the CO_2 comes from root respiration, which is not easily separated from CO_2 resulting from the mineralization of soil organic matter. The mean CO_2 emission from the four fallow sites was 5910kg CO_2-C ha^{-1} yr^{-1}. Annual mineralization of N estimated from the average C:N ratio of these sites (21.6) was 274kg N ha^{-1} yr^{-1}. Measured mean N_2O emission was 16.7kg N_2O-N ha^{-1} yr^{-1}, or 6 per cent of the mineralized N. While N_2O emissions from these disturbed ecosystems may not decline in the foreseeable future, their share of the greenhouse gas budget declines because emissions from other sources increase. Still, they can constitute a surprisingly large fraction of a country's greenhouse gas inventory. For Finland in 2004, for example, N_2O emissions from the cultivation of organic agricultural soils constituted 1.7 per cent of the total net greenhouse gas emissions in terms of global warming potential (Lapveteläinen et al, 2007).

Reduced tillage in annual field crops

In this last section, we look at reduced tillage as a land-management practice currently seen as a possible way to withdraw some CO_2 from the atmosphere and store it in the form of organic C in soil. Reduced tillage is, in principle, applicable to all types of annual cropping in any part of the world. Initially, cultivation methods with reduced tillage were developed for purely agronomic reasons, such as the preservation of soil moisture, for reducing wind erosion, or for reducing the energy demand in field preparations. Commonly these methods are termed 'no-till', although soil disturbance cannot be completely avoided, for example during the seeding of crops. No-till also tends to increase soil organic C contents, so additional interest in promoting this practice has been raised in the light of climate change.

Data in Smith et al (1998) indicate a potential for C sequestration through a change to no-till systems of about 350kg C ha^{-1} yr^{-1}. Results of 76 long-term experiments in the US provided an almost identical number (337kg C ha^{-1} yr^{-1}) (West and Marland, 2002). Tan et al (2006) estimate that if all croplands in the east-central US under conventional tillage in 1992 were converted to no-till, the soil organic C pool would increase by 16.8 per cent by 2012, equivalent to the sequestration of 0.5Pg CO_2 over 20 years. The scale and permanence of such measures are debatable but the short-term effect on atmospheric CO_2 concentrations is certainly welcome. However, concerning the effect of reduced tillage on N_2O emissions, we have to look carefully at different soil and environmental conditions.

Initially, we remain with the cultivated organic soils discussed in the previous section. Kasimir Klemedtsson et al (2009) observed large N_2O emission peaks, the largest >15.9mg N_2O m^{-2} hour^{-1}, after ploughing and harrowing of a highly fertile drained organic soil in southern Sweden. Elevated emissions continued for about four days before returning to previous levels. They presumed the mechanical perturbation had created and exposed new soil surfaces where rapid mineralization could take place. Since there were no growing plants at the time to provide a sink for released mineral N, the latter was entirely available to the nitrifier and denitrifier communities. Tillage was timed to coincide with soil conditions being favourable to sowing the crop. The same conditions, warm and moist, are also favourable for N_2O production. To reduce emissions, Kasimir Klemedtsson et al (2009) suggested that soil perturbations should be minimized. Increasing the water table was seen as an additional option, although annual crops may no longer be cultivated under such conditions. Consequently, annual crops would need to be replaced with long-standing grassland and tillage would cease completely. On organic soil in north-central Ohio, Elder and Lal (2008) investigated whether minimizing soil perturbation may indeed reduce N_2O emissions from cultivated organic soils. They compared mouldboard ploughing with direct seeding or planting of maize, associated with minimal soil disturbance (no-till). As usual in N_2O studies, observed emissions were notoriously variable. However, differences between no-till and ploughed cultivation were statistically significant; no-till soil management resulted in a reduction of N_2O emission by approximately 63 per cent. Assuming all cultivated organic soils in the US are currently being ploughed, Elder and Lal (2008) estimated a potential reduction in N_2O emissions by introducing no-till to all organic soils in the US of equivalent to 1.1Tg CO_2 yr^{-}.

In a number of studies on mineral soils, the contrary has been observed. Here, no-till may increase N_2O emissions (Smith and Conen, 2004; Rochette, 2008). Available oxygen often limits denitrification in mineral soils (Smith and Tiedje, 1979). Since no-till commonly leads to a reduced soil-pore volume, the fraction of water-filled pore space (WFPS) tends to increase when tillage ceases. Consequently, conditions for denitrification improve and more N_2O can be produced and emitted (Smith et al, 2003). Thus reduced aeration of soil through no-till has two counteracting effects on N_2O emissions, and it is a shift in the balance between these effects that determines whether more or less N_2O is emitted under the new cultivation practice. In organic soils, reduced aeration seems to have a greater effect on limiting mineralization and supply of mineral N than it does on improving conditions for denitrification by restricting oxygen supply. In mineral soils with an already poor aeration under tilled conditions, it is the other way round. Where aeration is good, little difference is caused by a change in cultivation. Rochette (2008) has summarized results of 25 studies on N_2O emissions (approximately 45 site-years of data) involving comparisons of tilled and no-till soils at the same site (Figure 7.4). He grouped the studies into three soil aeration categories, which we now analyse.

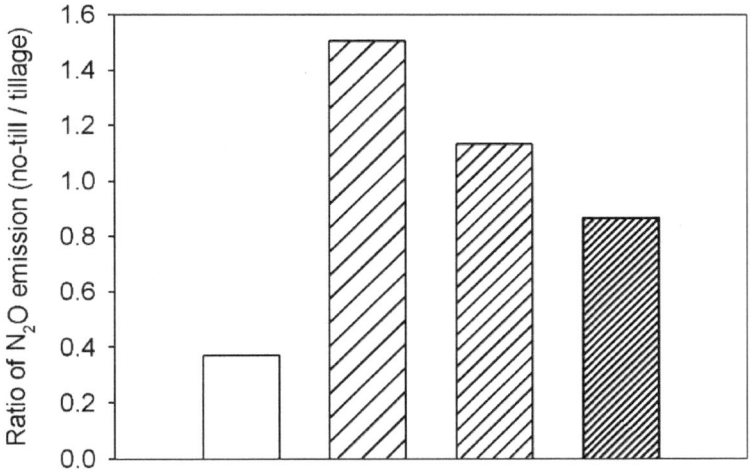

Figure 7.4 Relative N_2O emission from no-till compared to tilled cultivation on (white bar) organic soil; and on mineral soils with (hatched bars from left to right) poor, intermediate and good aeration status

Source: Data for white bar from Elder and Lal (2008); data for hatched bars from Rochette (2008)

On poorly aerated sites, no-till resulted in annual N_2O emissions being enhanced by 2.00kg N_2O-N (equivalent to 3.14kg N_2O). Let us assume an average CO_2 sequestration of 350kg C ha^{-1} yr^{-1} (equivalent to 1283kg CO_2 ha^{-1} yr^{-1}) and consider that 1kg of N_2O has 310 times the global warming potential (GWP) of 1kg of CO_2. Consequently, three-quarters of the benefit from C sequestration would be offset by the enhanced N_2O emissions on poorly aerated soils. Soils with a medium aeration status would have on average only about 5 per cent offset by additional N_2O emissions. On well-aerated soils, N_2O emissions from no-till may on average (difference between geometric means) be even lower than from tilled plots and increase the effect of C sequestration by another 2 per cent in terms of GWP. While rates of C sequestration will decline within a few decades and eventually approach zero, emissions of N_2O from poorly aerated mineral soils may continue to be enhanced. If so, the cumulative offset would grow with time and the greenhouse gas balance could become negative in the longer term.

Summary and concluding remarks

Land-use change always affects the cycling of N in soil and the exchange of N_2O with the atmosphere. Emissions tend to increase for a number of years following the conversion of tropical forest to grassland, or the conversion of temperate grassland to cropland. In the longer term, shrinking pools of soil organic matter and turnover rates of N may reverse the situation. Smaller emissions after a land-use change have been observed, often as a consequence of reduced ecosystem productivity, such as in degraded pastures on previously forested land. Where productivity, and with it the turnover of N, are maintained, such as in agroforestry systems, emissions of N_2O often tend to continue at an elevated level. Where N pools are large, such as in organic soils, emissions increase dramatically following drainage and cultivation. For a long time they continue to remain substantially larger than before. Even abandoning cultivation may not reduce emissions for decades. Reduced soil tillage, as a more recent development within mainly arable crop production systems, may either reduce or enhance N_2O emissions, the outcome depending on soil type and aeration status.

In summarizing all these findings, we can say that any particular type of land-use or management change does not always lead to the same result in terms of N_2O emissions and that the result may change with the time horizon. Also, it strongly depends on the specific soil conditions, in particular on the balance between factors limiting and factors promoting N_2O emission, and how this balance is shifted by the land-use or management change. Motivations for changing land use or farming practices are diverse, involving decisions by settlers, farmers, regulators, institutions and others, often with contrasting interests. As a consequence, arguments are plentiful. One of them should be about the effect on parameters affecting N_2O emissions, especially where land-use or management change is promoted as a way of mitigating climate change by increasing CO_2 sequestration.

References

Alm, J., Shurpali, N. J., Minkinen, K., Aro, L., Hytönen, J., Laurila, T., Lohila, A., Maljanen, M., Martikainen, P. J., Makiranta, P., Penttilä, P., Saarnio, S., Silvan, N., Tuittala, E.-S. and Laine, J. (2007) 'Emission factors and their uncertainty for the exchange of CO_2, CH_4 and N_2O in Finnish managed peatlands', *Boreal Environment Research*, vol 12, pp191–209

Arai, S., Ishizuka, S., Ohta, S., Ansori, S., Tokuchi, N., Tanaka, N. and Hardjono, A. (2008) 'Potential N_2O emissions from leguminous tree plantation soils in the humid tropics', *Global Biogeochemical Cycles*, vol 22, no 2, article number GB2028

Bateman, E. J. and Baggs, E. M. (2005) 'Contributions of nitrification and denitrification to N_2O emissions from soils at different water-filled pore space', *Biology and Fertility of Soils*, vol 41, no 6, pp379–388

Batjes, N. H. (1996) 'Total carbon and nitrogen in the soils of the world', *European Journal of Soil Science*, vol 47, no 2, pp151–163

Bothe, H., Ferguson, S. J. and Newton, W. E. (2007) *Biology of the Nitrogen Cycle*, Elsevier, Amsterdam

Bowman, R. A., Reeder, J. D. and Lober, R. W. (1990) 'Changes in soil properties in a central plains rangeland soil after 3, 20 and 60 years of cultivation', *Soil Science*, vol 150, pp851–858

Brumme, R., Borken, W. and Finke, S. (1999) 'Hierarchical control on nitrous oxide emission in forest ecosystems', *Global Biochemical Cycles*, vol 13, pp1137–1148

Chapin, F. S., III., Matson, P. and Mooney, H. (2002) *Principles of Terrestrial Ecosystem Ecology*, Springer-Verlag, New York

Cleveland, C. C. and Liptzin, D. (2007) 'C:N:P stoichiometry in soil: Is there a "Redfield ratio" for the microbial biomass?', *Biogeochemistry*, vol 85, no 3, pp235–252

Davidson, E. A., de Carvalho, C. J. R., Figueira, A. M., Ishida, F. Y., Ometto, J. P. H. B., Nardoto, G. B., Saba, R. T., Hayashi, S. N., Leal, E. C., Vieira, I. C. G. and Martinelli, L. A. (2007) 'Recuperation of nitrogen cycling in Amazonian forests following agricultural abandonment', *Nature*, vol 447, no 7147, pp995–998

Davidson, E. A., Sa, T. D. D., Carvalho, C. J. R., Figueiredo, R. D., Kato, M. D. A., Kato, O. R. and Ishida, F. Y. (2008) 'An integrated greenhouse gas assessment of an alternative to slash-and-burn agriculture in eastern Amazonia', *Global Change Biology*, vol 14, no 5, pp998–1007

Davies, M. G., Smith, K. A. and Vinten, A. J. A. (2001) 'The mineralisation and fate of N following ploughing of grass and grass-clover swards', *Biology and Fertility of Soils*, vol 33, pp423–434

do Carmo, J. B. and Cerri, C. E. P. (2007) 'Nitrogen dynamics in forestry and grassland soils in the Amazon: a review', *Outlook on Agriculture*, vol 36, issue 1, pp41–48

Elder, J. W. and Lal, R. (2008) 'Tillage effects on gaseous emissions from an intensively farmed organic soil in North Central Ohio', *Soil and Tillage Research*, vol 98, pp45–55

Gorham, E. (1991) 'Northern peatlands – role in the carbon-cycle and probable responses to climatic warming', *Ecological Applications*, vol 1, pp82–195

Hedin, L. O., Vitousek, P. M. and Matson, P. A. (2003) 'Nutrient lossess over four million years of tropical forest development', *Ecology*, vol 84, no 9, pp2231–2255

Hirsch, A. I., Little, W. S., Houghton, R. A., Scott, N. A. and White, J. D. (2004) 'The net carbon flux due to deforestation and forest re-growth in the Brazilian Amazon: analysis using a process-based model', *Global Change Biology*, vol 10, no 5, pp908–924

Hobbie, E. A. and Hobbie, J. E. (2008) 'Natural abundance of N-15 in nitrogen-limited forests and tundra can estimate nitrogen cycling through mycorrhizal fungi: A review', *Ecosystems*, vol 11, no 5, pp815–830

Iowa Department of Natural Resources (2000) *IOWA – Portrait of the Land*, Iowa Department of Natural Resources, Iowa, pp20–21, www.igsb.uiowa.edu/Portrait/3change/change.htm

IPCC (Intergovernmental Panel on Climate Change) (2000) *Special Report, Land-use, Land-use Change, and Forestry*, www.ipcc.ch/pdf/special-reports/spm/srl-en.pdf,

IPCC (2006) *2006 IPCC Guidelines for National Greenhouse Gas Inventories*, vol 4, prepared by the National Greenhouse Gas Inventories Programme, S. Eggelston, L. Buendia, K. Miwa, T. Ngara and K. Tanabe (eds), IGES, Hayama, Japan, www.ipcc-nggip.iges.or.jp/public/2006gl/vol4.html

Ishizuka, S., Iswandi, A., Nakajima, Y., Yonemura, S., Sudo, S., Tsuruta, H. and Murdiyarso, D. (2005) 'The variation of greenhouse gas emissions from soils of various land-use/cover types in Jambi province, Indonesia', *Nutrient Cycling in Agroecosystems*, vol 71, no 1, pp17–32

Melillo, J. M., Steudler, P. A., Feigl, B. J., Neill, C., Garcia, D., Piccolo, M. C., Cerri, C. C. and Tain, H. (2001) 'Nitrous oxide emissions from forests and pastures of various ages in the Brazilian Amazon', *Journal of Geophysical Research*, vol 106, pp34179–34188

Kasimir Klemedtsson, A., Klemedtsson, L., Berglund, K., Martikainen, P., Silvola, J. and Oenema, O. (1997) 'Greenhouse gas emissions from farmed organic soils: A review'. *Soil Use and Management*, vol 13, pp245–250

Kasimir Klemedtsson, A., Weslien, P. and Klemedtsson, L. (2009) 'Methane and nitrous oxide fluxes from a farmed Swedish Histosol', *European Journal of Soil Science*, vol 60, no 3, pp321–331

Keller, M., Veldkamp, E., Weitz, A. M. and Reiners, W. A. (1993) 'Effect of pasture age on soil trace-gas emissions from a deforested area of Costa Rica', *Nature*, vol 365, pp244–246

Keller, M., Varner, R., Dias, J. D., Silva, H., Crill, P. and de Oliveira, R. C. (2005) 'Soil-atmosphere exchange of nitrous oxide, nitric oxide, methane, and carbon dioxide in logged and undisturbed forest in the Tapajos National Forest, Brazil', *Earth Interactions*, vol 9, article number 23

Lapvetelänen, T., Regina, K. and Perälä, P. (2007) 'Peat-based emissions in Finland's national greenhouse gas inventory', *Boreal Environment Research*, vol 12, pp225–236

Lobe, I., Amelung, A. and Du Preez, C. C. (2001) 'Losses of carbon and nitrogen with prolonged arable cropping from sandy soils of the South African Highveld', *European Journal of Soil Science*, vol 52, pp93–101

Mäkiranta, P., Hytönen, J., Aro, L., Maljanen, M., Pihlatie, M., Potila, H., Shurpali, N. J., Laine, J., Lohila, A., Martikainen, P. J. and Minkkinen, K. (2007) 'Soil greenhouse gas emissions from afforested organic soil croplands and cutaway peatlands', *Boreal Environment Research*, vol 12, pp159–175

Maljanen, M., Liikanen, A., Silvola, J. and Martikainen, P. J. (2003) 'Nitrous oxide emissions from boreal organic soil under different land-use', *Soil Biology and Biochemistry*, vol 35, pp689–700

Maljanen, M., Hytönen, J., Mäkiranta, P., Alm, J., Minkkinen, K., Laine, J. and Martikainen, P. J. (2007) 'Greenhouse gas emissions from cultivated and abandoned organic croplands in Finland', *Boreal Environment Research*, vol 12, pp133–140

Manzoni, S., Jackson, R. B., Trofymow, J. A. and Porporato, A. (2008) 'The global stoichiometry of litter nitrogen mineralization', *Science*, vol 321, no 5889, pp684–686

Martikainen, P. J., Nykänen, H., Crill, P. and Silvola, J. (1993) ,'Effect of a lowered water table on nitrous oxide fluxes from northern peatlands', *Nature*, vol 366, pp51–53

Melillo, J. M., Steudler, P. A., Feigl, B. J., Neill, C., Garcia, D., Piccolo, M. C., Cerri, C. C. and Tian, H. (2001) 'Nitrous oxide emissions from forests and pastures of various ages in the Brazilian Amazon', *Journal of Geophysical Research*, vol 106, no D24, pp34179–34188

Melling, L., Hatano, R. and Goh, K. J. (2007) 'Nitrous oxide emissions from three ecosystems in tropical peatland of Sarawak, Malaysia', *Soil Science and Plant Nutrition*, vol 53, no 6, pp792–805

Millar, N. (2002) 'The effect of improved fallow residue quality on nitrous oxide emissions from tropical soils', PhD thesis, University of London, London

Mutuo, P. K., Cadisch, G., Albrecht, A., Palm, C. A. and Verchot, L. (2005) 'Potential of agroforestry for carbon sequestration and mitigation of greenhouse gas emissions from soils in the tropics', *Nutrient Cycling in Agroecosystems*, vol 71, pp43–54

Neill, C., Piccolo, M. C., Melillo, J. M., Steudler, P. A. and Cerri, C. C. (1999) 'Nitrogen dynamics in Amazon forest and pasture soils measured by ^{15}N pool dilution', *Soil Biology and Biochemistry*, vol 31, pp567–572

Neill, C., Steudler, P. A., Garcia-Montiel, D. C., Melillo, J. M., Feigl, B. J., Piccolo, M. C. and Cerri, C. C. (2005) 'Rates and controls of nitrous oxide and nitric oxide emissions following conversion of forest to pasture in Rondonia', *Nutrient Cycling in Agroecosystems*, vol 71, no 1, pp1–15

Prasad, V. K., Kant, Y., Gupta, P. K., Elvidge, C. and Badarinath, K. V. S. (2002) 'Biomass burning and related trace gas emissions from tropical dry deciduous forests of India: A study using DMSP-OLS data and ground-based measurements', *International Journal of Remote Sensing*, vol 23, no 14, pp2837–2851

Rochette, P. (2008) 'No-till only increases N_2O emissions in poorly-aerated soils', *Soil and Tillage Research*, vol 101, pp97–100

Rogiers, N., Conen, F., Furger, M., Stöckli, R. and Eugster, W. (2008) 'Impact of past and present land-management on the C-balance of a grassland in the Swiss Alps', *Global Change Biology*, vol 14, pp2613–2625

Rydin, H. and Jeglum, J. (2006) *The Biology of Peatlands*, Oxford University Press, Oxford

Schimel, J. P. and Bennett, J. (2004) 'Nitrogen mineralization: Challenges of a changing paradigm', *Ecology*, vol 85, pp591–602

Serrão, E. A. (1992) 'Alternative models for sustainable cattle ranching on already deforested lands in the Amazon', *Annaes da Academia Brasileira de Sciencias*, vol 64 (suppl. 1), pp97–104

Simek, M., Jisova, L. and Hopkins, D. W. (2002) 'What is the so-called optimum pH for denitrification in soil?', *Soil Biology and Biochemistry*, vol 34, pp1227–1234

Smith, K. A. and Conen, F. (2004) 'Impacts of land management on fluxes of trace greenhouse gases', *Soil Use and Management*, vol 20, pp255–263

Smith, K. A., Ball, T., Conen, F., Dobbie, K. E., Massheder, J. and Rey, A. (2003) 'Exchange of greenhouse gases between soil and atmosphere: Interactions of soil physical factors and biological processes', *European Journal of Soil Science*, vol 54, pp779–791

Smith, M. S. and Tiedje, J. M. (1979) 'Phases of denitrification following oxygen depletion in soil', *Soil Biology and Biochemistry*, vol 11, pp261–267

Smith, P., Powlson, D. S., Glendining, M. J. and Smith, J. U. (1998) 'Preliminary estimates of the potential for carbon mitigation in European soils through no-till farming', *Global Change Biology*, vol 4, pp679–685

Springob, G. and Kirchmann, H. (2003) 'Bulk soil C to N ratio as a simple measure of net N mineralization from stabilized soil organic matter in sandy arable soils', *Soil Biology and Biochemistry*, vol 35, no 4, pp629–632

Stehfest, E., and Bouwman, L. (2006) 'N$_2$O and NO emission from agricultural fields and soils under natural vegetation: Summarizing available measurement data and modeling of global annual emissions', *Nutrient Cycling in Agroecosystems*, vol 74, pp207–228

Stevens, R. J. and Laughlin, R. J. (1998) 'Measurement of nitrous oxide and di-nitrogen emissions from agricultural soils', *Nutrient Cycling in Agroecosystems*, vol 52, pp131–139

Tan, Z. X., Lal, R. and Liu, S. G. (2006) 'Using experimental and geospatial data to estimate regional carbon sequestration potential under no-till management', *Soil Science*, vol 171, no 12, pp950–959

Templer, P. H., Silver, W. L., Pett-Ridge, J., DeAngelis, K. M. and Firestone, M. K. (2008) 'Plant and microbial controls on nitrogen retention and loss in a humid tropical forest', *Ecology*, vol 89, no 11, pp3030–3040

Tiessen, H., Stewart, J. W. B. and Bettany, J. R. (1982) 'Cultivation effects on the amounts and concentration of carbon, nitrogen, and phosphorus in grassland soils', *Agronomy Journal*, vol 74, pp831–835

Veldkamp, E., Davidson, E., Erickson, H., Keller, M. and Witz, A. (1999) 'Soil nitrogen cycling and nitrogen oxide emissions along a pasture chronosequence in the humid tropics of Costa Rica', *Soil Biology and Biochemistry*, vol 31, pp387–394

Veldkamp, E., Purbopuspito, J., Corre, M. D., Brumme, R. and Murdiyarso, D. (2008) 'Land-use change effects on trace gas fluxes in the forest margins of Central Sulawesi, Indonesia', *Journal of Geophysical Research – Biogeosciences*, vol 113, no G2, article number G02003

Verchot, L. V., Davidson, E. A., Cattânio, J. H., Ackerman, I. L., Erickson, H. E. and Keller, M. (1999) 'Land-use change and biogeochemical controls of nitrogen oxide emissions from soils in eastern Amazonia', *Global Biogeochemical Cycles*, vol 13, pp31–46

Verchot, L. V., Hutabarat, L., Hairiah, K. and van Noordwijk, M. (2006) 'Nitrogen availability and soil N$_2$O emissions following conversion of forests to coffee in southern Sumatra', *Global Biogeochemical Cycles*, vol 20, article GB4008

Vieten, B., Conen, F., Neftel, A. and Alewell, C. (2009) 'Respiration of N_2O in suboxic soil', *European Journal of Soil Science*, vol 60, no 3, pp332–337

Weitz, A. M., Veldkamp, E., Keller, M., Neff, J. and Crill, P. M. (1998) 'Nitrous oxide, nitric oxide, and methane fluxes from soils following clearing and burning of tropical secondary forest', *Journal of Geophysical Research*, vol 103, pp28047–28058

Werner, C., Zheng, X. H., Tang, J. W., Xie, B., Liu, C., Kiese, R. and Butterbach-Bahl, K. (2006) 'N_2O, CH_4 and CO_2 emissions from seasonal tropical rainforests and a rubber plantation in Southwest China', *Plant and Soil*, vol 289, no 1–2, pp335–353

Weslien, P., Kasimir Klemendtsson, A., Börjesson, G. and Klemedtsson, L. (2009) 'Strong pH influence on N_2O and CH_4 fluxes from forested organic soils', *European Journal of Soil Science*, vol 60, no 3, pp311–320

West, T. O. and Marland, G. (2002) 'A synthesis of carbon sequestration, carbon emissions, and net carbon flux in agriculture: Comparing tillage practices in the United States', *Agriculture, Ecosystems and Environment*, vol 91, pp217–232

Wick, B., Veldkamp, E., de Mello, W. Z., Keller, M. and Crill, P. (2005) 'Nitrous oxide fluxes and nitrogen cycling along a pasture chronosequence in Central Amazonia, Brazil', *Biogeosciences*, vol 2, no 2, pp175–187

8

Indirect Emissions of Nitrous Oxide from Nitrogen Deposition and Leaching of Agricultural Nitrogen

Reinhard Well and Klaus Butterbach-Bahl

Introduction

N_2O emissions originating from agricultural land use include direct emissions from the surface of crop fields as well as indirect emissions caused by nitrogen flows from agricultural fields into adjacent systems (Mosier et al, 1998). Indirect emissions resulting from N leaching into aquatic systems are considered a potentially important N_2O source. However, its magnitude is still under debate (Nevison, 2000; Groffman et al, 2002; Weymann et al, 2008), with an uncertainty associated with current estimates of almost two orders of magnitude, which is larger than the uncertainty for other N_2O sources (IPCC, 2006). The aquatic pathway of reactive N that originates from leaching (and runoff) from agricultural fields and that ends up in the oceans includes downstream flow through a chain of connected systems, i.e. aquifers, riparian areas, rivers and estuaries (Figure 8.1).

A major fraction of agricultural surplus N is leached as nitrate (NO_3^-) to the groundwater. N_2O produced in soil by nitrification and denitrification can also be leached (Heincke and Kaupenjohann, 1999; Russow et al, 2002). In denitrifying aquifers, NO_3^- leached from agricultural soils is partially or completely reduced (Hiscock et al, 1991; Korom, 1992; Böhlke, 2002; Weymann et al, 2008) (Figure 8.1). N_2O produced under these conditions can be transported to the atmosphere via upward diffusion (Deurer et al, 2008; von der Heide et al, 2008; Weymann et al, 2009) or groundwater discharge to wells, springs and streams (Mühlherr and Hiscock, 1998; Heincke and Kaupenjohann, 1999). Groundwater containing NO_3^- and eventually N_2O reaches streams by direct discharge or via tile drainage (Hack and Kaupenjohann, 2002; Reay et al, 2003, 2004a, 2004b, 2005) and further flows

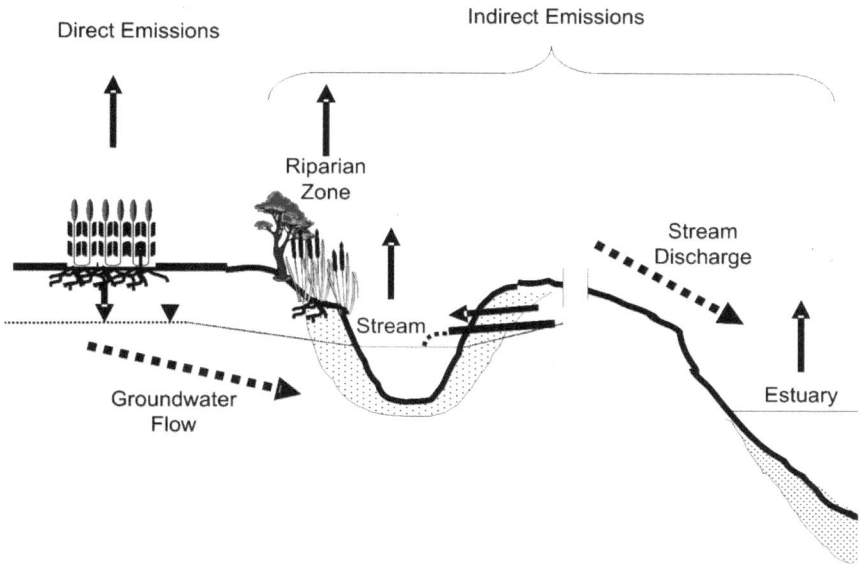

Figure 8.1 Nitrogen flows from crop fields to adjacent ecosystems and associated indirect N_2O emissions

Source: Well et al (2005b)

to downstream systems, i.e. lakes (Boontanon et al, 2000; Xiong et al, 2006), rivers and estuaries (Garnier et al, 2006), and finally to the open sea (Bange, 2006a). Once discharged to surface water bodies, dissolved N_2O may partially or completely degas to the atmosphere (Reay et al, 2004b).

During transport in streams, rivers and estuaries, NO_3^- can be denitrified (Laursen and Seitzinger, 2004) or assimilated by the biota (Mulholland et al, 2008). Within the N cycle in the open water bodies, mineral N species can be produced or retained, and N_2O can be produced by nitrification of ammonium (NH_4^+) as well as produced and reduced by denitrification. Point sources such as sewage treatment plants may significantly increase the load of reactive N. Estimates of fluxes at the various stages of the downstream flow chain are based on direct or indirect measurements, on process-based models or empirical emission factors (EFs).

Estimating indirect agricultural N_2O emission is complicated by the fact that it is often difficult to differentiate between fluxes originating from agricultural and other N sources. For example, a riparian buffer between agricultural land and a stream receives N via subsurface groundwater flow, atmospheric deposition from industrial, agricultural and natural sources, biological N_2 fixation and eventually N from various sources in the stream during flooding events (Figure 8.1). N_2O emitted at the soil surface is then a mixture of groundwater-derived N_2O of mostly agricultural origin, and N_2O

produced in soil that originates from industrial, agricultural and natural sources. Robust estimates of indirect agricultural emissions need to take this complex interaction of pathways into account.

Another major pathway of N loss from agricultural systems is the volatilization of ammonia (NH_3). On a global scale NH_3 emissions from agricultural systems are in the range of 27–38Tg NH_3-N yr^{-1} (Beusen et al, 2008). Of applied fertilizer or N excreted by animals, 10 to 30 per cent may be volatilized as NH_3 (Bouwman et al, 2002), which itself is deposited somewhere in the surrounding region, relatively close to its source. In many regions of Europe, N deposition is dominated by reduced N, i.e. NH_3 or NH_4^+, and deposition rates to natural and semi-natural systems can vary from 5kg N ha^{-1} yr^{-1} for unpolluted areas to over 80kg N ha^{-1} yr^{-1} (Fowler et al, 2004) in regions of intensive animal farming such as northern Germany or The Netherlands. Nitrogen input into natural and semi-natural systems via atmospheric deposition will increase N availability in the plant soil system and is regarded as one of the main drivers for increased soil N_2O emissions from temperate forests (Butterbach-Bahl et al, 1998; Pilegaard et al, 2006).

In this chapter, each system is described with respect to the control of N_2O fluxes and reported flux data. Moreover, concepts of EFs are compared and discussed and reported EFs and fluxes are summarized for each system. For the aquatic pathway, a comparison of systems is given which finally leads to some comments on mitigation options.

Aquatic pathway

Emission factors

Upscaling of emission data is often accomplished using specific EFs. In many cases, EFs are derived from the ratio between N_2O flux to the atmosphere ($F_{N2O,out}$) and N flux into the system (F_{Nin}). This approach can be seen as one of several concepts for deriving EFs and is thus referred to as conceptual EF 1 (CEF_1) which is defined as follows:

$$CEF_1 = F_{N2O,out} / F_{Nin} \qquad (8.1)$$

This concept is used by the IPCC methodology for calculation of national inventories (Mosier et al, 1998; Groffman et al, 2002; IPCC, 2006). Indirect emissions consist of emissions associated with depositions of agricultural N that has been transferred to the atmosphere (N_2O_A), or to human sewage (N_2O_S), and N that leaves crop fields in leaching and runoff (N_2O_L), all of which end up in aquatic systems.

N_2O_L is calculated as follows:

$$N_2O_L = N_{Leach} \times EF5, \qquad (8.2)$$

where N_{Leach} is the amount of N that leaves crop fields in leaching and runoff, EF5 is the N_2O EF for N that leaves crop fields in leaching and runoff and is processed as it moves ultimately to the world ocean. This factor is partitioned as EF5g (groundwater), EF5r (rivers) and EF5e (estuaries), with a value of 0.0025 for each of the partial EFs and thus 0.0075 for the overall EF5 (IPCC 2006). An earlier EF5 estimate of 0.025 (Mosier et al, 1998; IPCC, 2001) has been revised since more recent results indicated that the previous EF5g (0.015) and EF5r (0.0075) were too high (Hiscock et al, 2002, 2003; Reay et al, 2004a, 2004b, 2005; Sawamoto et al, 2005; Clough et al, 2006).

Moreover, although the basic idea was to use Equation 8.1 for deriving EFs, EF5g was based on a different concept with the following ratio:

$$CEF_2 = cN_2O_{aq} / cNO3_{aq} \tag{8.3}$$

where cN_2O_{aq} and $cNO3_{aq}$ are the concentrations of N_2O and NO_3^- measured in agricultural drainage water or groundwater. This concept assumes that these aquatic systems act solely as a domain of transport without any processing of NO_3^- and N_2O. The conceptual basis of Equation 8.3 has been questioned (Groffman et al, 2000; Nevison, 2000; Well et al, 2005a) because reduction of NO_3^- as well as production and reduction of N_2O occur in denitrifying systems and N_2O loss to the atmosphere can occur before the water reaches the sampling spots. Thus $cNO3_{aq}$ is often much lower than the influx of N (N_{in}) and cN_2O_{aq} can both increase and decrease during transport, which implies that CEF_2 (Equation 8.3) is likely to yield lower values than CEF_1 (Equation 8.1) (Figure 8.2). Consequently, CEF_1 would yield more realistic estimates of EF5g. However, determining CEF_1 implies more difficult measurements as both F_{in}, i.e. NO_3^--leaching, and $F_{N2O,out}$, i.e. the total advective and diffusive N_2O-flux from the water leaving crop fields, have to be quantified (Figure 8.1).

Another concept for EFs is to compare N_2O flux with N loss via denitrification (Well et al, 2005a):

$$CEF_3 = F_{N2O,out} / [\text{N-loss by denitrification}] \tag{8.4}$$

It is known that in surface soils, the N_2O-to-(N_2+N_2O) ratio of emitted N gases, i.e. CEF_3, can vary between 0 and 1 (Granli and Bøckman, 1994). Applying CEF_3 to aquatic systems means to compare the beneficial and harmful effects of denitrification, i.e. improving water quality and polluting the atmosphere, respectively. Thus CEF_3 could be used to classify the environmental tolerance of denitrification in various systems. Determining CEF_3 for groundwater, rivers and estuaries would help to answer the question of how N_2O emission resulting from NO_3^- flow across these systems is influenced, if measures are taken to enhance upstream denitrification in order to reduce NO_3^- discharge to the oceans, for example by riparian and wetland restoration projects (Groffman et al, 2000).

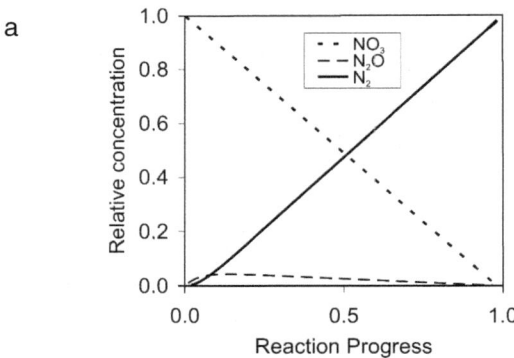

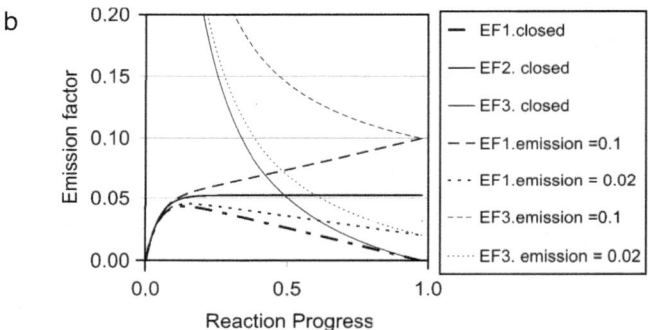

Figure 8.2 (a) Modelled NO_3^-, N_2O and N_2 during ongoing progress of denitrification; and (b) resulting EFs CEF_1, CEF_2 and CEF_3

Note: Reaction progress is the ratio between denitrified and initial NO_3^-. Concentrations of NO_3, N_2O and N_2 are modelled assuming a two-step reaction (NO_3^- to N_2O; N_2O to N_2) and using first-order kinetics. For the rate coefficients of the NO_3^--to-N_2O step and N_2O-to-N_2 steps, a 1:20 ratio was set to fit maximum N_2O concentration approximately to observations (see Table 8.1). N_2O emission to the atmosphere in relation to NO_3^- reduction during each time step is assumed to be 0 (closed systems) or >0 (0.02 or 0.1, semi-open systems).
Source: Adapted from Well et al (2005b)

Estimated emissions from various pathways

Groundwater

Due to the difficulty in measuring diffusive N_2O emission from the aquifer surface and convective discharge via streams, wells and springs, N_2O emission from aquifers has mostly been estimated from the N_2O and NO_3^- concentrations at groundwater monitoring wells. Most reported estimates of CEF_2 (Table 8.1) are within the range given in the review of Hiscock et al (2003). The lowest value of 0.0003 has been calculated for the central High Plains aquifer in the US, where the impacts of agricultural nitrogen inputs may be yet to be fully realized due to a thick unsaturated zone. Highest values of

Table 8.1 Indirect N_2O fluxes from riparian areas, aquifers and land drainage and associated conceptual emission factors

Location and description	N_2O flux (g N ha^{-1} d^{-1})	N_2O conc. (nM)	CEF$_1$	CEF$_2$	CEF$_3$	Reference
Riparian areas						
Eolian sand over glacial till; forested (*Alnus glutinosa*), The Netherlands	55			0.03 to 0.06	0.03 to 0.16[3]	Hefting et al, 2003
Eolian sand over glacial till; grassland (*Glyceria maxima*), The Netherlands	6 to 11			0.02 to 0.036	0.03 to 0.05[3]	Hefting et al, 2003
Sandy soils in Pleistocene deposits, (*Alnus glutinosa*), The Netherlands	2.6 to 13.6			0.01 to 0.08	0.05 to 2.6[3]	Hefting et al, 2006
Aquifers						
Review of seven aquifers		<1 to 21771	.	0.00003 to 0.042		Hiscock et al, 2003
Cretaceous chalk and weathered bedrock aquifers, UK		149 to 1928 (602)[2]	0.001	0.0019		Hiscock et al, 2003
Three Pleistocene sand and gravel aquifers in northern Germany		121 to 657[1]	0.0006 to 0.004[1,5]	0.001 to 0.04[1,5]	0.001 to 0.005[1,4,5]	Weymann et al, 2008
Pleistocene sand aquifer in northern Germany		7 to 45,428 (3178)[2]	0.00004 to 0.12 (0.011)[2,5]	0.00005 to 0.24 (0.024)[2,5]	0.000001 to 0.37 (0.0023)[2,4,5]	Deurer et al, 2008; Weymann et al, 2008
Near-surface groundwater of a pleistocene sand aquifer in northern Germany		3357 to 97,250		0.005 to 0.1		Well et al, 2005c
Tayhoo Valley, China, cropped with upland rice		15 to 571 (161)[2]		0.0001 (0.00003 to 0.005)		Xiong et al, 2006
Review of three sites		11 to 35,6571		0.00003 to 0.04		Hiscock et al, 2003
Mix of arable and grassland, Aberdeenshire, UK		<2500		0.002		Reay et al, 2003, 2004a, 2004b, 2005

Note: CEF$_1$, CEF$_2$ and CEF$_3$ were estimated using Equations 8.1, 8.3 and 8.4. Individual data sets are only listed if EFs were given or can be calculated from presented data in the reference. [1] Range of aquifer medians. [2] Total range (median in brackets) as in Well et al (2005a) based on data given by Hefting et al (2003, 2006). [3] Calculation of CEF$_3$ as in Well et al (2005a) based on data given by Hefting et al (2003, 2006). [4] CEF$_3$ was calculated for each sample in the basic data set of Weymann et al (2008). [5] Potential EFs measured at groundwater monitoring wells.

up to 0.1 were observed near the surface of sandy aquifers in northern Germany under agricultural management. A value in between these extremes (0.0019) was found in unconfined chalk groundwater in Cambridgeshire and Norfolk, eastern England.

An estimate of CEF_1 (Equation 8.1) at the aquifer scale was obtained for the same aquifers (Hiscock et al, 2003). $F_{N2O,out}$ of 0.04kg N ha^{-1} yr^{-1} was determined from groundwater concentrations and discharge by bore hole extractions and springs. Diffusive flux to the unsaturated zone was estimated to be negligible. F_{Nin} was derived from average N fertilizer applications to crop fields (130kg N ha^{-1} yr^{-1}) and the IPCC default fraction of leached N (0.3), giving a CEF_1 of 0.001.

CEF can also be determined more directly: summing excess N_2 from denitrification and residual NO_3^--N of individual groundwater samples reflects initial NO_3^--N discharge to the groundwater, i.e. F_{Nin} (Equation 8.1) (Green et al, 2008; Weymann et al, 2008). If this information is available in addition to N_2O, CEF_1 can be calculated using Equation 8.1. Excess N_2 can also be used to determine CEF_3, i.e. N_2O accumulation in relation to denitrification (Equation 8.4). However, it must be noted that CEF_1 and CEF_3 as determined for individual groundwater samples from monitoring wells are potential EFs, since the concentration of dissolved N species may further change during transport within the aquifer. Actual EFs could be obtained from excess N_2 and N_2O in groundwater abstraction wells or monitoring wells close to groundwater discharge zone, but this was not reported in the studies mentioned above.

How large is the potential error caused by neglecting denitrification when using CEF_2? This can be seen by comparing CEF_2 and CEF_1 within the same data sets. For example, the data set given by Weymann et al (2008) yields CEF_1 site medians between 0.0006 and 0.01, which is much lower than the site medians of CEF_2 for the same sites (0.001–0.04). This demonstrates the importance of accounting for denitrified N in reduced aquifers.

EFs based on dissolved N in groundwater samples are estimates of the potential lateral N_2O flux to springs, streams or wells. But the total groundwater-derived N_2O flux also includes vertical diffusive flux at the aquifer surface. When taking this into account, care must be taken to avoid double-counting of total agricultural N_2O fluxes, since the diffusive groundwater emissions under agricultural land are included in the total emission at the soil surface. Strictly speaking, estimates of direct N_2O emission from agricultural land based on surface flux measurements thus include some of the indirect emissions. But the question is; is the magnitude of these diffusive fluxes significant? Extremely high N_2O concentrations of up to 100μM (Table 8.1) (Well et al, 2005c) result in significant N_2O fluxes of up to 3g N ha^{-1} day^{-1}, or approximately 1kg N ha^{-1} yr^{-1}. But such high values are rare individual observations. Recently, a large number of vertical N_2O profiles at 10cm resolution were determined in a sandy aquifer in northern Germany (Deurer et al, 2008; von der Heide et al, 2009b). Resulting diffusive fluxes were 0.0009–0.3kg N ha^{-1} yr^{-1}. It could be shown that these values were related to aquifer properties such as dissolved and particulate organic C, O_2, pH, and NO_3^- (von der Heide et al, 2008, 2009a). In the same aquifer, vertical fluxes from the groundwater to the soil surface were studied using an in situ ^{15}N-tracer experiment (Weymann et al, 2009) in which an $8m^2$ area of the aquifer surface was amended with ^{15}N-labelled NO_3^- solution. Groundwater-derived fluxes estimated from $^{15}N_2O$ in flux chambers at the soil surface were very small (0.0022 to 0.0207μg N_2O-N m^{-2} hour^{-1}, equivalent to 0.0002–0.0018kg N_2O-N ha^{-1} yr^{-1}). When comparing locations with similar concentration gradients among different studies, direct flux estimates (Weymann et al, 2009) were similar to the indirect estimates obtained from concentration gradients (Deurer et al, 2008). Thus the gradient-derived fluxes were confirmed. Peak $^{15}N_2O$ fluxes during falling groundwater levels suggested that a release of N_2O from entrapped gas bubbles or from residual soil water in drained layers might be a fast path for vertical N_2O fluxes (Weymann et al, 2009). However, the magnitude of this emission path still needs to be investigated.

Overall, the latest studies on CEF_1, CEF_2 and diffusive groundwater fluxes confirm that the downward revision of EF5g was adequate.

Riparian area including constructed wetland buffers

A variety of recent studies supplies data on N_2O emission from riparian buffer zones (Blicher-Mathiesen and Hoffmann, 1999; Bernal et al, 2003; Hefting et al, 2003, 2006; Dhondt et al, 2004; Machefert and Dise, 2004), demonstrating that these systems are potential locations of substantial indirect N_2O emissions. In addition to N_2O fluxes, Hefting et al (2003) investigated NO_3^- loading in grassland and forested buffer zones along first-order streams in The Netherlands, and could thus determine EFs for these systems. EF5g determined using Equation 8.1 (CEF_1) ranged between 0.03 and 0.06 in the forested and between 0.02 and 0.03 in the grassland buffer zone (Table 8.1). The influx of nitrate to these zones was 467 and 192g N m^{-2} yr^{-1}, respectively. N_2O emissions were 20 and 2–4kg N ha yr^{-1}, respectively. The authors concluded that the IPCC EF5g underestimates riparian buffer zones of this type. This supports previous suggestions that a separate EF for riparian systems should be defined (Groffman et al, 1998). Using the nitrate removal and N_2O fluxes reported for riparian systems (Hefting et al, 2003, 2006), CEF_3 can be calculated, giving 0.021–0.042. This range is about one order of magnitude higher than the site medians of CEF_3 of the aquifer studies (0.001–0.005). This demonstrates that, in comparison with aquifers, NO_3^- removal within these riparian buffers can be more harmful to the atmosphere. However, care must be taken before generalization, since the reported fluxes were much higher compared to the other riparian studies cited above.

Constructed wetlands

Constructed wetlands can be used for reducing N discharge from wastewater or leaching from agricultural soils. Hernandez and Mitsch (2006) investigated

N_2O fluxes induced by N retention in constructed riverbank marshes. Higher N_2O fluxes occurred in higher marsh areas with periodical flooding than in the lower areas with permanent flooding. Wetland plants in the lower marshes favoured N_2O fluxes due to aerenchyma transport. N_2O emission in relation to denitrification, estimated using the static-core acetylene-inhibition method, was 0.005–0.03. This is about one order lower than the CEF_3 of riparian wetlands receiving high agricultural N loads (Hefting et al, 2003, 2006). However care must be taken when comparing these data since different methods were used to estimate denitrification. N_2O fluxes from constructed wetlands for sewage treatment (Mander et al, 2003; Teiter and Mander, 2005) exhibited high N_2O emissions per unit area (up to 590g N ha^{-1} day^{-1}) but due to their relatively small surface area, the impact of these systems at larger scales is small. Because N_2 fluxes had been determined in sediment cores incubated under N_2-free atmospheres, CEF_3 could be derived from the N_2O to (N_2+N_2O) ratio, which ranged approximately from 0.001 to 0.01. Similar to the river bank marshes, this ratio and the total N_2O fluxes were inversely related to the water level.

Until now, studies in riparian areas and constructed wetlands have focused on surface emissions, with few reports of N_2O in the saturated zone of these systems (Blicher-Mathiesen and Hoffmann, 1999). It is thus not clear to what extent emissions originate from the saturated or unsaturated zone, respectively. Moreover the convective flux from riparian areas to streams is poorly known. Elevated N_2O levels in river water draining arable land with intensive upland rice production in the Yangtze River delta in China are interpreted as resulting from convective N_2O fluxes from riparian areas (Xiong et al, 2006). The significance of this path needs to be further investigated.

Land drainage

Land drainage systems (whether traditional tiles or modern perforated plastic pipe drains) accelerate the discharge of temporary or permanent near-surface groundwater to streams. This practice thus affects the convective flux of groundwater N_2O. Reay et al (2003, 2004a, 2004b) conducted a study on N_2O emission from a drainage system serving crop fields and discharging into a ditch. They demonstrated that N_2O discharged by 14 field drain outfalls was rapidly emitted to the atmosphere during transport in the open ditch, which was enhanced by turbulence in the stream flow. An EF (CEF_2) of 0.002 resulted from the N_2O to NO_3^- mass ratio at the drain outfalls. The total range of CEF_2 for artificial drainage is similar to CEF_2 that was reported for groundwater (Table 8.1). However, this range is based on only a few studies and it is thus not clear if artificial land drainage leads to elevated convective N_2O fluxes to streams in comparison with natural drainage.

Until now there have been no direct comparisons between N_2O measurement at groundwater monitoring wells and corresponding field drainage outlets. Moreover, the effect of accelerated groundwater discharge via field drains on convective N_2O fluxes to streams has not yet been studied. By comparing excess N_2 and N_2O in denitrifying aquifers, Weymann et al (2008) found typical patterns showing that, with increasing excess N_2, i.e. ongoing denitrification progress during passage through the aquifer, N_2O levels initially increased to a certain level, but then gradually decreased and finally disappeared after NO_3^- was completely consumed by denitrification. Highest N_2O levels were generally observed at intermediate reaction stages, when 20–60 per cent of the NO_3^- was denitrified. It was concluded that low N_2O fluxes via artificial drainage might be due to short groundwater residence times in aquifers. However, one might expect that field drainage fluxes can be high under conditions when denitrification at the groundwater surface is rapid, causing substantial reaction progress even during the short residence time of groundwater passing through drainage systems. But this needs to be confirmed.

So far, there are no estimates of CEF_1 for artificial drainage systems. Because significant denitrification is a frequent phenomenon in near-surface groundwater (Well et al, 2005b), it can be expected that in many cases some of the leached NO_3^- is denitrified before groundwater is discharged via this path. Consequently, CEF_1 must be lower than CEF_2 (see Equations 8.3 and 8.4 and section above). But due to the relatively short groundwater residence time, this discrepancy is probably lower compared to groundwater without artificial drainage. It would also be interesting to determine CEF_3 for drainage in order to check its impact on the overall N_2O emission of the total aquatic pathway including rivers and the ocean. Estimating CEF_1 and CEF_3 using excess N_2 (see above) will not be possible by direct measurement at drainage outlets, since the water equilibrates with the atmosphere during passage through the drains. Thus excess N_2 in the water would be lost before sampling. To measure excess N_2 it would thus be necessary to measure it in groundwater monitoring wells installed within the groundwater body of the drained area, which has not been done up to now.

Rivers and estuaries

According to model estimates (Dumont et al, 2005), the total global load of dissolved inorganic N transported by rivers to the oceans is $25Tg \, yr^{-1}$, of which $16Tg \, N \, yr^{-1}$ comes from anthropogenic sources, including sewage point sources (0.4Tg), diffuse agricultural emissions from mineral fertilizer (5.3Tg), biological N_2 fixation (4.5Tg) and manure (3.8Tg). In addition, 12Tg organic N is transported by rivers to coastal waters (Harrison et al, 2005), plus 39Tg of particulate N (Beusen et al, 2005). The latter may not be all biologically available. N_2O discharged by rivers to the ocean or emitted from river surfaces originates from various sources. Both nitrification and denitrification can occur in the water column, in the sediments and in the interior of suspended particles (Bange et al, 2006b). In addition to this in situ production, N_2O can also be discharged to streams from groundwater or sewage plants (Seitzinger and Kroeze, 1998). The contribution from these sources is variable and difficult to estimate. However, Toyoda et al (2009) were able to show by using stable isotope signatures of N_2O that peak N_2O concentrations in an urban river were always dominated by sewage sources.

Model estimates of global N_2O emissions from rivers and estuaries have been based on river loads of dissolved inorganic nitrogen (DIN) and the assumption that 50 per cent of the DIN is lost through denitrification and that all DIN is nitrified once (Seitzinger and Kroeze, 1998). It was also assumed that 0.3 per cent of the nitrified and denitrified N is lost as N_2O, or 3 per cent when the average watershed N loading is relatively high (>10kg N ha^{-1} yr^{-1}). This is similar to the basic concept of CEF_3, but it includes nitrification and N-fluxes from natural sources. A direct comparison with CEF_3 from other systems is thus not possible (Table 8.1). Further calculations with this model resulted in estimates for N_2O emissions from rivers and estuaries of 1.26 and 0.25Tg N yr^{-1}, respectively, totalling 1.5Tg N yr^{-1} (Kroeze et al, 2005). More recently, they estimated emissions from rivers and estuaries in 2000 to be 0.3–2.1Tg N_2O-N, using updates of the global nutrient models (Kroeze et al, 2009).

Principally, the N-budget of rivers and estuaries might be used to derive EFs using the concepts of Equations 8.3 and 8.4. Assuming an average N retention in rivers of 50 per cent (Seitzinger and Kroeze, 1998) or 30 per cent (van Drecht et al, 2003) and using the total dissolved N export and N_2O fluxes of about 40Tg N yr^{-1} and 1Tg N yr^{-1}, respectively, yields F_{Nin} of 80Tg N yr^{-1} and 56Tg N yr^{-1}, respectively, and CEF_1 of 0.013 and 0.019, respectively. The lower value is close to the sum of IPCC-EF5r and IPCC-EF5e in the 1996 IPCC Guidelines (0.01). Both values are clearly higher than the mean values of the current IPCC Guidelines, where EF5r + EF5e = 0.005 (IPCC, 2006), but close to the upper limit of the respective uncertainty range (0.0004–0.017). An approximate CEF_3 can be obtained from the ratio between N_2O flux and N retention (16–40Tg yr^{-1}) since the major fraction of N retention in rivers can probably be attributed to denitrification (Seitzinger and Kroeze, 1998). This would give a CEF_3 in the range 0.025–0.063, but the actual value must be higher since some of the retention is not denitrification but burial of N in sediments. This estimate of river CEF_3 is thus relatively large compared to the CEF_3 of other systems (Table 8.2).

Reported N_2O concentrations levels in rivers and estuaries range from 9 to 201nM, with medians of 25 and 83nM, respectively (Table 8.2). These levels are clearly lower than most of the reported groundwater concentrations (up to 100,000nM, with site medians always >100nM). There are several factors explaining the higher groundwater levels. Because degassing from aquifers is small (see above), N_2O concentrations result mainly from the balance between production and reduction to N_2. In the rivers, there is degassing of the groundwater-derived load and of N_2O produced in situ during river transport. Another factor might be the fact that most groundwater data come from aquifers under agricultural land, whereas river data also include catchments without agricultural land use.

Table 8.2 Indirect N_2O fluxes from constructed wetlands, rivers and estuaries and associated conceptual emission factors

Location and description	N_2O flux (g N ha^{-1} d^{-1})	N_2O conc. (nM)	CEF_1	CEF_2	CEF_3	Reference
Constructed wetlands						
Constructed river bank marshes, USA	1 to 3.5				0.005 to 0.03[4]	Hernandez and Mitsch, 2006
Constructed wetlands for wastewater treatment, Estonia	1 to 590				approx. 0.001 to 0.01[5]	Mander et al, 2003
Rivers and estuaries						
Model estimates of global rivers and estuaries					0.003 or 0.03[1]	Seitzinger and Kroeze, 1998
N-budget model of global rivers and estuaries			0.013 to 0.018[6]		0.033 to 0.063[6]	
Review of nine rivers		12.6 to 301 (83)[3]				Toyoda et al, 2009
Whole reaches of three small rivers in north-west Indiana and north-east Illinois, USA	3 to 403	8.9 to 18.1		0.00003 to 0.00017	0.003	Laursen and Seitzinger, 2004
Three lowland agricultural streams, New Zealand	4 to 151	39 to 167				Wilcock and Sorrell, 2008
LII river, New Zealand	3 to 120	14 to 68		0.00006		Clough et al, 2006, 2007
Seine river and estuary, France	12 to 22	Up to approx. 140		0.02 to 0.04[2]		Garnier et al, 2006
Review of 17 estuaries		8.6 to 220 (24.7)[3]				Toyoda et al, 2009

Note: CEF_1, CEF_2 and CEF_3 were estimated using Equations 8.1, 8.3 and 8.4. Individual data sets are only listed if EFs were given or can be calculated from presented data in the reference. [1] 0.03 if watershed N loading is relatively high (>10kg N ha^{-1} yr^{-1}), otherwise 0.003. [2] Estimated from the ratio between river-derived fluxes and estimated N-input within the river catchment. [3] Total range (median in brackets). [4] Denitrification estimated using C_2H_2 static core incubation, range of means. [5] Denitrification by core incubation under He/O_2. An approximate CEF_3 range was estimated from the ranges of N_2O and N_2 fluxes since the range of (N_2O/N_2+N_2O) ratios was not given. [6] Estimate for rivers plus estuaries derived from global budget calculations (see text).

N_2O dynamics in rivers are extremely variable since process-related factors such as concentrations of O_2 and suspended or dissolved organic carbon depend on factors like flow velocity, depth, path length, sediment properties and nutrient load. For example, extremely low EF5r values were reported for a short, shallow, fast and well-oxygenated river in New Zealand (Clough et al, 2006, 2007), whereas high values were obtained for the deep, slow, partially sub-oxic Seine river system in France (Garnier et al, 2006) (Table 8.2).

Some studies report on the contribution of the N_2O influx from groundwater and on in situ production. Xiong et al (2006) observed high N_2O concentrations in river water similar to those in the groundwater wells in the Taihu river catchment and concluded that river N_2O mostly originated from groundwater discharge in riparian areas. Laursen and Seitzinger (2004) measured N_2O emission and denitrification in three rivers at the whole-reach scale by measuring dissolved N_2, Ar and N_2O, as well as concentrations of injected gas tracers. Denitrification and N_2O emission ranged from 0.4 to 60µmol N m^{-2} hour^{-1} (median 6.9µmol N m^{-2} hour^{-1}) and from 0.31 to 15.91mmol N m^{-2} hour^{-1} (median 2.62mmol N m^{-2} hour^{-1}), respectively. Using Equation 8.4, a CEF$_3$ of 0.0026 (ratio of median N_2O flux to median denitrification) can be calculated, which is similar to the median CEF$_3$ of the reduced aquifers (Table 8.1). To fully characterize the fate of inorganic N during passage through aquifers, rivers and estuaries to the ocean it would be necessary to combine such studies with investigation of N_2O loads and associated process dynamics in the aquifers of the catchment, for example using excess N_2.

Principally, N_2O dynamics in estuaries tend to be different from those in rivers, due to smaller flow velocities. This implies lower gas exchange and thus lower oxygen concentrations and higher residence times for N_2O. Theoretically, there should be more N_2O reduction to N_2 and CEF$_3$ should thus be lower than in rivers. Reported median N_2O concentrations of reviewed estuaries (Table 8.2) are indeed lower than those of rivers (Table 8.2). Since gas exchange rates are also lower in the former locations, resulting N_2O fluxes must be clearly smaller.

Comparing emission factors among systems and mitigation options for the aquatic pathway

To illustrate the impact of system properties such as residence time, N-retention, production and consumption of N_2O and gas exchange on EFs, a simple process simulation is shown (Figure 8.2). During closed conditions, i.e. when there is no exchange of N species between a water parcel with adjacent domains, there is an ongoing progress of retention reactions leading eventually to complete consumption of reactive N, for example by reduction of NO_3^- to N_2 via denitrification (Figure 8.2a). During this process, N_2O is produced and reduced, resulting in the typical pattern described above (Weymann et al, 2008). The magnitude of maximum N_2O concentration depends on the balance between production and reduction and can thus be highly variable. The curves in Figure

8.2b represent EFs at a given stage of reaction progress, assuming that all N_2O is emitted to the atmosphere at this stage, for example by discharge of groundwater via land drains or springs. By definition, CEF_1 then follows the time course of N_2O and thus depends on the reaction progress (RP) as described above. Low values of CEF_1 can thus be found at early as well as late stages of the reaction. During the later phase, CEF_3 must inevitably decrease with RP as the amount of produced N_2 increases, while N_2O decreases. This demonstrates that extended residence times under closed conditions in denitrifying systems lead to decreasing EFs. In semi-closed systems such as rivers and estuaries, there is emission of a certain fraction of produced N_2O during ongoing reaction progress. This leads to final values of CEF_1 and CEF_3 >0 and both EFs may increase or decrease towards the end (Figure 8.2b).

The complete pathway of reactive N from the aquifer surface to the sea might be seen as a chain of connected systems where N_2O dynamics follow varying patterns similar to the examples in Figure 8.2. Can the observed EFs in the various components of the aquatic pathway (Tables 8.1 and 8.2) be explained with these simple models of closed or semi-closed system? In riparian areas and constructed wetlands potential denitrification is often high due to richness in organic C. Residence time is very short at the interface between saturated and unsaturated layers. This might explain why the highest values of CEF_1 and/or CEF_3 are reported for riparian areas and constructed wetlands, i.e. semi-closed systems where reactive zones are in direct contact with or close to the atmosphere. The high EFs are thus in agreement with the EF pattern of semi-closed systems with high gas exchange (Figure 8.2b). The potential values of CEF_1 and CEF_3 of aquifers exhibit a large range, but EFs effective at the aquifer scale are probably much lower than maximum values of up to 0.5 (Table 8.1), which can also be seen from the effective CEF_1 of 0.001 reported by Hiscock et al (2003). Due to almost closed conditions during aquifer transport, EFs can be expected to be at the low end.

Artificial land drainage decreases groundwater residence time. CEF_3 may thus be larger compared to non-drained aquifers (Figure 8.2a). CEF_1 of tile drainage might be lower or higher, depending on denitrification rates: if reaction progress is below the N_2O maximum, artificial drainage decreases CEF_1 and vice versa (Figure 8.2a).

For rivers and estuaries, CEF_3 values derived from global budget calculations (0.03 and 0.06, respectively) are much higher than values based on measurements (Laursen and Seitzinger, 2004) or the default EFs that had been used in a river N_2O model (Seitzinger and Kroeze, 1998). This discrepancy needs to be clarified. The river CEF_3 of 0.003 derived from Laursen and Seitzinger (2004) is lower compared to CEF_3 of riparian areas and constructed wetland and within the range of potential CEF_3 of aquifers. From the process dynamics in rivers and estuaries one might expect that EFs are in between these groups, since the water column is more open compared to aquifers but there is no direct contact between active zones and the atmosphere such as at the saturated/unsaturated interface zone of wetlands and riparian areas.

What are the options for mitigating N_2O fluxes from aquatic systems? Because anthropogenic discharge of reactive N is the driver for indirect N_2O fluxes from aquatic systems, each measure to reduce this discharge certainly yields some mitigation effect. High mitigation efficiency can be expected from lowering discharge of reactive N to systems with high EFs, which might be riparian areas and constructed wetlands. Conversely, little effect will result, if systems with small EFs such as deep denitrifying aquifers receive lower N loads.

Principally, mitigation might also be achieved using end-of-the-pipe measures. Mathematically, N_2O emission during the aquatic pathway is the product of CEF_3 and N loss by denitrification within any time interval (Figure 8.2). Total N_2O flux from denitrification during downstream movement of leached agricultural N through aquifers, wetlands, streams and estuaries is thus defined by the time course of denitrification and CEF_3. Properties enhancing CEF_3 thus increase total N_2O fluxes. It might be possible to lower the high CEF_3 values induced by near-surface processes in wetlands by assuring permanent flooding of these systems (Hernandez and Mitsch, 2006). Land drainage increases aquifer N_2O fluxes since it enhances CEF_3. Control of land drainage systems to achieve high water tables and thus longer residence times can be used to enhance denitrification within the drained groundwater zone. If CEF_3 of this additional denitrification was lower than CEF_3 of downstream systems, this measure would reduce overall N_2O fluxes from the aquatic pathway. This needs to be investigated.

In rivers, measures to reduce flow velocity and increase depth, such as damming or restoration of natural meandering, might decrease the exchange of dissolved gases with the atmosphere. This might eventually reduce N_2O fluxes, but care must be taken before generalizing because reduced gas exchange can lower oxygen concentrations and might thus favour N_2O production by denitrification.

N_2O from groundwater abstraction wells might be trapped, for example by stripping the water and injecting the exhaust air into wastewater treatment plants. Because some drinking water plants use gas stripping to remove excess CO_2 from the water, this measure might be realized with moderate effort. This might be an effective measure in cases of extremely high N_2O levels as observed by Well et al (2005c).

Overall, each of these end-of-the-pipe measures can only mitigate a small fraction at various points within the rather long hydrological pathway, and each requires costly construction work. Moreover, there is a need to investigate the magnitude of potential mitigation effects. It is thus highly questionable whether these measures will be competitive compared to other more cost-effective greenhouse gas mitigation measures.

Nitrogen-deposition pathway

The magnitude of terrestrial N deposition largely depends on the height of vegetation, for example forests may receive a two- to three-fold higher N deposition than shorter vegetation types such as grasslands (Fowler et al, 2004) and the proximity of the ecosystem to the pollutant source, for example being in the vicinity of a poultry farm (Skiba et al, 2006). In regions with intensive agriculture, reduced N compounds dominate N deposition, whereas in areas unaffected by agricultural activities oxidized N compounds, originating mainly from combustion processes (industry, vehicles; see Chapter 9), are the most important. An example is the Höglwald Forest in southern Bavaria, situated in a landscape with intermixed forest and agricultural areas, where the NH_3-to-NO_3^- ratio in the throughfall is 2:1, while N deposition, depending on forest type, is in the range of 20–35kg N ha^{-1} yr^{-1} (Kreutzer and Weis, 1998). N input of such a magnitude significantly exceeds the N demand of a growing forest, which is approximately 5–10kg N ha^{-1} yr^{-1} (Scarascia-Mugnozza et al, 2000). Thus N deposition leads to increased N availability in the soil–plant system, reflected by, for example, a narrowing of the C:N ratio of the litter, forest floor or mineral soil and increased concentrations of nitrate and ammonium in the soil solution (Kristensen et al, 2004; Mannig et al, 2008). Several studies show that N deposition and N_2O, as well as NO emissions, from forest soils are positively correlated (for example Brumme and Beese, 1992; Brumme et al, 1999; Papen and Butterbach-Bahl, 1999; van Dijk and Duyzer, 1999; Butterbach-Bahl et al, 2002a, 2002b; Pilegaard et al, 2006; Skiba et al, 2006) and that the observed stimulation of fluxes may mainly be attributed to the increased availability of N (as NH_4^+ and NO_3^-) for the microbial processes of nitrification and denitrification (Rennenberg et al, 1998; Corre et al, 1999), i.e. the key microbial processes responsible for N trace-gas production in soils (see Chapter 2). In a study involving microbial process studies along an N-enrichment gradient, Corre et al (2007) demonstrated that gross N mineralization, NH_4^+ immobilization, gross nitrification and NO_3^- immobilization rates increased up to intermediate N-enrichment levels and somewhat decreased in highly N-enriched conditions. With regard to N trace-gas emissions the authors concluded that decreasing turnover rates of the NO_3^- pool, due to a slowdown of microbial nitrate immobilization, may be the reason for increased gaseous emissions as well as increased nitrate leaching.

A possible further explanation for increased N_2O emissions due to ecosystem N enrichment was recently provided by Conen and Neftel (2007). They speculated that increased N availability may have reduced N_2O reduction in soils via denitrification, i.e. that the ratio of N_2O to N_2 increases with increasing N availability. Since increased N deposition also affects nitrate N leaching and runoff (Dise et al, 1998; Borken and Matzner, 2004), one also needs to consider indirect N_2O emissions from water bodies due to N deposition to natural systems. However, a thorough evaluation and quantification of N-deposition effects on soil N trace-gas emissions also remains difficult, since environmental conditions such as meteorology or soil and plant properties do significantly affect the magnitude, temporal course and composition of the emitted N gases ($NO/N_2O/N_2$) (for example Figures 8.3 and 8.4).

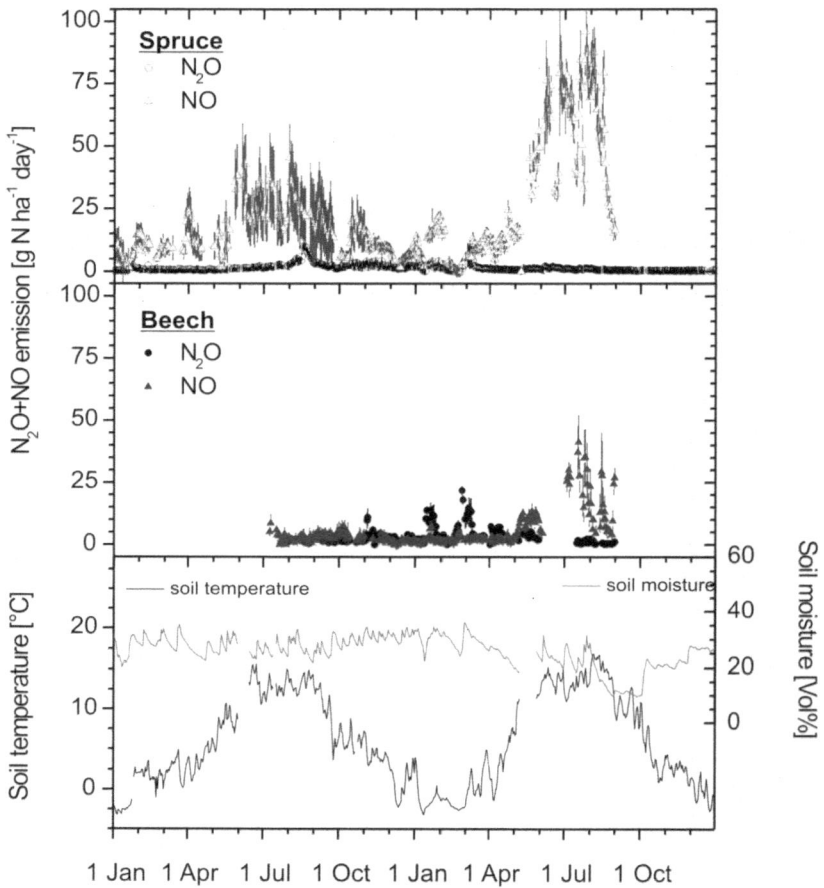

Figure 3.3 Soil N₂O and NO fluxes (±standard error (SE)) at two directly adjacent beech and spruce sites of the Höglwald Forest in the years 2002 and 2003

Note: Fluxes are daily mean values of sub-daily observations for five different chambers per site and trace gas as measured with automated measuring systems. The lower panel shows simultaneous observed temporal variations in forest floor temperature and moisture at the spruce site. Noteworthy are the pronounced differences in N₂O and NO emissions between both sites, i.e. with NO emissions dominating at the spruce site and with N₂O and NO emissions being balanced at the beech site. These differences in N trace-gas emissions are mainly due to differences in atmospheric N deposition and effects of tree species on soil hydrological and chemical properties. For further details see Papen and Butterbach-Bahl (1999) and Pilegaard et al (2006).
Source: Butterbach-Bahl et al (1997)

Having these difficulties in mind, there have been several attempts to estimate the stimulating effect of N deposition on N₂O emissions from forest soils. Skiba et al (2006) used a gradient approach, with measuring sites being located at increasing distances from a poultry farm, i.e. a strong NH₃ source. They estimated that >3 per cent of the N deposited to the woodland sites was released as N₂O. Butterbach-Bahl et al (1998) used a regression-type approach, time

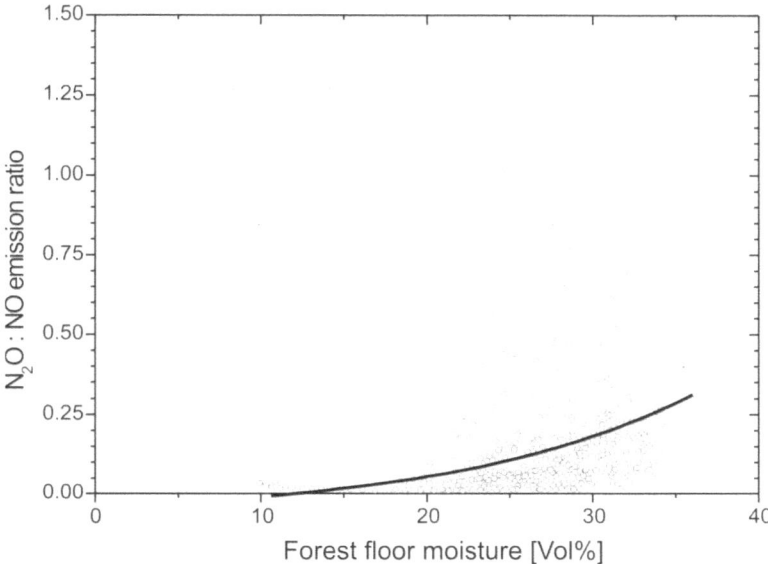

Figure 8.4 Effect of changes in soil moisture on the ratio of N_2O to NO

Note: The figure is based on daily mean N_2O and NO fluxes at the spruce site in the Höglwald Forest (see Plate 8.1) and time domain reflectometry (TDR) measurements of volumetric moisture in the forest floor. At all water contents NO is the dominant N trace gas emitted. However, at volumetric water contents >30 per cent (approximately >65 per cent WFPS), N_2O emissions increase exponentially at the cost of NO.
Source: Butterbach-Bahl et al (1997)

series of nitrogen deposition throughfall data, and continuous N_2O and NO emission measurements (Figure 8.3; Plate 8.1) at the long-term monitoring site at Höglwald Forest for estimating N-deposition-driven N_2O losses. Their estimate is comparable to that in the study by Skiba et al (2006), i.e. 1.4 per cent for coniferous forests and 5.4 per cent for deciduous forest. Also, a literature review by Denier van der Gon and Bleeker (2005) showed that N deposition to forests stimulates N_2O emissions within the same range; they concluded that the stimulating effect was higher for deciduous forests (5.7 per cent of deposited N being lost as N_2O) than for coniferous forests (3.7 per cent).

This marked forest-type effect (Figure 8.3 and Table 8.3) may be due to differences in canopy structure and resulting effects on soil moisture, and in acidity of the forest floor, as well as differences in soil C storage and distribution, which favour nitrification rather than denitrification activity in the soils of coniferous forests as compared to soils of deciduous forests (Butterbach-Bahl et al, 2002a; Pilegaard et al, 2006; de Vries et al, 2007). In a scenario-type study at the European Union scale, Kesik et al (2005) estimated N deposition effects on forest soil N_2O emissions by running the biogeochemical model Forest-DNDC either with actual values of atmospheric N deposition or by assuming that N deposition was zero. The results showed that across Europe 1.8 per cent of atmospheric N deposition was returned to the atmosphere as N_2O. All published estimates, therefore, show that the default N_2O EF of 1 per cent used by IPCC for indirect emissions from soils following N deposition (Mosier et al, 1998; IPCC, 2006) is most likely too low by at least a factor of two.

Table 8 3 Summary of published N_2O emission data for deciduous forests and coniferous forests and derived emission factor as a function of N input

Number of observations[a]	N_2O emission (kg N ha^{-1} yr^{-1})	Emission factor[b]	Reference(s)
Deciduous forest			
3	0.49	0.023	Ambus et al (2001); Beier et al (2001)
1	0.23	0.023	Bowden et al (2000)
8	1.67	0.053	Brumme et al (1999)
2	2.98	0.111	Butterbach-Bahl et al (2002b)
1	1.45	0.072	Butterbach-Bahl et al (1997)
2	0.02	0.001	Corre et al (1999)
2	2.65	0.044	Mogge et al (1998)
1	0.20	0.013	Oura et al (2001)
2	4.44	0.222	Papen and Butterbach-Bahl (1999)
6	0.65	0.035	Skiba et al (1998)
3	4.03	0.115	Zechmeister-Boltenstern et al (2002)
Indicative average[c]		0.065	
Coniferous forest			
3	0.31	0.016	Borken et al (2002)
4	0.58	0.034	Brumme et al (1999)
2	0.70	0.020	Butterbach-Bahl et al (2002b)
4	1.59	0.073	Butterbach-Bahl et al (1997)
6	0.11	0.005	
4	0.36	0.005	
2	0.39	0.013	
2	3.20	0.032	
6	0.85	0.028	Papen and Butterbach-Bahl (1999)
14	0.37	0.016	Skiba et al (1998, 1999)
Indicative average[c]		0.024	

Note: [a] 'Number of observations' (n) counts separately the different years of each study and the various plots with different N treatments and/or tree species within the class 'deciduous forest' or 'coniferous forest'; thus for example for an experiment in which a beech plot and an alder plot were each monitored for two consecutive years, n = 4. [b] The EF (i.e. the fraction of N input that is emitted as N_2O) is calculated for individual plots on the original data for N_2O emission and N input, and then averaged. [c] The average EF presented here is only indicative, as it is not corrected for the number of observations or N input levels, and weights each location and/or study equally.
Source: Based on Denier van der Gon and Bleeker (2005)

Global estimates of indirect emissions

Based on IPCC default EFs of 1996 (Mosier et al, 1998), indirect emissions from atmospheric deposition had been estimated to be 0.3 Tg N yr^{-1} with an uncertainty range of 0.06 to 0.6 Tg N. The respective indirect flux from leaching and runoff had been 1.6 Tg N yr^{-1} (0.13–7.7 Tg N yr^{-1}). Indirect emissions had thus been close to the direct emission from agricultural soils of 2.1 Tg N yr^{-1}.

Using 2006 IPCC data, del Grosso et al (2008) calculated much lower indirect emissions of 0.76Tg N yr^{-1}, which are also much lower than the current estimates of direct soil emission of 3.77Tg N yr^{-1}. Indirect emissions from leaching and runoff are now much lower than before, with a total of 0.4Tg yr^{-1} or 0.13Tg yr^{-1} for emissions from groundwater, rivers and estuaries, respectively (S. del Grosso, personal communication, 2009). The downward revision of the EF for N_2O from groundwater (IPCC-EF5g) is supported by recent groundwater data (Weymann et al, 2008, Deurer et al, 2008). Indirect emissions from atmospheric deposition slightly increased to 0.36Tg N yr^{-1}. However, the flux from atmospheric N deposition is probably higher than this value because IPCC-EF4 is likely to be underestimated by at least a factor of two (see previous section).

Based on river N loads, global fluxes from rivers and estuaries have recently been modelled, giving 0.3 to 2.1Tg N yr^{-1} (Kroeze et al, 2009), which is clearly higher than the current IPCC-based estimate for rivers and estuaries (0.13 + 0.13 = 0.26Tg N yr^{-1}). This inconsistency needs further attention. However, it must be noted that the extent of uncertainty for all estimates of indirect N_2O fluxes is still huge and much larger than estimates of direct fluxes.

Conclusions

Indirect agricultural N_2O emissions via the aquatic pathway occur while reactive nitrogen leached from agricultural fields is transported downstream through aquifers, riparian areas, wetlands, rivers and estuaries. Estimates of indirect N_2O fluxes from leached N are derived from direct or indirect measurements, process-based models or empirical EFs. EFs for each system are based on a relatively large database for rivers, estuaries and aquifers. Few data are available for land drainage, constructed wetlands or riparian areas. There are different concepts of EFs and various methods have been used to estimate them. Despite the resulting difficulty in comparing the contribution of each system to total indirect N_2O fluxes, some clear conclusions can be drawn:

- The highest EFs occur in more open systems such as riparian areas and constructed wetlands, where a relatively large fraction of N_2O production can rapidly be released to the atmosphere.
- The lowest EFs occur in deeper aquifers, where produced N_2O is almost trapped until groundwater is discharged to wells springs or streams.
- Current estimates of the total N_2O emission from this pathway are lower

than previously assumed, with a global total of 0.4Tg N per year or 11 per cent of total agricultural N$_2$O fluxes.

- In view of the large variation within the limited data sets, these estimates are still uncertain.
- Mitigating indirect N$_2$O fluxes by reducing emissions of reactive N to the environment is thus most effective in systems with the highest EFs, and vice versa. Principally there are some possible end-of-the-pipe measures to reduce indirect fluxes, but their effectiveness is probably relatively low.

Agricultural activities affect N$_2$O emissions from natural and semi-natural ecosystems via volatilization of reactive nitrogen (NH$_3$, NO/NO$_2$) and following re-deposition. In many regions in Europe, Asia and North America, atmospheric N deposition to natural/semi-natural ecosystems is already exceeding the ecosystem net N demand. Part of the surplus nitrogen is emitted as N$_2$O. Estimated EFs for N deposition are in a range of 1.4–5.4 per cent, significantly higher than suggested by IPCC (1 per cent). Published reports suggest that N$_2$O emissions due to atmospheric N deposition are strongly affected by vegetation type, for example indirect N$_2$O fluxes due to N deposition tend to be higher for deciduous forests as compared to coniferous forests. Existing estimates of nitrogen deposition effects on N$_2$O emissions from natural and semi-natural terrestrial ecosystems are still uncertain and more long-term studies for contrasting environmental settings are needed to improve our understanding of mechanisms and environmental consequences of nitrogen deposition with regard to N$_2$O, but also with regard to nitrate leaching, acidification or ecosystem carbon storage. The only promising mitigation strategy is a strategy to reduce nitrogen deposition. Finally, this means that losses of reactive nitrogen from agriculture, being one of the main sources of atmospheric NH$_3$ and NO$_x$, need to be minimized as far as possible.

To improve and validate the current knowledge of indirect agricultural N$_2$O fluxes, it would be useful to combine process-based models with direct flux measurements. Moreover, to optimize mitigation strategies, there is a need to validate systems-specific EFs. This is necessary to better identify the stages of downstream flow where the retention of each unit of reactive N causes highest N$_2$O fluxes.

Acknowledgements

We acknowledge the contributions from Stephen del Grosso and Carolien Kroeze to the estimates of global fluxes. We also thank Carolien Kroeze, Keith Smith and Daniel Weymann for helpful comments on the manuscript.

References

Ambus, P., Jensen, J. M., Priemé, A., Pilegaard, K. and Kjøller, A. (2001) 'Assessment of CH$_4$ and N$_2$O fluxes in a Danish beech (*Fagus sylvatica*) forest and an adjacent N-fertilised barley (*Hordeum vulgare*) field: Effects of sewage sludge amendments', *Nutrient Cycling in Agroecosystems*, vol 60, pp15–21

Bange, H. W. (2006a) 'New Directions: The importance of the oceanic nitrous oxide emissions', *Atmospheric Environment*, vol 40, no 1, pp198–199

Bange, H. W. (2006b) 'Nitrous oxide and methane in European coastal waters', *Estuarine, Coastal and Shelf Science*, vol 70, pp361–374

Beier, C., Rasmussen, L., Pilegaard, K., Ambus, P., Mikkelsen,T., Jensen, N. O., Kjøller, A., Priemé, A. and Ladekarl, U. L. (2001) 'Fluxes of NO_3^-, NH_4^+, NO, NO_2, and N_2O in an old Danish beech forest', *Water, Air, and Soil Pollution: Focus*, vol 1–2, pp187–195

Bernal, S., Butturini, A., Nin, E., Sabater, F. and Sabater, S. (2003) 'Leaf litter dynamics and nitrous oxide emission in a Mediterranean riparian forest: Implications for soil nitrogen dynamics', *Journal of Environmental Quality*, vol 32, no 1, pp191–197

Beusen, A. H. W., Dekkers, A. L. M., Bouwman, A. F., Ludwig, W. and Harrison, J. (2005) 'Estimation of global river transport of sediments and associated particulate C, N, and P', *Global Biogeochemical Cycles*, vol 19, GB4S05

Beusen, A. H. W., Bouwman, A. F., Heuberger, P. S. C., van Drecht, G. and van der Hoek, K. W. (2008) 'Bottom-up uncertainty estimates of global ammonia emissions from agricultural production systems', *Atmospheric Environment*, vol 42, pp6067–6077

Blicher-Mathiesen, G. and Hoffmann, C. C. (1999) 'Denitrification as a sink for dissolved nitrous oxide in a freshwater riparian fen', *Journal of Environmental Quality*, vol 28, pp257–262

Böhlke, J. K. (2002) 'Groundwater recharge and agricultural contamination', *Hydrogeology Journal*, vol 10, pp153–179

Boontanon, N., Ueda, S., Kanatharana, P. and Wada, E. (2000) 'Intramolecular stable isotope ratios of N_2O in the tropical swamp forest in Thailand', *Naturwissenschaften*, vol 87, pp188–192

Borken, W. and Matzner, E. (2004) 'Nitrate leaching in forest soils: An analysis of long-term monitoring sites in Germany', *Journal of Plant Nutrition and Soil Science*, vol 167, pp189–196

Borken, W., Beese, F., Brumme, R. and Lamersdorf, N. (2002) 'Long-term reduction in nitrogen and proton inputs did not affect atmospheric methane uptake and nitrous oxide emission from a German spruce forest soil', *Soil Biology and Biochemistry*, vol 34, pp1815–1819

Bouwman, A. F., Boumans, L. J. M. and Batjes, N. H. (2002) 'Estimation of global NH_3 volatilization loss from synthetic fertilizers and animal manure applied to arable lands and grasslands', *Global Biogeochemical Cycles*, vol 16, no 2, article no 1024

Bowden, R. D., Rullo, G., Stevens, G. R. and Steudler, P. A. (2000) 'Soil fluxes of carbon dioxide, nitrous oxide, and methane at a productive temperate deciduous forest', *Journal of Environmental Quality*, vol 29, pp268–276

Brumme, R. and Beese, F. (1992) 'Effects of liming and nitrogen fertilization on emissions of CO_2 and N_2O from a temperate forest', *Journal of Geophysical Research*, vol 97, pp851–858

Brumme, R., Borken, W. and Finke, S. (1999) 'Hierarchical control on nitrous oxide emission in forest ecosystems', *Global Biogeochemical Cycles*, vol 13, pp1137–1148

Butterbach-Bahl, K., Gasche, R., Breuer, L. and Papen, H. (1997) 'Fluxes of NO and N$_2$O from temperate forest soils: Impact of forest type, N deposition and of liming on the NO and N$_2$O emissions', *Nutrient Cycling in Agroecosystems*, vol 48, pp79–90

Butterbach-Bahl, K., Gasche, R., Huber, C., Kreutzer, K. and Papen, H. (1998) 'Impact of nitrogen input by wet deposition on N-trace gas fluxes and CH$_4$-oxidation in spruce forest ecosystems of the temperate zone in Europe', *Atmospheric Environment*, vol 32, pp559–564

Butterbach-Bahl, K., Breuer, L., Gasche, R., Willibald, G. and Papen, H. (2002a) 'Exchange of trace gases between soils and the atmosphere in Scots pine forest ecosystems of the northeastern German Lowlands; 1. Fluxes of N$_2$O, NO/NO$_2$ and CH$_4$ at forest sites with different N-deposition', *Forest Ecology and Management*, vol 167, pp123–134

Butterbach-Bahl, K., Rothe, A. and Papen, H. (2002b) 'Effect of tree distance on N$_2$O- and CH$_4$-fluxes from soils in temperate forest ecosystems', *Plant and Soil*, vol 240, pp91–103

Clough, T. J., Bertram, J. E., Sherlock, R. R., Leonard, R. L. and Nowicki, B. L. (2006) 'Comparison of measured and EF5-r-derived N$_2$O fluxes from a spring-fed river', *Global Change Biology*, vol 12, pp352–363

Clough, T. J., Buckthought, L. E., Kelliher, F. M. and Sherlock, R. R. (2007) 'Diurnal fluctuations of dissolved nitrous oxide (N$_2$O) concentrations and estimates of N$_2$O emissions from a spring-fed river: Implications for IPCC methodology', *Global Change Biology*, vol 13, no 5, pp1016–1027

Conen, F. and Neftel, A. (2007) 'Do increasingly depleted δ^{15}N values of atmospheric N$_2$O indicate a decline in soil N$_2$O reduction?, *Biogeochemistry*, vol 82, pp321–326

Corre, M. D., Pennock, D. J., van Kessel, C. and Elliott, D. K. (1999) 'Estimation of annual nitrous oxide emissions from a transitional grassland-forest region in Saskatchewan, Canada', *Biogeochemistry*, vol 44, pp29–49

Corre, M. D., Brumme, R., Veldkamp, E. and Beese F. O. (2007) 'Changes in nitrogen cycling and retention processes in soils under spruce forests along a nitrogen enrichment gradient in Germany', *Global Change Biology*, vol 13, pp1509–1527

Del Grosso, S. J., Wirth, T., Ogle, S. M. and Parton, W. J. (2008) 'Estimating agricultural nitrous oxide emissions', *Eos (Transactions of the American Geophysical Union)*, vol 89, pp529–540

Denier van der Gon, H. A. C. and Bleeker, A. (2005) 'Indirect N$_2$O emission due to atmospheric N deposition for the Netherlands', *Atmospheric Environment*, vol 39, pp5827–5838

Deurer, M., von der Heide, C., Böttcher, J., Duijnisveld, W. H. M., Weymann, D. and Well, R. (2008) 'The dynamics of N$_2$O in the surface groundwater and its transfer into the unsaturated zone: A case study from a sandy aquifer in Germany', *Catena*, vol 72, pp362–373

De Vries, W., Oenema O., Butterbach-Bahl, K. and Denier van der Gon, H. (2007) 'The impact of atmospheric nitrogen deposition on the exchange of carbon dioxide, nitrous oxide and methane from European forests', in D. Reay, N. Hewitt, K. Smith and J. Grace (eds), *Greenhouse Gas Sinks*, CABI, Wallingford, pp249–283

Dhondt, K., Boeckx, P., Hofman, G. and van Cleemput, O. (2004) 'Temporal and spatial patterns of denitrification enzyme activity and nitrous oxide fluxes in three adjacent vegetated riparian buffer zones', *Biology and Fertility of Soils*, vol 40, no 4, pp243–251

Dise, N. B., Matzner, E. and Gundersen, P. (1998) 'Synthesis of nitrogen pools and fluxes from European forest ecosystems', *Water, Air and Soil Pollution*, vol 105, pp143–154

Dumont, E, Harrison, J. A., Kroeze, C., Bakker, E. J. and Seitzinger S. P. (2005) 'Global distribution and sources of dissolved inorganic nitrogen export to the coastal zone: Results from a spatially explicit, global model', *Global Biogeochemical Cycles*, vol 19, article no GB4S02, doi:10.1029/2005GB002488

Fowler, D., O'Donoghue, M., Muller, J. B. A., Smith, R. I., Dragosits, U., Skiba, U., Sutton, M. and Brimblecombe, P. (2004) 'A chronology of nitrogen deposition in the UK between 1900 and 2000', *Water, Air and Soil Pollution*, vol 4, pp9–23

Garnier, J., Cébron, A., Tallec, G., Billen, G., Sebilo, M. and Martinez, A. (2006) 'Nitrous oxide emission in the Seine River estuary (France): Comparison with upstream sector of the Seine basin', *Biogeochemistry*, vol 77, pp305–326

Granli, T. and Bøckman, O. C. (1994) 'Nitrous oxide from agriculture', *Norwegian Journal of Agricultural Science*, Supplement no 12, pp1–128

Green, T. C., Puckett, L. J., Böhlke, J. K., Bekins, B. A., Phillips, S. P., Kauffman, L. J., Denver, J. M. and Johnson, H. M. (2008) 'Limited occurrence of denitrification in four shallow aquifers in agricultural areas of the United States', *Journal of Environmental Quality*, 37, pp994–1009

Groffman, P. M., Gold, A. J. and Jacinthe, P. A. (1998) 'Nitrous oxide production in riparian zones and groundwater', *Nutrient Cycling in Agroecosystems*, vol 52, pp179–186

Groffman, P. M., Gold, A. J. and Addy, K. L. (2000) 'Nitrous oxide production in riparian zones and its importance to national emission inventories', *Chemosphere-Global Change Science*, vol 2, pp291–299

Groffman, P. M., Gold, A. J., Kellogg, D. Q. and Addy, K. (2002) 'Mechanisms, rates and assessment of N_2O production in groundwater, riparian zones and rivers', in J. Van Ham, A. P. M. Baede, R. Guicherit and J. G. F. M. Williams-Jacobse (eds) *Non-CO$_2$ Greenhouse Gases: Scientific Understanding, Control Options and Policy*, Millpress, Rotterdam, pp159–166

Hack, J. and Kaupenjohann, M. (2002) 'N_2O discharge with drain water from agricultural soils of the upper Neckar region in Southern Germany', in J. Van Ham, A. P. M. Baede, R. Guicherit and J. G. F. M. Williams-Jacobse (eds) *Non-CO$_2$ Greenhouse Gases: Scientific Understanding, Control Options and Policy*, Millpress, Rotterdam, pp185–190

Harrison, J. A., Caraco, N. and Seitzinger, S. P. (2005) 'Global patterns and sources of dissolved organic matter export to the coastal zone: Results from a spatially explicit, global model', *Global Biogeochemical Cycles*, vol 19, article no GB4S04

Hefting, M. M., Bobbink, R. and de Caluwe, H. (2003) 'Nitrous oxide emission and denitrification in chronically nitrate-loaded riparian buffer zones', *Journal of Environmental Quality*, vol 32, no 4, pp1194–1203

Hefting, M. M., Bobbink, R. and Janssens, M. P. (2006) 'Spatial variation in denitrification and N₂O emission in relation to nitrate removal efficiency in a N-stressed riparian buffer zone', *Ecosystems*, vol 9, pp550–563

Heincke, M. and Kaupenjohann, M. (1999) 'Effects of soil solution on the dynamics of N₂O emissions: A review', *Nutrient Cycling in Agroecosystems*, vol 55, pp133–157

Hernandez, M. E. and Mitsch, W. J. (2006) 'Influence of hydrologic pulses, flooding frequency, and vegetation on nitrous oxide emissions from created riparian marshes', *Wetlands*, vol 26, no 3, pp862–877

Hiscock, K. M., Lloyd, J. W. and Lerner, D. N. (1991) 'Review of natural and artificial denitrification of groundwater', *Water Research*, vol 25, pp1099–1111

Hiscock, K. M., Bateman, A. S., Fukada, T. and Dennis, P. F. (2002), 'The concentration and distribution of groundwater N₂O in the chalk aquifer of eastern England', in J. van Ham, A. P. M. Baede, R. Guicherit and J. G. F. M. Williams-Jacobs (eds), *Non-CO₂ Greenhouse Gases: Scientific Understanding, Control Options and Policy Aspects*, Millpress, Rotterdam, The Netherlands, pp179–184

Hiscock, K. M., Bateman, A. S., Mühlherr, I. H., Fukada, T. and Dennis, P. F. (2003) 'Indirect emissions of nitrous oxide from regional aquifers in the United Kingdom', *Environmental Science and Technology*, vol 37, no 16, pp3507–3512

IPCC (Intergovernmental Panel on Climate Change) (2001) *Climate Change 2001: The Scientific Basis*, Cambridge University Press, Cambridge

IPCC (2006) *2006 IPCC Guidelines for National Greenhouse Gas Inventories*, Prepared by the National Greenhouse Gas Inventories Programme, S. Eggelston, L. Buendia, K. Miwa, T. Ngara and K. Tanabe (eds), IGES, Hayama, Japan

Kesik, M , Ambus, P., Baritz, R., Brüggemann, N., Butterbach-Bahl, K., Damm, M., Duyzer J., Horváth, L., Kiese, R., Kitzler, B., Leip, A., Li, C., Pihlatie, M., Pilegaard, K., Seufert, G., Simpson, D., Skiba, U., Smiatek, G., Vesala, T. and Zechmeister-Boltenstern, S. (2005) 'Inventories of N₂O and NO emissions from European forest soils', *Biogeosciences*, vol 2, pp353–375

Korom, S. F. (1992) 'Natural denitrification in the saturated zone: A review', *Water Resources Research*, vol 28, pp1657–1668

Kreutzer, K. and Weis, T. (1998) 'The Höglwald field experiments – aims, concept and basic data', *Plant and Soil*, vol 199, no 1, pp1–10

Kristensen, H. L., Gundersen, P., Callesen, I. and Reinds, G. J. (2004) 'Throughfall nitrogen deposition has different impacts on soil solution nitrate concentration in European coniferous and deciduous forests', *Ecosystems*, vol 7, no 2, pp180–192

Kroeze, C., Dumont, E. and Seitzinger, S. P. (2005) 'New estimates of global emissions of N₂O from rivers, estuaries and continental shelves', *Environmental Sciences*, vol 2, nos 2–3, pp159–167

Kroeze, C., Dumont, E. and Seitzinger, S. (2009) 'Future trends in emissions of N₂O from rivers, estuaries and continental shelves', Abstract for the 4th International Symposium on Non-CO₂ Greenhouse Gases (NCGG-5), 30 June–3 July 2009, Wageningen, The Netherlands

Laursen, A. E. and Seitzinger, S. P. (2004) 'Diurnal patterns of denitrification, oxygen consumption and nitrous oxide production in rivers measured at the whole-reach scale', *Freshwater Biology*, vol 49, no 11, pp1448–1458

Machefert, S. E. and Dise, N. B. (2004) 'Hydrological controls on denitrification in riparian ecosystems', *Hydrology and Earth System Sciences*, vol 8, no 4, pp686–695

Mander, U., Kuusemets, V., Lohmus, K., Mauring, T., Teiter, S. and Augustin, J. (2003) 'Nitrous oxide, dinitrogen, and methane emission in a subsurface flow constructed wetland', *Water Science and Technology*, vol 48, no 5, pp135–142

Mannig, P., Saunders, M., Bardgett, R. D., Bonkowski, M., Bradford, M. A., Ellis, R. J., Kandeler, E., Marhan, S. and Tscherko, D. (2008) 'Direct and indirect effects of nitrogen deposition on litter decomposition', *Soil Biology and Biochemistry*, vol 40, pp688–698

Mogge, B., Kaiser, E. A. and Munch, J. C. (1998) 'Nitrous oxide emissions and denitrification N-losses from forest soils in the Bornhöved lake region (northern Germany)', *Soil Biology and Biochemistry*, vol 30, pp703–771

Mosier, A., Kroeze, C., Nevison, C., Oenema, O., Seitzinger, S. and van Cleemput, O. (1998) 'Closing the global N_2O budget: Nitrous oxide emissions through the agricultural nitrogen cycle, OECD/IPCC/IEA Phase II: Development of IPCC guidelines for national greenhouse gas inventory methodology', *Nutrient Cycling in Agroecosystems*, vol 52, pp225–248

Mühlherr, I. H. and Hiscock, K. M. (1998) 'Nitrous oxide production and consumption in British limestone aquifers', *Journal of Hydrology*, vol 211, pp126–139

Mulholland, P. J., Helton, A. M., Poole, G. C., Hall, R. O., Hammilton, S. K., Peterson, B. J.,Tank, J. L., Ashkenas, L. R., Cooper, L. W., Dahm, C. N., Dodd, W. K., Findlay, S. E. G., Gregory, S. V., Grimm, N. B., Johnson, S. L., McDowell, W. H., Meyer, J. L., Valett, H. M., Webster, J. R., Arango, C. P., Beaulieu, J. J., Bernot, M. J., Burgin, A. J., Crenshaw, C. L., Johnson, L. T., Niederlehner, B. R., O'Brien, J. M., Potter, J. D., Sheibley, R. W., Sobota, D. J. and Thomas, S. M. (2008) 'Stream denitrification across biomes and its response to anthropogenic nitrate loading', *Nature*, vol 452, pp202–205

Nevison, C. (2000) 'Review of the IPCC methodology for estimating nitrous oxide emissions associated with agricultural leaching and runoff', *Chemosphere*, vol 2, pp493–500

Oura, N., Shindo, J., Fumoto, T., Toda, H. and Kawashima, H. (2001) 'Effects of nitrogen deposition on nitrous oxide emissions from the forest floor', *Water, Air, and Soil Pollution*, vol 130, pp673–678

Papen, H. and Butterbach-Bahl, K. (1999) 'A 3-year continuous record of nitrogen trace gas fluxes from untreated and limed soil of a N-saturated spruce and beech forest ecosystem in Germany: 1. N_2O emissions', *Journal of Geophysical Research*, vol 104, pp18487–18503

Pilegaard, K., Skiba, U., Ambus, P., Beier, C., Brüggemann, N., Butterbach-Bahl, K., Dick, J., Dorsey, J., Duyzer, J., Gallagher, M., Gasche, R., Horvath, L., Kitzler, B., Leip, A., Pihlatie, M. K., Rosenkranz, P., Seufert, G., Vesala, T., Westrate, R. and Zechmeister-Boltenstern, S. (2006) 'Factors controlling regional differences in forest soil emission of nitrogen oxides (NO and N_2O)', *Biogeosciences*, vol 3, pp615–661

Reay, D. S., Smith, K. A. and Edwards, A. C. (2003) 'Nitrous oxide emission from agricultural drainage waters', *Global Change Biology*, vol 9, pp195–203

Reay, D. S., Edwards, A. C. and Smith, K. A. (2004a) 'Determinants of nitrous oxide emission from agricultural drainage waters', *Water, Air and Soil Pollution: Focus*, vol 4, pp107–115

Reay, D. S., Smith, K. A. and Edwards, A. C. (2004b) 'Nitrous oxide in agricultural drainage waters following field fertilisation', *Water, Air and Soil Pollution: Focus*, vol 4, pp437–451

Reay, D. S., Smith, K. A., Edwards, A. C., Hiscock, K. M., Dong, L. F. and Nedwell, D. B. (2005) Indirect nitrous oxide emissions: Revised emission factors', *Environmental Sciences*, vol 2, pp153–158

Rennenberg, H., Kreutzer, K., Papen, H. and Weber, P. (1998) 'Consequences of high loads of nitrogen for spruce (*Picea abies*) and beech (*Fagus sylvatica*) forests', *New Phytologist*, vol 139, pp71–86

Russow, R., Knappe, S. and Neue, H.-U. (2002) 'The N$_2$O content of soil air at different depths as well as its related content in and transport by seepage in lysimeter soils', in J. van Ham, A. P. M. Baede, R. Guicherit and J. G. F. M. Williams-Jacobs (eds) *Non-CO$_2$ Greenhouse Gases: Scientific Understanding, Control Options and Policy Aspects*, Millpress, Rotterdam, pp151–152

Sawamoto, T., Nakajima, Y., Kasuya, M., Tsuruta, H. and Yagi, K. (2005) 'Evaluation of EFs for indirect N$_2$O emission due to nitrogen leaching in agro-ecosystems', *Geophysical Research Letters*, vol 32, no 3, doi:10.1029/2004GL021625

Seitzinger, S. P. and Kroeze, C. (1998) 'Global distribution of nitrous oxide production and N inputs in freshwater and coastal marine ecosystems', *Global Biogeochemical Cycles*, vol 12, pp93–113

Scarascia-Mugnozza, G., Bauer, G. A., Persson, H., Matteucci, G. and Masci, A. (2000) 'Tree biomass, growth and nutrient pools', in E. D. Schulze (ed) *Carbon and Nitrogen Cycling in European Forest Ecosystems*, Springer, Berlin, pp49–62

Skiba, U., Sheppard, L. J., MacDonald, J. and Fowler, D. (1998), 'Some key environmental variables controlling nitrous oxide emissions from agricultural and seminatural soils in Scotland', *Atmospheric Environment*, vol 32, pp3311–3320

Skiba, U., Sheppard, L. J., Pitcairn, C. E. R., van Duk, S. and Rossal, M. J. (1999) 'The effect of N deposition on nitrous oxide and nitric oxide emissions from temperate forest soils', *Water, Air, and Soil Pollution*, vol 116, pp89–104

Skiba, U., Dick, J., Storeton-West, R., Lopez-Fernandez, S., Woods, C., Tang, S. and van Di k, N. (2006) 'The relationship between NH$_3$ emissions from a poultry farm and soil NO and N$_2$O fluxes from a downwind forest', *Biogeosciences*, vol 3, pp375–382

Teiter, S. and Mander, Ü. (2005) 'Emission of N$_2$O, N$_2$, CH$_4$ and CO$_2$ from constructed wetlands for wastewater treatment and from riparian buffer zones', *Ecological Engineering*, vol 25, pp528–541

Toyoda, S., Iwai, H., Koba, K. and Yoshida, N. (2009) 'Isotopomeric analysis of N$_2$O dissolved in a river in the Tokyo metropolitan area', *Rapid Communications in Mass Spectrometry*, vol 23, pp809–821

van Dijk, S. M., and Duyzer, J. H. (1999) 'Nitric oxide emissions from forest soils', *Journal of Geophysical Research*, vol 104, pp15955–15961

van Drecht, G., Bouwman, A. F., Knoop, J. M., Beusen, A. H. W. and Meinardi, C. R. (2003) 'Global modeling of the fate of nitrogen from point and nonpoint sources in soils, groundwater, and surface water', *Global Biogeochemical Cycles*, vol 17, article no 1115 von der Heide, C., Böttcher, J., Deurer, M., Weymann, D., Well, R. and Duijnisveld, W. H. M. (2008) 'Spatial variability of N₂O concentrations and of denitrification-related factors in the surficial groundwater of a catchment in Northern Germany', *Journal of Hydrology*, vol 360, pp230–241

von der Heide, C., Böttcher, J., Deurer, M., Weymann, D., Well, R. and Duijnisveld, W. H. M. (2009a) 'Spatial and temporal variability of N₂O in the surface groundwater: A detailed analysis from a sandy aquifer in northern Germany', *Nutrient Cycling in Agroecosystems*, doi:10.1007/s10705-009-9310-7

von der Heide, C., Böttcher, J., Deurer, M., Duijnisveld, W. H. M., Weymann, D. and Well, R. (2009b) 'Estimation of indirect nitrous oxide emissions from a shallow aquifer in northern Germany', *Journal of Environmental Quality*, doi:10.2134/jeq2008.0320

Well, R., Weymann, D. and Flessa, H. (2005a) 'Recent research progress on the significance of aquatic systems for indirect agricultural N₂O emissions', *Environmental Sciences – Journal of Integrative Environmental Research*, vol 2, pp143–152

Well, R., Höper, H. and Mehranfar, O. (2005b) 'Denitrification in the saturated zone of hydromorphic soils – laboratory measurement, regulating factors and stochastic modeling', *Soil Biology and Biochemistry*, vol 10, pp1822–1836

Well, R., Flessa, H., Jaradat, F., Toyoda, S. and Yoshida, N. (2005c) 'Measurement and simulation of isotopomer signatures of N₂O in groundwater', *Journal of Geophysical Research – Biogeosciences*, vol 110, G02006, doi:10.1029/2005JG000044

Weymann, D., Well, R., Flessa, H., von der Heide, C., Deurer, M., Meyer, K., Konrad, C. and Walther, W. (2008) 'Denitrification and nitrous oxide accumulation in nitrate-contaminated aquifers estimated from dissolved nitrate, dinitrogen, argon and nitrous oxide', *Biogeosciences*, vol 5, pp1215–1226

Weymann, D., Well, R., von der Heide, C., Böttcher, J., Flessa, H. and Duijnisveld, W. H. M. (2009) 'Recovery of groundwater N₂O at the soil surface and its contribution to total N₂O emissions' *Nutrient Cycling in Agroecosystems*, vol 85, pp 299–312

Wilcock, R. J. and Sorrell, B. K. (2008) 'Emissions of greenhouse gases CH₄ and N₂O from low-gradient streams in agriculturally developed catchments', *Water, Air, and Soil Pollution*, vol 188, no1–4, pp155–170

Xiong, Z. Q., Xing, G. X. and Zhu, Z. L. (2006) 'Water dissolved nitrous oxide from paddy agroecosystem in China', *Geoderma*, vol 136, pp524–532

Zechmeister-Boltenstern, S., Hahn, M., Meger, S. and Jandl, R. (2002) 'Nitrous oxide emissions and nitrate leaching in relation to biomass dynamics in a beech forest soil', *Soil Biology and Biochemistry*, vol 34, pp823–832

9
Abiotic Nitrous Oxide Sources: Chemical Industry and Mobile and Stationary Combustion Systems

Peter Wiesen

Introduction

For many years the potential impact of anthropogenic N_2O emissions on the global N_2O budget has been a controversial subject of discussion. In the 1980s there was a tendency to attribute the increase in atmospheric N_2O concentration primarily to anthropogenic emissions associated with biomass burning, fossil fuel combustion, industrial production of adipic and nitric acids, and the use of N fertilizers. The discovery of an artefact in the procedures used to evaluate the emissions of N_2O from fossil fuel combustion (involving also the evaluation of the emission from biomass burning) has drastically changed this situation and the strength of the combustion source has decreased significantly. Linak and co-workers (1990) re-examined the direct role of combustion sources on the observed global rise of N_2O in the atmosphere and concluded that direct N_2O emissions from conventional fossil fuel combustion are very low. As a consequence, the global N_2O flux from this source was reported to be lower by a factor of 50 to 100 than that inferred from previous work and accounted for less than 3 per cent of the entire anthropogenic N_2O flux, which was less than the contribution from fertilizers at that time. In this chapter the current knowledge on N_2O formation during mobile and stationary combustion, as well as N_2O emissions from industrial processes, is reviewed.

Kinetics of nitrous oxide formation during combustion

The formation of N_2O as a by-product during combustion has been well known for many years. The gas-phase formation/destruction of nitrous oxide is strongly linked to gas-phase NO kinetics and more specifically to the transformation of cyanide species into NO and N_2 in the so-called 'fuel-NO mechanism'.

The presence of amines or other organic N compounds in fossil fuels – 'fuel-bound nitrogen' – is of paramount importance for the formation of N_2O. Kramlich et al (1989) reported a significant increase in the N_2O concentration by adding nitrogen-containing compounds such as ammonia (NH_3), hydrogen cyanide (HCN) and acetonitrile (CH_3CN) to a gas flame in the temperature range 1050–1400K. The results of this study are shown in Figure 9.1. It is quite evident that N_2O formation is limited to a relatively small temperature range. These results, together with the findings of Miller and Bowman (1989) and Hayhurst and Lawrence (1992), provide further evidence that N_2O formation in combustion occurs by homogeneous gas-phase reactions through the so-called NH_i or HCN pathways.

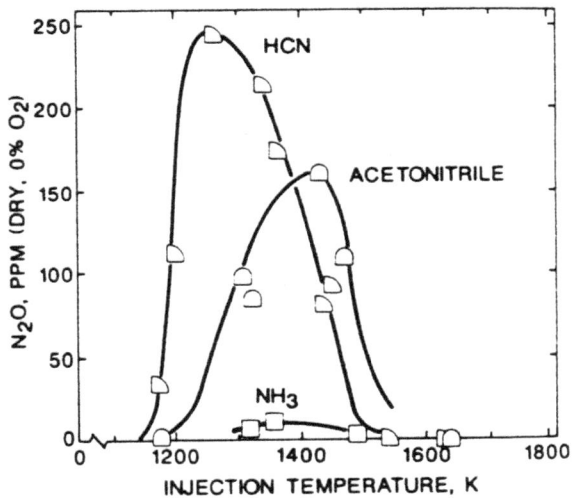

Figure 9.1 N_2O formation after adding NH_3, HCN and acetonitrile into the exhaust of a gas flame as a function of temperature

Source: Kramlich et al (1989)

The NH_i pathway

In the presence of amine-like nitrogen compounds in combustion a reservoir of NH_i radicals is generated. Depending on factors such as temperature, pressure and the air:fuel ratio, reaction products such as N_2, NO and N_2O are formed. In Figure 9.2 the most important reactions of NH_i radicals in flames are summarized.

The NH_2 radical is the dominant specie under lean-burning conditions. By increasing the fuel:air ratio more hydrogen atoms are formed, by which more

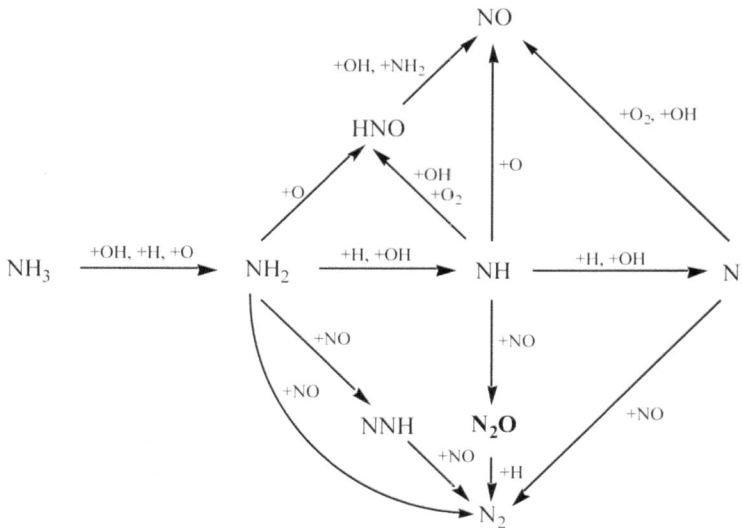

Figure 9.2 Reaction scheme showing the oxidation of NH_i radicals in flames

Source: Adapted from Miller and Bowman (1989)

NH radicals and nitrogen atoms are generated. The most important reaction leading to N_2O formation under flame conditions is the reaction:

$$NH + NO \rightarrow N_2O + H \tag{9.1}$$

However, under these conditions N_2O is destroyed very quickly by its reaction with H atoms and/or OH radicals. Only if the local flame temperature is in the region of 1000K is a significant formation of N_2O observed.

The HCN pathway

In the presence of fuel-bound nitrogen in the form of pyrrole- or pyridine-type compounds HCN is formed; this compound is also generated to some extent through the reaction of CH radicals with molecular nitrogen (prompt NO). The consecutive reactions in the flame system are very complex, as shown in Figure 9.3.

N_2O is formed mainly by the reaction of NCO radicals with NO:

$$NCO + NO \rightarrow N_2O + CO \tag{9.2}$$

and to a lesser extent by the reaction:

$$NH + NO \rightarrow N_2O + H \tag{9.3}$$

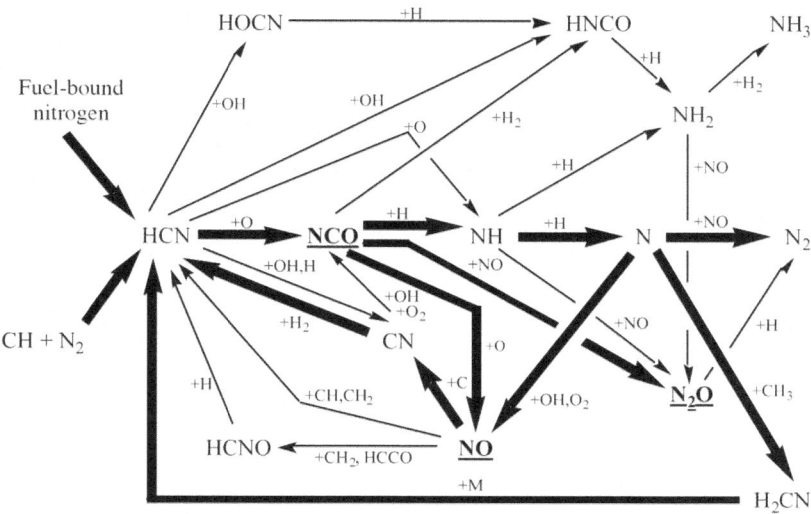

Figure 9.3 Scheme of the most important reactions during the formation of NO and N_2O through fuel-bound nitrogen and prompt NO through the reaction $CH + N_2$

Source: After Becker et al (2000a)

The formation of nitrous oxide is, however, counterbalanced by its very fast destruction by hydrogen radicals, according to the reaction:

$$N_2O + H \rightarrow N_2 + OH \tag{9.4}$$

Calculations performed by Kilpinen and Hupa (1991) demonstrated a decrease in the N_2O concentration with increasing temperature. Above 1200K, NCO radicals are converted almost completely into NO, whereas with decreasing temperature N_2O formation through NCO radicals increases.

From the competition between N_2O formation and destruction reactions, it follows that any changes in conditions that substantially decrease the hydrogen atom concentration in the N_2O formation zone can be expected to increase gas-phase N_2O emissions. Examples are (1) lowering the gas-phase combustion temperature, which might well provide a partial explanation of relatively high N_2O emissions measured from fluidized bed combustors; and (2) air or fuel staging, i.e. where secondary air or air/fuel is introduced into low-temperature combustion zones. For the same reasons, an increase of N_2O emissions should be expected when – as mentioned above – ammonia, urea or other amine or cyanide species are injected into combustors, especially at relatively low temperatures.

In conclusion, non-negligible N_2O emissions may arise from gas-phase combustion when:

- mixture inhomogeneities are created;
- temperatures of oxidation zones are low;
- the oxygen concentration is increased.

In addition to gas-phase chemistry, N_2O can be formed or destroyed during combustion by heterogeneous reactions, defined here as reactions either occurring between a gaseous reactant and a solid reactant, or between two gaseous reactants but taking place at the gas/solid interface, where the solid may play the role of a catalyst.

As far as formation and destruction of nitrous oxide is concerned, the following heterogeneous reaction mechanisms have been identified and, at least, partly described:

- In reactions taking place during combustion:
  - reduction of N_2O on char- and soot-bound carbon atoms;
  - N_2O formation from char-bound nitrogen atoms.
- In reactions taking place during catalytic after-treatment of combustion products:
  - formation/destruction of N_2O during selective catalytic reduction (SCR);
  - formation/destruction of N_2O during NO reduction in engine exhaust by three-way catalysts.

Nitrous oxide emissions from stationary combustion

As already mentioned above, the combustion of any fuel leads to the emission of varying quantities of nitrous oxide. The N_2O emission from stationary combustion is strongly dependent on the temperature in the combustion vessels. The emission rates are highest when the combustion temperature ranges from 800 to 1200K, while the emissions are negligible below 800K and above 1200K.

However, compared to N_2O emissions from conventional stationary combustion units – which are generally quite low – those from either bubbling, recirculating or pressurized fluidized bed combustors were found to reach values around 200ppm (De Soete, 1997), even if obtained by adequate analysis and sampling techniques.

It is worth mentioning here that in the late 1980s artificial N_2O formation in grab samples from combustion processes that contained sulphur dioxide (SO_2) led to an overestimation of these emissions from combustion processes because the sampling procedure and storage was not adequate (see Introduction, above). This artificial N_2O formation has been investigated in detail and can be explained on the basis of an oxidation/reduction system that is composed of the nitrogen oxides N_2O, NO and NO_2 on the one hand and the sulphur oxides SO_2 and SO_3 (or the corresponding HSO_3^- and SO_3^{2-} ions) on the other hand.

With respect to SO_2, both NO_2 and NO behave as oxidants, oxidizing SO_2 to SO_3 and thereby being reduced to NO and N_2O, respectively. On the contrary, with respect to SO_3, NO behaves as a reductant, reducing SO_3 again to SO_2 and thereby being oxidized to NO_2. In this way, a recycling of SO_2 becomes possible when an excess of NO is present, which explains some experimental findings that the amount of N_2O formed may exceed the initial amount of SO_2 being present in the sample (De Soete, 1989).

The role played by NO in the reduction of the higher sulphur oxide into the lower one can be played as well by any other reducing agent, for example bound carbon. During dense-phase combustion of coal and char, for example in fluidized beds, in the presence of calcium sulphate ($CaSO_4$), which is being formed from limestone added for sulphur capture, the latter will be reduced by this way into $CaSO_3$ and CaS at temperatures well below those where thermal decomposition of $CaSO_4$ is observed. Subsequent reaction of NO with these sulphur compounds yields small but measurable amounts of nitrous oxide. Some measurements obtained from fluidized bed combustors, showing an increase of N_2O emissions when limestone is added, support this interpretation and may help to explain the higher N_2O emissions from this type of combustor.

However, the relatively high N_2O emission from fluidized bed combustors is primarily due to the much lower temperature levels of combustion, i.e. 800–900K (Leckner and Åmand, 1992, and references therein) and is supported by the following findings: (1) it is commonly observed that N_2O emissions increase when the bed temperature is decreased; and (2) results from laboratory test rigs, in which conventional combustion has been realized at temperatures significantly below usual burner-flame temperatures, also exhibit a pronounced increase in N_2O emission (Leckner and Åmand, 1992, and references therein).

Nitrous oxide emissions from mobile combustion

Vehicular emissions

It is well established that vehicle exhaust contains N_2O (Smith and Carey, 1982; Metz, 1984; Prigent and De Soete, 1989; Dasch, 1992; Berges et al, 1993; Sjödin et al, 1995). Furthermore, it is widely accepted that new vehicles equipped with three-way catalysts generally emit more N_2O than older vehicles without catalysts (Dasch, 1992; Berges et al, 1993; Hupa and Matinlinna, 1994; Sjödin et al, 1995; Siegl et al, 1996). In the early 1990s concern was expressed that N_2O emissions would rise substantially as the global fleet of old vehicles was replaced with modern vehicles equipped with three-way catalysts to reduce urban air pollution (Berges et al, 1993).

In 1992, Berges et al (1993) studied N_2O emissions from motor traffic in tunnels in Stockholm, Sweden, and Hamburg, Germany, and concluded that catalyst-equipped vehicles emit 106mg N_2O km^{-1} (170mg N_2O mile^{-1}).

Extrapolation to the global car fleet led Berges et al (1993) to predict that global N_2O emissions from vehicles could reach 6–32 per cent of the atmos - pheric growth rate. Sjödin et al (1995) studied N_2O emissions from motor traffic passing though a different tunnel in Sweden in 1992 (in Gothenburg) and reported an average emission rate of 25mg N_2O km^{-1}. Sjödin et al (1995) estimated that the traffic passing through the tunnel consisted of approximately 10 per cent heavy-duty vehicles, 45 per cent catalyst-equipped and 45 per cent non-catalyst-equipped cars. If one takes the extreme assumption that N_2O emission from heavy-duty vehicles and non-catalyst equipped cars is zero, then an EF of approximately 56mg N_2O km^{-1} (90mg N_2O mile^{-1}) for catalyst-equipped vehicles could be derived, which is significantly less than that reported by Berges et al (1993).

To better understand the environmental impact of vehicle exhaust, Becker et al (1999) conducted field measurements of traffic passing through the Kiesbergtunnel in Wuppertal, Germany, and this was followed by laboratory studies at the Ford Motor Company using a chassis dynamometer facility to measure N_2O emissions from vehicles (Becker et al, 2000b). Figure 9.4 shows as an example of N_2O and CO_2 mixing ratios from the measurements in the Kiesbergtunnel and their good correlation, which indicates that high concentrations of N_2O measured in the tunnel resulted from vehicle exhaust. The combined database from the laboratory studies at the Ford Motor Company provided emission rates for non-methane hydrocarbons (NMHC), CO, CO_2, NO_x, NO, NO_2, N_2O and NH_3 from 26 different vehicles powered by six different fuels and represents a comprehensive survey of the emissions of N-containing compounds from individual modern vehicles. The EF (g N_2O g^{-1} CO_2) measured in the 'real world' of the tunnel was 4.1 ± 1.2 × 10^{-5} while that measured in the 'laboratory' was 4.3 ± 1.2 × 10^{-5}. The consistency between these EFs was rather surprising. Typically, 'real world' studies report emission rates that are significantly higher (perhaps by a factor of two to three) than those observed in the laboratory. This has been ascribed to the fact that a small fraction of the 'real world' vehicle fleet is poorly maintained and contributes disproportionately to the observed pollutant levels. The consistency between the 'real world' and 'laboratory' results reported in the papers by Becker et al (1999, 2000b) may reflect either: (1) the contribution of heavy-duty and non-catalyst vehicles in the tunnel that are believed to emit less N_2O than catalyst-equipped vehicles; or (2) that N_2O emissions are not very sensitive to the maintenance condition of the vehicle.

In the light of the results of Jiménez et al (1997, 2000), the former explanation seems more likely. In any case, the results suggest that a reasonable estimate of the EF is (4 ± 2) × 10^{-5} when calculating traffic-related N_2O emissions for emission inventory purposes. This range is consistent with the overall EF measured by Berges et al (1993) in 1992 in a tunnel study in Germany of (6 ± 3) × 10^{-5} but is substantially lower than the range of (1.4 ± 0.9) × 10^{-4} derived from a similar study in Sweden (Berges et al, 1993).

It is useful to place these emission rates into perspective in terms of vehicle contributions to the global N_2O budget and to radiative forcing of climate change. The contribution of global vehicular traffic to the global N_2O budget can be estimated using two different approaches. First, one could assume that the N_2O observed in the tunnel studies is attributable solely to catalyst-equipped passenger cars, calculate the EF for such vehicles, and combine this result with an estimate of their global fuel consumption. Second, one could assume that the vehicle mix travelling through the Wuppertal tunnel is representative of the global vehicle population and simply multiply the measured EF by the global vehicle fuel consumption. For simplicity, and because the tunnel measurements do not support the assumption that the N_2O is attributable solely to catalyst-equipped passenger cars, Becker et al (1999) adopted the second approach. Using values of 964Tg for the annual global vehicle fuel consumption in 1995 (637Tg of gasoline, 327Tg of automotive diesel (Associated Octel, 1996), 0.855 for the average carbon content of gasoline by mass (Marland and Rotty, 1984), 44 for the molecular weights of N_2O and CO_2, 12 for the atomic weight of carbon, and $4 \pm 2 \times 10^{-5}$ for the EF of N_2O, the authors estimated the contribution of vehicular traffic to the global N_2O budget to be $964 \times 0.855 \times (44/12) \times (4 \pm 2) \times 10^{-5} = 0.12 \pm 0.06$Tg. Atmospheric levels of N_2O are increasing at a rate of 4.7 ± 0.9Tg yr^{-1} (3.0 ± 0.6Tg N yr^{-1}) (Berges et al, 1993). Hence, emissions from the global vehicle fleet represent approximately 1–4 per cent of the atmospheric growth rate of N_2O. The global warming potential of N_2O is 298 times that of CO_2 (IPCC, 2007). Using an emission factor (g N_2O g^{-1} CO_2) of $(4 \pm 2) \times 10^{-5}$ it follows that N_2O emissions from vehicles have a global warming impact of 1–2 per cent of that of the CO_2 emitted from vehicles.

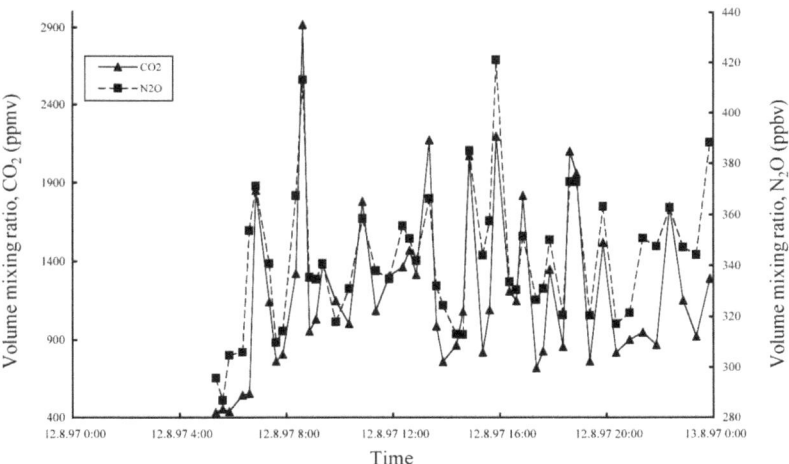

Figure 9.4 Correlation between CO_2 and N_2O emitted from vehicular traffic in the Wuppertal Kiesbergtunnel

Source: After Becker et al (2000b)

Becker et al (1999) concluded that N_2O emissions from vehicles make minor (though non-negligible) contributions to the global atmospheric N_2O budget and to anthropogenic radiative forcing of global climate change. These findings are in disagreement with the conclusions of the previous study by Berges et al (1993) described above, i.e. that if the entire fleet of cars were to be equipped with catalysts the global N_2O emissions from vehicles would double and reach 6–32 per cent of the atmospheric growth rate. This work of Berges et al (1993) took place in 1992, when catalyst-equipped passenger cars represented only a small fraction (20–30 per cent) of the total traffic. In contrast, in 1997 catalyst-equipped passenger cars dominated the traffic volume. Berges et al (1993) assumed that N_2O emissions from non-catalyst cars and trucks were zero and derived EFs by attributing all of the measured N_2O to the relatively small number of catalyst-equipped cars. The results from the work of Becker et al (1999, 2000b) indicate that this assumption is not valid and that N_2O emissions from catalyst cars operating under 'real world' conditions are substantially lower than previously reported. Becker et al (1999, 2000b) recommended the use of an EF of $(4 \pm 2) \times 10^{-5}$ in emission inventory calculations for the modern vehicle fleet.

Since 2000, several publications have appeared reporting N_2O emissions from various type of vehicles (for example Durbin et al, 2003; Huai et al, 2003, 2004; Behrentz et al, 2004; Karlson, 2004; Winer and Behrentz, 2005). Data from these publications were added to the body of data reported in a review (US EPA, 2004) by the United States Environmental Protection Agency (US EPA).

It is nowadays commonly accepted that N_2O emissions from light-duty gasoline-powered vehicles are very dependent on the type of pollution control technology and the age of this technology. Emissions also vary with the vehicle in question and with operating conditions such as fuel sulphur level, driving cycle, ambient temperature and catalyst operating temperature (Jobson et al, 1994; Laurikko and Aakko, 1995; Michaels, 1998; Odaka et al, 1998, 2000, 2002; Koike et al, 1999; Baronick et al, 2000; Meffert et al, 2000). The most up-to-date EFs for light-duty gasoline- and diesel-powered vehicles have been published very recently by Graham et al (2009), who investigated the N_2O emission from 200 of those vehicles between April 2004 and June 2007. This work was mainly triggered by the fact that major changes in both vehicle and fuel specifications had occurred during the previous ten years. Taken together with data from the same group going back to 2004, information on a test fleet of 467 vehicles is available, representing a wide range of emission standards.

The data analysis has shown that the distinction between light-duty auto - mobiles and light-duty trucks within a given emission standard is not significant, whereas the distinction between new and aged catalysts remains. The N_2O emission rates appeared to be well correlated with the numerical NMHC or non-methane organic gases (NMOG) emission standard to which the vehicles were certified, but less so with NO_x or CO emission standards. Of the vehicles investigated, those with aged catalysts showed a greater rate of increase of N_2O emissions with NMHC or NMOG emission standard than did new vehicles.

To what extent new NO_x reduction techniques such as the AdBlue® may alter the N_2O emission in particular from heavy-duty vehicles is still an open question and needs further clarification. The basic principle of AdBlue®, which is used in heavy-duty vehicles, is very similar to the thermal DeNOx process used in stationary combustion; an aqueous solution of urea is injected into a SCR catalyst, leading to almost 90 per cent NO_x reduction in the exhaust.

Nitrous oxide emissions from aircraft

Only a few data are available up to now on the potential impact of nitrous oxide emissions from commercial aero engines on the global N_2O budget, particularly with respect to different flight, i.e. engine operating, conditions. Wiesen et al published in 1994 and 1996 emission data from two modern commercial aero engines, namely a Rolls Royce RB211 engine, which is still being used, for example in Boeing B747 aeroplanes, and a Pratt & Whitney Canada PW305 engine, which is used in smaller business jets. Both engines were tested in an altitude test facility, which was used to simulate real-flight conditions. The authors used gas sampling – taking into account sampling line artefacts – and off-line analyses of the gas samples by tunable-diode laser spectrometry. Within the experimental errors the N_2O emission data obtained from the altitude test of the PW305 engine showed neither a dependence on the altitude nor on the engine speed. For the RB211 engine, the N_2O concentrations were found to increase slightly with increasing engine power, with a much higher N_2O emission index than for the PW305. The authors proposed an average N_2O emission index of 0.15 g N_2O kg^{-1} fuel. Taking into account the global annual fuel consumption for air transportation, one can conclude that the present contribution of this transport sector to the global N_2O emission is small if not negligible.

Atmospheric processes leading to nitrous oxide formation

The chemical transformation and removal of atmospheric trace gases by heterogeneous processes involving water droplets in clouds and fog, sulphuric acid and particulates is of great importance for atmospheric chemistry, since it is known that these processes can lead, for example, to major shifts in gas-phase photochemistry. For the removal of atmospheric nitrogen compounds, heterogeneous reactions leading to the final end-product nitric acid (HNO_3) are of special interest. However, these reactions may also lead, at least in part, to the formation of other nitrogen species such as nitrous acid (HONO) and nitrous oxide.

In a study in the mid-1990s, Wiesen et al (1995) studied the heterogeneous conversion of NO_2 on different surfaces in both an 11-litre Pyrex glass reactor (surface to volume (s:v) ratio ~22m^{-1}) under simulated atmospheric conditions and in a 64-litre quartz glass reactor (s:v ratio ~21m^{-1}). The quartz glass reactor was equipped with a White mirror system (base length 2.0m), and operated at reduced pressure, which allowed in situ concentration measurements of the different species by long-path tunable-diode laser absorption spectroscopy (TD-LAS).

In a first set of experiments the authors studied the heterogeneous conversion of NO_2 in the quartz glass reactor (White cell) in the dark. Concentration–time profiles of NO_2, HONO and N_2O were determined for initial NO_2 concentrations, which were varied from 1.5 to 26ppm in dry synthetic air. Figure 9.5 shows a typical concentration–time profile for NO_2, HONO and N_2O.

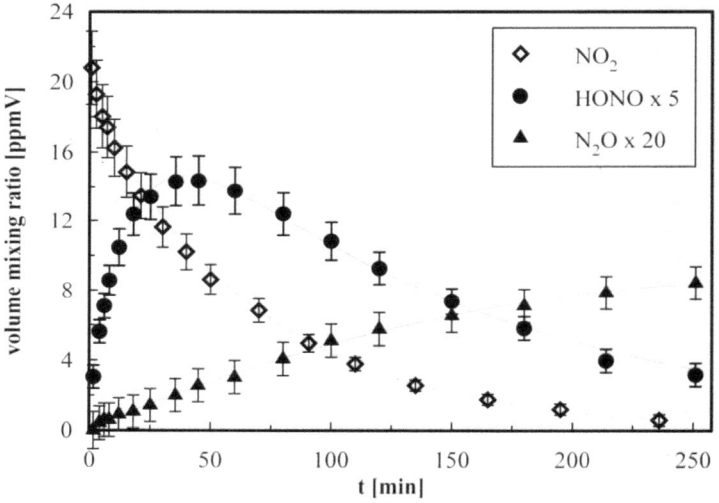

Figure 9.5 NO_2, N_2O and HONO concentration–time profiles in the quartz glass reactor (White cell) in the dark; $[NO_2]_0$ = 21.2ppm, p_{tot} = 6.5torr (c. 870Pa) synthetic air

Source: Wiesen et al (1995)

Since it proved to be very difficult to control the surface activity of the different reactors the experimental set-up was modified. Instead of the White cell, a long-path absorption Herriott cell was used for monitoring the different species. In addition, the gas mixtures in the 11-litre Pyrex glass reactor were pumped continuously through a bubbler containing different solutions, using well-defined parameters such as concentration, temperature, volume, flow velocity and bubble diameter. It was observed that N_2O formation took place in acid solutions such as H_2SO_4, HCl, $HClO_4$ and H_3PO_4, whereas for water or different salt solutions no N_2O formation was observed (Figure 9.6).

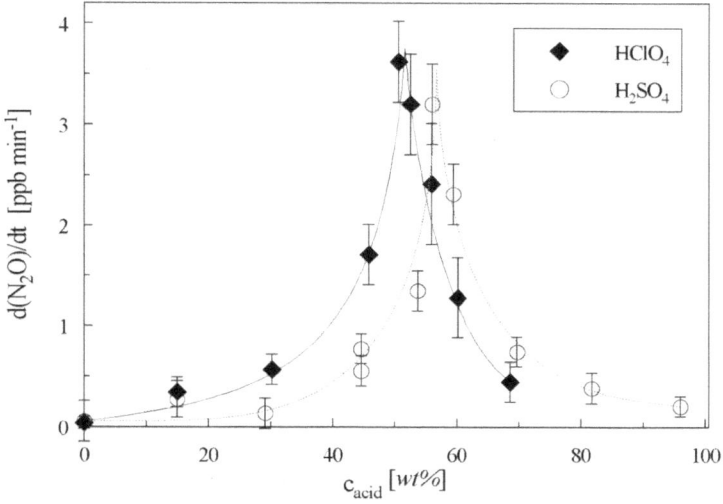

Figure 9.6 Dependence of the N_2O formation rate on the concentration of sulphuric and perchloric acid measured in the modified experimental set-up (Pyrex glass reactor, bubbler, Herriott cell); $[NO_2]_0 = 25ppm$, $p_{tot} = 740torr$ (c. 99kPa), $V_{H_2SO_4} = 200mL$

Source: Kleffmann et al (1998b)

Based on the experimental observations the authors proposed the following reaction mechanism for sulphuric acid as reactant:

$$8NO_2 + 4H_2O \overset{(surface)}{\rightleftharpoons} 4HONO + 4HNO_3 \qquad (9.5)$$

$$2HONO + 2H_2SO_4 \rightleftharpoons 2H_2O + 2NO^+HSO_4^- \qquad (9.6)$$

$$2NO^+HSO_4^- + 2HONO + 2H_2O \rightarrow (HON)_2 + 2HNO_3 + 2H_2SO_4 \quad (9.7)$$

$$(HON)_2 \rightarrow N_2O + H_2O \qquad (9.8)$$

$$8NO_2 + 3H_2O \rightarrow N_2O + 6HNO_3 \quad \text{(net reaction, 9.9)}$$

In a first step, Reaction 9.5, NO_2 is converted on the surface into HONO and HNO_3. In the reaction system the NO_2 decay constant was found to be independent of the NO_2 concentration and the volume of the sulphuric acid solution but was found to be directly proportional to the surface area. Accordingly, the observed NO_2 decay should mainly be caused by an uptake of NO_2 in the sulphuric acid solution.

In a second step, Reaction 9.6, dissolved HONO reacts with sulphuric acid to form $NO^+HSO_4^-$. It is proposed that the nitrosyl cation (NO^+), which is a strong oxidizer, oxidizes HONO to form HNO_3, Reaction 9.7. Through this reaction NO^+ is reduced to $(HON)_2$, which can easily decompose into N_2O and water, Reaction 9.8 (Greenwood and Earnshaw, 1984).

For H_2SO_4 concentrations >55 wt per cent HONO is converted almost completely into $NO^+HSO_4^-$. Accordingly, no HONO can be observed in the gas phase and Reaction 9.7 slows down because no HONO is available. As a consequence, less $(HON)_2$ can be formed, resulting in less N_2O formation. For low H_2SO_4 concentrations, the N_2O yield is also small. This is probably caused by the low formation yields of $NO^+HSO_4^-$, Reaction 9.6, since it is known that $NO^+HSO_4^-$ is not stable in an excess of water (Coleman et al, 1939). Based on the proposed mechanism, the observed maxima in the N_2O formation rate must occur for H_2SO_4 and $HClO_4$ concentrations for which NO^+ and HONO concentrations are equal. Previous studies in which the equilibrium between NO^+ and HONO was investigated as a function of acid concentration reported equal amounts of NO^+ and HONO for 56 wt per cent H_2SO_4 (Seel and Winkler, 1960; Bayliss et al, 1963; Becker et al, 1996) and 52 wt per cent $HClO_4$ (Singer and Vamplew, 1956; Turney and Wright, 1958; Bayliss et al, 1963), which is in excellent agreement with the result from the present study.

It should be pointed out that the proposed reaction mechanism can also explain the experimentally observed formation of N_2O in the presence of other acid solutions such as HCl and H_3PO_4.

In order to estimate the possible influence of the conversion process investigated on the global N_2O budget, a simple estimation was performed for a s:v of the atmospheric aerosol of 3×10^{-4} cm^2 cm^{-3}, which is often observed in heavily polluted air masses. Furthermore, it was assumed that 5 per cent of the aerosol consists of 56 wt per cent H_2SO_4 with an uptake coefficient of 3×10^{-7} (Kleffmann et al, 1998a). A global source strength for NO_x of 160Mt yr^{-1} (as NO_2) was used (Lammel and Graßl, 1995) with an atmospheric NO_2 conversion rate of 10 per cent hour^{-1} (Finlayson-Pitts and Pitts, 1986). Based on these assumptions a global N_2O source strength from NO_2 conversion of 100kt N_2O/yr was calculated.

If one takes into account that N_2O formation occurs also on aerosols with c_{H2SO4} >56 wt per cent and that other sources, for example NO_2 conversion processes on soils, contribute to atmospheric HONO formation, an upper limit for the global N_2O source strength from NO_2 conversion of 0.3Mt N_2O yr^{-1} can be roughly estimated. This result can be compared with the deficit of ~3Mt N_2O/yr in the global N_2O budget. Accordingly, only 10 per cent of the deficit can be explained by heterogeneous NO_2 conversion on acid surfaces.

Another possible source of N_2O formation in the atmosphere that has been discussed is the heterogeneous reaction of N_2O_5 with ammonia on wet aerosols, proposed by Behnke et al (1997). Yet another possible source was proposed by Zellner et al (1992), Maric et al (1992) and Kummer and Zellner (1997): the photochemical formation of N_2O in the atmosphere through the gas-phase reaction of electronically excited NO_2 and NO_3 radicals by reactive collisional quenching with molecular nitrogen. Adema et al (1990) suggested heterogeneous formation of N_2O in air containing NO_2, O_3 and ammonia. The authors suggested a rather complex reaction mechanism; however, they did not provide an estimation of how much N_2O may be formed in the atmosphere by the suggested mechanism.

Nitrous oxide emissions from industrial processes

Nitrous oxide from nitric acid production

Nitric acid is a key ingredient in N-based fertilizers. As a raw material, it also is used for the production of adipic acid (another important N_2O source – see next section) and explosives, metal etching and in the processing of ferrous metals.

Nitric acid production involves the oxidation of ammonia (NH_3) using a platinum catalyst (Ostwald process). Nitrous oxide forms during the catalytic oxidation of ammonia over platinum/rhodium gauzes, the major product being NO. It is estimated that around 5g of nitrous oxide are produced in this way for every kilogram of nitric acid that is produced. Nitric acid plants now represent the single largest industrial process source of nitrous oxide and there has been a strong need for technologies to lower these emissions. The installation of N_2O abatement technologies is obligatory for nitric acid plants in the European Union because of regulations entering into force in 2013.

In principle there are three different ways to achieve this goal, all of which are based on the use of various ceramic catalysts. These may be placed at different points in a nitric acid plant, leading to various advantages and drawbacks of the respective measure. In the US the nitric acid industry currently controls N_2O emissions by using both non-selective catalytic reduction (NSCR) and SCR technologies to reduce N_2O to molecular nitrogen. While NSCR is more effective than SCR at controlling N_2O, these units are not generally preferred in today's plants because of high energy costs and associated high gas temperatures.

Recently much progress has been made in reducing nitrous oxide emissions from the ammonia oxidation process by using more efficient oxidation conditions, with lower reaction temperatures preventing so much of the ammonia ending up as nitrous oxide. The commercially tried and tested EnviNOx® process for the abatement of N_2O and NO_x emissions in nitric acid plants provides a highly viable solution to this problem, as it achieves almost complete removal of N_2O and NO_x. This is accomplished by using a special iron zeolite catalyst, which showed under laboratory conditions a diverse reactivity towards N_2O. The catalyst either decomposes N_2O into N_2 and O_2 – an effect that increases significantly in the presence of NO_x in the tail gas – or by reducing N_2O using various reducing agents, such as hydrocarbons. In addition, the iron zeolites were proven to be excellent DeNOx catalysts, allowing N_2O abatement to be ideally combined with NO_x reduction. Currently four different process variants are available, allowing either the catalytic reduction of NO_x alone, or the combined catalytic decomposition of N_2O and catalytic reduction of NO_x, the catalytic reduction of N_2O and NO_x, or, finally, the catalytic decomposition of N_2O alone.

Several nitric acid plants have been equipped with these systems, which achieve N_2O removal rates of 98–99 per cent, with NO_x emissions being reduced to as low as 1ppmv. The longest-operating EnviNOx® unit has been running since 2003 to the complete satisfaction of its owner and still with the first charge of catalyst.

Basing their judgement on this successful track record and the exceptionally high rates of N_2O and NO_x abatement, the responsible European Union body has declared the EnviNOx® process to be the best available technique (BAT) for N_2O and NO_x abatement in nitric acid plants.

Nitrous oxide from adipic acid production

Nylon production is responsible for nitrous oxide emission through its requirement for adipic acid, $(CH_2)_4(COOH)_2$, as a precursor. Adipic acid is a fine powder also used for some plastics, clothing, carpets and tyres, and in the production of dyes and insecticides.

Nitrous oxide arises from adipic acid production during the oxidation of a ketone-alcohol mixture with nitric acid. It is estimated that for each kilogram of adipic acid made, around 30g of nitrous oxide is also produced.

Although global adipic acid demand is expected to increase, N_2O emissions from this source are expected to continue to decrease substantially, as they have since 1996. As reported by industrial sources, the decrease is a result of the installation by nearly all adipic acid producers of N_2O abatement technologies. The main technologies used to reduce nitrous oxide in the adipic acid industry are catalytic decomposition and thermal destruction. These methods convert nitrous oxide into nitrogen and oxygen. Catalytic decomposition operates at about 500°C and thermal destruction operates at and above 1000°C (Shimizu et al, 2000). Other than these processes, there are reports of a method that consumes N_2O as an oxidant for phenol synthesis. Shimizu et al (2000) give a detailed description of the technology for each adipic acid-producing company that has been made public through their patent applications. Briefly, the *catalytic* decomposition of N_2O uses, for example, a spinel-type $CuAl_2O_4$, a catalyst with Ag and CuO supported on Al_2O_3, a catalyst with Ag supported on Al_2O_3, or CoO and NiO supported on ZrO_2. When the catalysis is around 500°C, using the off-gas from the adipic acid production process that contains 23 per cent N_2O, all of these catalysts give a decomposition rate of 99 per cent or higher. The *thermal* decomposition of N_2O leading to N_2 and O_2 is exothermic; however, it leads also to the formation of NO, which can be recovered as nitric acid.

Using these reduction technologies allows the adipic acid manufacturers to reduce N_2O emissions by 90 per cent or more. Figure 9.7 shows as an example the estimated US nitrous oxide emissions from industry by source from 1990 to 2007. For adipic acid production, N_2O abatement is estimated to have improved from approximately 32 per cent in 1990 to approximately 90 per cent in 2000. The current efficiency range of abatement technologies is 90–99 per cent reduction of N_2O emissions.

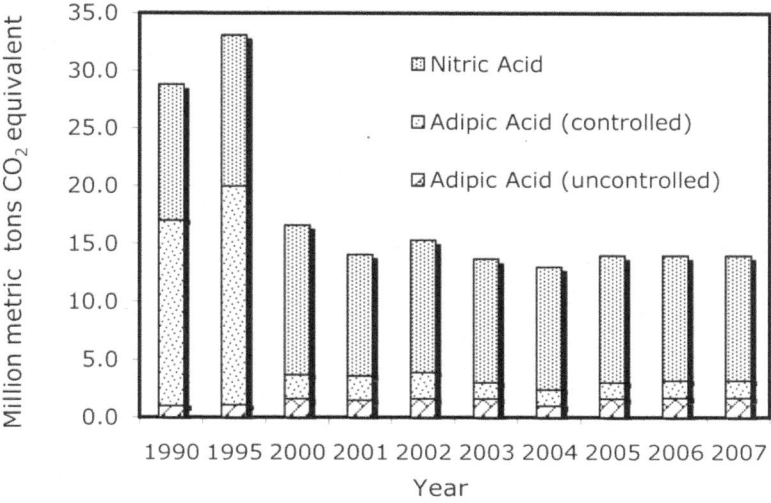

Figure 9.7 US nitrous oxide emissions from industrial sources (nitric acid and adipic acid production)

Source: data taken from the US Energy Information Administration (see www.eia.doe.gov/olaf/1605/ggrpt/nitrous.html)

References

Adema, E. H., Ybema, J. R., Heeres, P. and Wegh, H. C. P. (1990) 'The heterogeneous formation of N_2O in air containing NO_2, O_3 and NH_3', *Journal of Atmospheric Chemistry*, vol 11, pp255–269

Associated Octel (1996) *World Gasoline and Diesel Fuel Survey 1995*, Associated Octel Ltd, London

Baronick, J., Heller, B., Lach, G. and Ramacher, B. (2000) 'Impact of sulfur in gasoline on nitrous oxide and other exhaust gas components', Society of Automotive Engineers (SAE) technical paper no. 2000-01-0857, SAE, Warrendale, PA

Bayliss, N. S., Dingle, R., Watts, D. W. and Wilkie, R. J. (1963) 'The spectrophotometry of sodium nitrite solutions in aqueous sulphuric and perchloric acids and the equilibrium between nitrosonium ion and nitrous acid', *Australian Journal of Chemistry*, vol 16, pp933–942

Becker, K. H., Kleffmann, J., Kurtenbach, R. and Wiesen, P. (1996) 'Solubility of nitrous acid (HONO) in sulfuric acid solutions', *Journal of Physical Chemistry*, vol 100, pp14984–14990

Becker, K. H., Lörzer, J. C., Kurtenbach, R., Wiesen, P., Jensen, T. and Wallington, T. J. (1999) 'Nitrous oxide (N_2O) emissions from vehicles', *Environmental Science and Technology*, vol 33, pp4134–4139

Becker, K. H., Kurtenbach, R., Schmidt, F. and Wiesen, P. (2000a) 'Kinetics of NCO radical reactions with atoms and selected molecules', *Combustion and Flame*, vol 120, pp570–577

Becker, K. H., Lörzer, J. C., Kurtenbach, R., Wiesen, P., Jensen, T. and Wallington, T. J. (2000b) 'Contribution of vehicle exhaust to the global N_2O budget', Chemosphere: Global Change Science, vol 2, pp387–395

Behnke, W., Elend, M., Krüger, U. and Zetzsch, C. (1997) 'The formation of N_2O from the heterogeneous reaction of N_2O_5 with NH_3 on wet aerosols', in K. H. Becker and P. Wiesen (eds) Proceedings of 7th International Workshop on Nitrous Oxide Emissions, Cologne, 21–23 April 1997, Bericht No 41, Bergische Universität Gesamthochschule, Wuppertal, Germany, pp135–141

Behrentz, E., Ling, R., Rieger, P. and Winer, A. (2004) 'Measurements of nitrous oxide emissions from light duty motor vehicles: A pilot study', Atmospheric Environment, vol 38, pp4291–4303

Berges, M. G. M., Hofmann, R. M., Scharffe, D. and Crutzen, P. J. (1993) 'Nitrous oxide emissions from motor vehicles in tunnels and their global extrapolation', Journal of Geophysical Research, vol 98, pp18527–18531

Coleman, G. H., Lillis, G. A. and Goheen, G. E. (1939) 'Nitrosyl chloride', in H. S. Booth ed) Inorganic Syntheses, Vol. 1, McGraw-Hill, New York, p55–59

Dasch, J. M. (1992) 'Nitrous oxide emissions from vehicles', Journal of the Air and Waste Management Association, vol 42, pp63–67

De Soete, G. (1989) 'Updated evaluation of nitrous oxide emissions from industrial fossil fuel combustion', Report, Institute Francais du Petrole (IFP), Paris

De Soete, G. (1997) 'N_2O emissions from combustion and industry', in K. H. Becker and P. Wiesen (eds) Proceedings of 7th International Workshop on Nitrous Oxide Emissions, Cologne, 21–23 April 1997, Bericht No 41, Bergische Universität Gesamthochschule, Wuppertal, Germany, pp11–13

Durbin, T. D., Miller, J. W., Pisano, J. T., Younglove, T., Sauer, C. G., Rhee, S. H. and Huai, T. (2003) 'The effect of fuel on NH_3 and other emissions from 2000–2001 model year vehicles', Final Report, CRC Project E-60, University of California, Riverside, CA

Finlayson-Pitts, B. J. and Pitts, Jr., J. N. (1986) Atmospheric Chemistry: Fundamentals and Experimental Techniques, John Wiley & Sons, New York

Graham, L. A., Belisle, S. L. and Rieger, P. (2009) 'Nitrous oxide emissions from light duty vehicles', Atmospheric Environment, vol 43, pp2031–2044

Greenwood, N. N. and Earnshaw, A. (1984) Chemistry of Elements, Pergamon Press, Oxford

Hayhurst, A. N. and Lawrence, A. D. (1992) 'Emissions of nitrous oxide from combustion sources', Progress in Energy and Combustion Science, vol 18, pp529–552

Huai, T., Durbin, T. D. and Norbeck, J. M. (2003) 'Analysis of nitrous oxide and ammonia emissions from motor vehicles', Final Report for California Air Resources Board, University of California, Riverside, CA

Huai, T., Durbin, T. D., Miller, W. and Norbeck, J. M. (2004) 'Estimates of emission rates of nitrous oxide from light duty vehicles using different dynamometer test cycles', Atmospheric Environment, vol 38, pp6621–6629

Hupa, M. and Matinlinna, J. (eds) (1994) Proceedings of 6th International Workshop on Nitrous Oxide Emissions, Turku, Finland, 7–9 June

IPCC (Intergovernmental Panel on Climate Change) (2007) *Climate Change 2007: The Physical Science Basis*. Contribution of Working Group I to the Fourth Assessment Report of the Intergovernmental Panel on Climate Change, Cambridge University Press, Cambridge

Jiménez, J. L., Nelson, D. D., Zahniser, M. S., McManus, J. B., Kolb, C. E., Koplow, M. D. and Schmidt, S. E. (1997) 'Remote sensing measurements of on-road vehicle nitric oxide emissions and of an important greenhouse gas: Nitrous oxide', in *Proceedings of the 7th CRC On-Road Vehicle Emissions Workshop*, San Diego, CA

Jiménez J. L., McManus, J. B., Shorter, J. H., Nelson, D. D., Zahniser, M. S., Koplow, M., McRae, G. J. and Kolb, C. E. (2000) 'Cross road and mobile tunable infrared laser measurements of nitrous oxide emissions from motor vehicles', *Chemosphere: Global Change Science*, vol 2, pp397–412

Jobson, E., Smedler, G., Malmberg, P., Bernler, H., Hjortsberg, O., Gottberg, I. and Rosen, A. (1994) 'Nitrous oxide formation over three-way catalyst', SAE technical paper no. 940926, SAE International, Warrendale, PA, pp219–230

Karlson, H. L. (2004) 'Ammonia, nitrous oxide and hydrogen cyanide emissions from five passenger vehicles', *Science of the Total Environment*, vol 334–335, pp125–132

Kilpinen, P. and Hupa, M. (1991) 'Homogeneous N_2O chemistry at fluidized bed combustion conditions: A kinetic modeling study', *Combustion and Flame*, vol 85, pp94–104

Kleffmann, J., Becker, K. H. and Wiesen, P. (1998a) 'Heterogeneous NO_2 conversion processes on acid surfaces: Possible atmospheric implications', *Atmospheric Environment*, vol 32, pp2721–2729

Kleffmann, J., Becker, K. H. and Wiesen, P. (1998b) 'Investigation of the heterogeneous NO_2 conversion on perchloric acid surfaces', *Journal of the Chemical Society, Faraday Transactions*, vol 94, pp3289–3292

Koike, N., Odaka, M. and Suzuki, H. (1999) 'Reduction of N_2O from automobiles equipped with three-way catalyst – analysis of N_2O increase due to catalyst deactivation', SAE technical paper no. 1999-0-1081, SAE International, Warrendale, PA

Kramlich, J. C., Cole, J. A., McCarthy, J. M. and Lanier, S. (1989) 'Mechanisms of nitrous oxide formation in coal flames', *Combustion and Flame*, vol 77, pp375–384

Kummer, W. and Zellner, R. (1997) 'In-situ photochemical formation processes on N_2O in the atmosphere', in K. H. Becker and P. Wiesen (eds) *Proceedings of 7th International Workshop on Nitrous Oxide Emissions*, Cologne, 21–23 April 1997, Bericht No 41, Bergische Universität Gesamthochschule, Wuppertal, Germany, pp127–134

Lammel, G. and Graßl, H. (1995) 'On the greenhouse effect of NO_x', *Environmental Science and Pollution Research*, vol 2, pp40–45

Laurikko, J. and Aakko, P. (1995) 'The effect of ambient temperature on the emissions of some nitrogen compounds: A comparative study on low-, medium- and high mileage three-way catalyst vehicles', SAE technical paper no. 950933, SAE International, Warrendale, PA

Leckner, B. and Åmand, L.-E. (1992) 'N_2O emissions from combustion in circulating fluidized bed', in *Proceedings of 5th International Workshop on Nitrous Oxide Emissions*, Tsukuba, Japan, 1–3 July 1992, ppKL-6-1–KL-6-15

Linak, W. P., McSorley, J. A., Hall, R. E., Ryan, J. V., Srivastava, R. K., Wendt, J. O. L and Mereb, J. B. (1990) 'Nitrous oxide emissions from fossil fuel combustion', *Journal of Geophysical Research*, vol 95, pp7533–7541

Maric, D., Burrows, J. P. and Moortgat, G. K. (1992) 'A study of the formation of N_2O in the reaction of $NO_3(A^2E')$ with N_2', *Journal of Atmospheric Chemistry*, vol 15, pp157–169

Marland, G. and Rotty, R. M. (1984) 'Carbon dioxide emissions from fossil fuels: A procedure for estimation and results for 1950–1982', *Tellus*, vol 36B, pp232–261

Meffert, M. W., Lenane, D. L., Openshaw, M. and Roos, J. W. (2000) 'Analysis of nitrous oxide emissions from light duty passenger cars', SAE technical paper no. 2000-01-1952, SAE International, Warrendale, PA

Metz, N. (1984) 'Trace gas emissions in the exhausts of passenger cars', *Automobiltechnische Zeitschrift*, vol 86, pp425–430

Michaels H. (1998) 'Emission of nitrous oxide from highway mobile sources', Report EPA420-R-98-009, US EPA, Washington, DC

Miller, J. A. and Bowman, C. T. (1989) 'Mechanism and modeling of nitrogen chemistry in combustion', *Progress in Energy and Combustion Science*, vol 15, pp287–338

Odaka, M., Koike, N. and Suzuki, H. (1998) 'Deterioration effect of three-way catalyst on nitrous oxide emission', SAE technical paper no. 980676, SAE International, Warrendale, PA

Odaka, M., Koike, N. and Suzuki, H. (2000) 'Influence of catalyst deactivation on N_2O emissions from automobiles', *Chemosphere – Global Change Science*, vol 2, pp413–423

Odaka, M., Koike, N., Ishii, H., Suzuki, H. and Goto, Y. (2002) 'N_2O emissions from vehicles equipped with three-way catalysts in a cold climate', SAE technical paper no. 2002-01-1717, SAE International, Warrendale, PA

Prigent, M. and De Soete, G. (1989) 'Nitrous oxide in engines exhaust gases – A first appraisal of catalyst impact', SAE technical paper no. 890492, SAE International, Warrendale, PA

Seel, F. and Winkler, R. (1960) 'Das Gleichgewicht salpetrige Säure – Stickoxydkation im System Schwefelsäure – Wasser', *Zeitschrift für Physikalische Chemie Neue Folge*, vol 25, pp217–232

Shimizu, A., Tanaka, K. and Fujimori, M. (2000) 'Abatement technologies for N_2O emissions in the adipic acid industry', *Chemosphere – Global Change Science*, vol 2, pp425–434

Siegl, W. O., Korniski, T. J., Richert, J. F. O., Chladek, E., Weir, J. E. and Jensen, T. E. (1996) 'A comparison of the emissions from a vehicle in both normal and selected malfunctioning operation modes', SAE technical paper no. 961903, SAE International, Warrendale, PA

Singer, K. and Vamplew, P. A. (1956) 'Oxidation by nitrous and nitric acid. Part IV. Spectroscopic investigation of the equilibrium between NO+ and nitrous acid in aqueous perchloric acid', *Journal of the Chemical Society*, pp3971–3974

Sjödin, A., Cooper, D. A. and Andréasson, K. (1995) 'Estimations of real-world N_2O emissions from road vehicles by means of measurements in a traffic tunnel', *Journal of the Air and Waste Management Association*, vol 45, pp186–190

Smith, L. R. and Carey, P. M. (1982) 'Characterization of exhaust emissions from high mileage catalyst-equipped automobiles', SAE technical paper no. 820783, SAE International, Warrendale, PA

Turney, T. A. and Wright, G. A. (1958) 'Nitrous acid equilibria in perchloric acid', *Journal of the Chemical Society*, pp2415–2418

US EPA (United States Environmental Protection Agency) (2004) 'Update of methane and nitrous oxide emission factors for on-highway vehicles', Report EPA420-P-04-016, US EPA, Washington, DC, www.epa.gov/otaq/models/ngm/420p04016.pdf

Wiesen, P., Kleffmann, J., Kurtenbach, R. and Becker K. H. (1994) 'Nitrous oxide and methane emissions from aero engines', *Geophysical Research Letters*, vol 21, pp2027–2030

Wiesen, P., Kleffmann, J., Kurtenbach, R. and Becker, K. H. (1995) 'Mechanistic study of the heterogeneous conversion of NO_2 into HONO and N_2O on acid surfaces', *Faraday Discussions of the Chemical Society*, vol 100, pp121–127

Wiesen, P., Kleffmann, J., Kurtenbach, R. and Becker, K. H. (1996) 'Emission of nitrous oxide and methane from aero engines: Monitoring by tunable diode laser spectroscopy', *Infrared Physics and Technology*, vol 37, pp75–81

Winer, A. M. and Behrentz, E. (2005) 'Estimates of nitrous oxide emissions from motor vehicles and the effects of catalyst composition and aging', Final Report on Contract No. 02-313, California Air Resources Board, Sacramento, CA

Zellner, R., Hartmann, D. and Rosner, I. (1992) 'N_2O formation in the reactive collisional quenching of NO_3^* and NO_2^* by N_2', *Berichte der Bunsengesellschaft für Physikalisch Chemie*, vol 96, pp385–390

10

Conclusions and Future Outlook

Keith Smith

Nitrous oxide's contribution to radiative forcing

Our knowledge of the sources and sinks of atmospheric nitrous oxide has been expanded greatly in the last two or three decades by the results of pure and applied research in many countries, much of it documented in the bibliographies of the preceding chapters. The major driver has been the concern about the potential consequences of anthropogenic global warming. To improve predictions about what might happen to the climate we need to quantify better the rates of emissions of the important greenhouse gases and the likely future trends in those emissions under various scenarios, and to improve ways of minimizing the emissions; all of these desires have required additional information from research.

The most recent assessment of climate change (IPCC, 2007) shows the relatively minor role played by N_2O at present in the warming ('radiative forcing') process, compared with the other long-lived greenhouse gases (Figure 10.1), but the picture varies greatly between countries. Where agriculture is the dominant industry, the contribution of N_2O (and methane) to total national emissions can be very much greater than that implied by Figure 10.1. Among the developed countries, this is well illustrated by the data for New Zealand: there in 2007 N_2O (almost entirely from livestock-based agriculture) contributed 17 per cent, in CO_2-equivalent terms, (and methane 35.2 per cent), to emissions from all sectors. Thus the trace gases together exceeded the contribution from CO_2, and contributed more than half of all national emissions (New Zealand Government, 2009). However, leaving aside such special cases, there is the likelihood that in many countries, if major curbs on CO_2 emissions are forthcoming over the period 2010–2050 (see below), the trace gases will contribute a significant and ever-increasing proportion of emissions. This will mean that they are likely to receive more and more attention concerning mitigation options.

The implications of such a trend will be particularly important in agriculture. The technology for controlling industrial and vehicular emissions already exists, and many major sources have already been dealt with (see

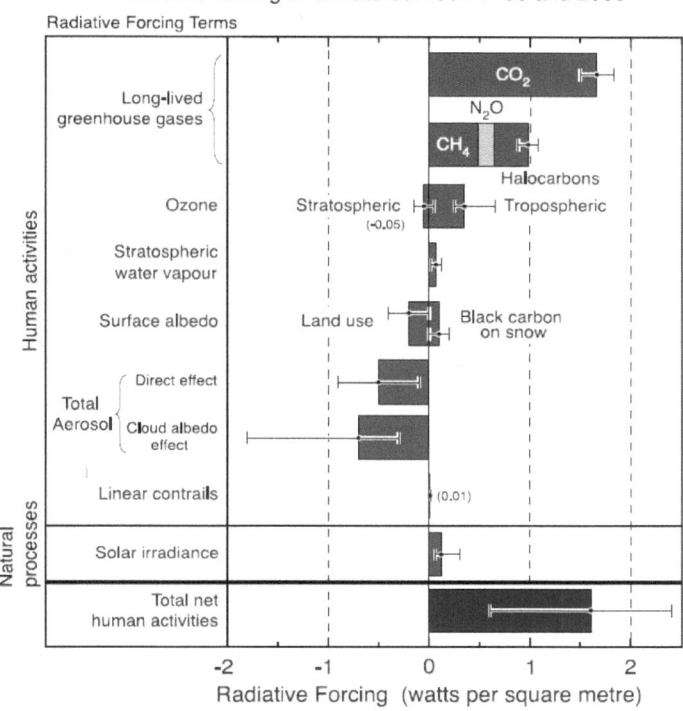

Figure 10.1 Summary of the principal components of the radiative forcing of climate change

Source: IPCC (2007)

Chapter 9). Step-changes in national N_2O emissions have occurred in a number of industrialized countries when large adipic acid-manufacturing plants have been modified to include the available abatement technology. For example, the voluntary agreements in the 1990s between the German government and the two German adipic acid producers to capture and transform N_2O emissions reduced total national emissions by 8 per cent, or approximately 26Mt CO_2-eq (Eichhammer et al, 2001). When any remaining industrial sources have been similarly abated, emissions from agriculture will become, more and more, the principal concern.

Potential impact of population trends

The growth of world population, to around 6.7 billion in 2008 according to the World Bank (2010), is expected to continue for some decades into the future, reaching perhaps 9 billion by 2040. An inevitable consequence of such growth is more intensification of food production. Erisman et al (2008) estimate that, by 2008, nitrogen fertilizers manufactured by the Haber-Bosch process made possible the feeding of virtually half of all humanity. Nevertheless the conversion of fertilizer N to protein N is low. For example, in 2005, approximately 100Tg of N from the Haber-Bosch process was used in global agriculture, whereas only 17Tg of this was consumed by humans in crop, dairy and meat products (UNEP and Woods Hole Research Center, 2007). As pointed out by Erisman et al (2008), although livestock provide many other benefits in addition to meat and milk (for example transport, hides and wool), this highlights the extremely low nitrogen-use efficiency in agriculture.

Erisman et al, on the basis of work by Tilman et al (2002) and the International Fertilizer Association (2007), conclude that the global nitrogen-use efficiency in the production of cereals decreased from ~80 per cent in 1960 (when N application rates were generally low by modern standards) to ~30 per cent in 2000. This latter value is even lower than those arrived at elsewhere in this book: around 50 per cent by Bouwman et al in Chapter 5, and 40 per cent by the present author and colleagues in Chapter 4, based on different sources in the literature. In spite of these uncertainties, it seems safe to conclude that more than half of all fertilizer N is currently lost by denitrification to N_2O and N_2, volatilization as ammonia and leaching of nitrate. All these loss mechanisms lead to N_2O emissions, whether direct or indirect, and it is evident that an improvement in the efficiency with which fertilizer N is used in the future would mean that some of the extra food required in the future could be produced without a matching increase in fertilizer N use, thus avoiding additional N_2O emissions. Smil (2001) and Erisman et al (2008) both suggest that a 50 per cent increase in N use efficiency is achievable.

From Kyoto to Copenhagen

Additional impetus to investigation of biogeochemical pathways and processes, and to development of mitigation measures to reduce emissions, has been provided by the adoption of the Kyoto Protocol. The Protocol was signed in 1997 but only came legally into force in 2005. By November 2009, 187 nations had ratified it.

Under the Protocol, 37 industrialized countries (the 'Annex I countries') committed themselves to a reduction of a 'basket' of four greenhouse gases (carbon dioxide, methane, nitrous oxide, sulphur hexafluoride) and two groups of gases (hydrofluorocarbons and perfluorocarbons) produced by them, by an average of 5.2 per cent below the 1990 level, by 2008–2012 (8 per cent by the 15 countries then members of the European Union). The most notable non-member of the Kyoto Protocol has been the US, although it is a signatory of the UNFCCC and was responsible for an estimated 36 per cent of the 1990 emission levels; however, the US does prepare an annual greenhouse gas inventory and report it to UNFCCC. Latest estimates by the European Commission show that the European Union as a whole is set to overshoot its reduction target, predicting a 13 per cent drop in greenhouse gas emissions below the base year of 1990, whereas some individual Union member states and other Annex I countries are failing to meet their targets.

There were high hopes that the recent climate change conference in Copenhagen (December 2009) would agree new, and much more drastic, reduction targets, and include for the first time commitments to lessen greenhouse gas emissions by major developing nations such as Brazil, China and India. In the event, the conference resulted in a document called the Copenhagen Accord, devised by a small group of countries – including the world's two biggest greenhouse gas polluters, China and the US. The conference as a whole did not adopt the accord, but voted to 'take note' of it. The accord unites the US, China and other major developing countries in an effort to curb global greenhouse gas emissions. There is a commitment 'to reduce global emissions so as to hold the increase in global temperature below 2°C' and to achieve 'the peaking of global and national emissions as soon as possible'. This is a step forward, compared with the Kyoto Protocol, which imposed no obligations on developing countries to restrain the growth of their emissions, and to which the US never acceded. The US is now committed to cut its absolute carbon emissions by about 17 per cent below 2005 levels by 2020, China to reduce its emissions growth by up to 45 per cent from 2005 levels by 2020, and India to reduce its corresponding emissions growth by up to 25 per cent. However, on the negative side, the summit did not result in a legally binding deal or any commitment to reach one in future. Furthermore, there is no global target for emissions reductions by 2050 and the accord is vague as to how its goals – such as the proposed US$100 billion of funds annually for developing countries – will be achieved.

Under the Copenhagen Accord, the EU has pledged to cut its carbon emissions by 20 per cent below 1990 levels by 2020, and to cut by 30 per cent if other nations agree to greater reductions. In fact decisions had already been taken in some individual countries, before the conference took place, that will sharpen the focus on each and every one of the greenhouse gases being emitted, and on possible mitigation methods that might be employed. For example, the UK Government adopted The Climate Change Act in 2008, which states:

The target for 2050:
(1) It is the duty of the Secretary of State [for Energy and Climate Change] to ensure that the net UK carbon account for the year 2050 is at least 80 per cent lower than the 1990 baseline.
(2) 'The 1990 baseline' means the aggregate amount of—
    (a) net UK emissions of carbon dioxide for that year, and
    (b) net UK emissions of each of the other targeted greenhouse gases for the year that is the base year for that gas.

Thus all the key greenhouse gases, including N_2O, will be included in the calculations on the basis of their CO_2 equivalents, derived from their GWP. The existence of the Act has already led to the commissioning of additional research, for example to improve quantification of agricultural emissions of methane and N_2O, and to develop robust and verifiable mitigation options. Because of the high GWP of N_2O – almost 300 times that of CO_2 (Chapter 4) – a saving of just over 3kg of N_2O will therefore be equivalent to a saving of 1 tonne of CO_2, providing a considerable incentive to reduce N_2O emissions as much as possible.

Dealing with emissions uncertainties

For N_2O, the observed growth in atmospheric concentration (amounting to 3.9Tg N_2O-N yr^{-1}), combined with the best assessment of how much new reactive N has become available for N_2O production (discussed in Chapter 4), point to an overall EF of 4 per cent, whereas summing the default values for individual emission pathways gives a *lower* value; in other words, estimated emissions are too small to account for observed atmospheric growth, unless the upper bounds of the uncertainty ranges for the default values (IPCC, 2006) are taken into account.

This situation is the inverse of that relating to sources and sinks of CO_2. The estimated net global CO_2 emission rate from fossil fuel burning and cement manufacture was 7.8Gt CO_2-C yr^{-1} in 2005 (up from 6.5Gt CO_2-C yr^{-1} in 1999), to which must be added somewhere between 0.5 and 2.7Gt C yr^{-1} from deforestation (IPCC, 2007), whereas the observed growth in the atmospheric concentration is much less than indicated by these figures. The difference of around 3Gt CO_2-C yr^{-1} is attributed to the combined effect of a terrestrial carbon sink in growing vegetation and an oceanic sink (for example Hymus and Valentini, 2007). Thus for CO_2 we need to improve quantification of the apparent sinks, while for N_2O we need to do the same for the sources, to achieve balanced global budgets.

A priority area for future investigation of N_2O EFs is indirect emissions caused by nitrogen transfer from land to waters. Chapter 8 documents the existing knowledge, and highlights the variability encountered in studies to date. For example, recent evidence is presented that emissions from riparian areas receiving water from agricultural land may be higher than those predicted by the IPCC default EF. Further work in a wider range of such environments is desirable, together with more studies of the biogeochemical processes going on in the major rivers of the world. There is potentially a big difference in the propensity for leached nitrate to denitrify to N_2O in the course of its passage to the sea, between river systems such as the Mississippi-Missouri and the Yangtze, where the distance may be thousands of kilometres, and the short rivers measured in tens or a few hundreds of kilometres where many recent studies have taken place. It may well be that, because of the sheer complexity of the systems needing study, only techniques integrating emissions over substantial areas will provide the data required. A new generation of sensors capable of determining changes in atmospheric N_2O concentration at the ppt (parts per trillion), rather than the ppb level, may be required.

Such devices would also have a complementary role (if ship- or air-borne) in determining fluxes over the estuaries and seas receiving river discharges and the wider ocean receiving atmospheric deposition of reactive N. According to Duce et al (2008), increasing quantities of atmospheric anthropogenic fixed nitrogen entering the open ocean could account for up to about a third of the ocean's external (non-recycled) nitrogen supply and up to around 3 per cent of the annual new marine biological production, around 0.3Pg of carbon per year. This input, they calculate, could account for the production of up to around 1.6Tg of N_2O per year. This obviously needs experimental verification.

One additional area that deserves priority for new work is in emissions from irrigated agriculture. Worldwide, about 70 per cent of all water abstracted from rivers and aquifers is used for agriculture, and although only 18 per cent of agricultural land is irrigated, this land provides 40 per cent of global food production (Siebert et al, 2007). In view of the forecast upward trend in population (see above), the demand for irrigation is likely to intensify, even though some areas are predicted to experience water shortages because of climate change. The diversity, and the variation in quality, of land management systems is great, and once again there is a need for targeted research to investigate the direct EF for, for example, flood irrigation systems in hot countries; in these environments, nitrification of urea fertilizer to nitrate can occur very rapidly, and the lack of sophisticated land-levelling equipment can result in variable ponding of water in parts of a field: ideal conditions for substantial denitrification to take place. Should it turn out that such environments are, indeed, hot spots of N_2O emission, the introduction of technology available elsewhere, particularly water-saving systems such as drip irrigation, should both mitigate the emissions and help to conserve scarce and/or expensive water – a good example of a 'win-win' strategy. A substantial part of the efforts of the developed world to minimize N_2O emissions from agriculture would be best devoted to challenges of this type in the developing world; for a long-lived, well-mixed greenhouse gas such as N_2O, reducing its emission anywhere on the planet is equally valuable, and making a large emission reduction abroad is a better strategy than reducing a smaller amount at home. It is important that the need for national governments to meet their Kyoto (and post-Copenhagen) domestic commitments do not make them lose sight of this important consideration.

N_2O and the ozone layer

As was mentioned in Chapter 1, before N_2O emissions began to receive attention because of their contribution to global warming, the main environmental concern relating to N_2O was its role in depleting the stratospheric ozone layer. This topic has been given prominence again; Ravishankara et al (2009) wrote recently:

> By comparing the ozone depletion potential-weighted anthropogenic emissions of N_2O with those of other ozone-depleting substances, we show

that N_2O emission currently is the single most important ozone-depleting emission and is expected to remain the largest throughout the 21st century. N_2O is unregulated by the Montreal Protocol. Limiting future N_2O emissions would enhance the recovery of the ozone layer from its depleted state and would also reduce the anthropogenic forcing of the climate system, representing a 'win-win' for both ozone and climate.

Thus although this book has focused on N_2O as a global warming issue, a similar one in the future may well need to be broadened to cover the role of this gas within a wider definition of global change, to embrace also its effect on the ozone layer.

References

Duce, R. A., LaRoche, J., Altieri, K., Arrigo, K. R., Baker, A. R., Capone, D. G., Cornell, S., Dentener, F., Galloway, J., Ganeshram, R. S., Geider, R. J., Jickells, T., Kuypers, M. M., Langlois, R., Liss, P. S., Liu, S. M., Middelburg, J. J., Moore, C. M., Nickovic, S., Oschlies, A., Pedersen, T., Prospero, J., Schlitzer, R., Seitzinger, S., Sorensen, L. L., Uematsu, M., Ulloa, O., Voss, M., Ward, B. and Zamora L. (2008) Impacts of atmospheric anthropogenic nitrogen on the open ocean', *Science*, vol 320, pp893–897

Eichhammer, W., Boede, U., Gagelmann, F., Jochem, E., Kling, N., Schleich, J., Schlomann, B., Chesshire, J. and Ziesing, H.-J. (2001) 'Greenhouse gas reductions in Germany and the UK – coincidence or policy induced?', Research Report 201 41 133 UBA-FB 000193, Federal Ministry of The Environment, Nature Conservation and Nuclear Safety, Berlin, www.umweltdaten.de/publikationen/fpdf-l/1987.pdf

Erisman, J. W., Sutton, M. A., Galloway, J., Klimont, Z. and Winiwarter, W. (2008) '100 years of ammonia synthesis: How a single patent changed the world', *Nature Geoscience*, vol 1, pp636–639

Hymus, G. and Valentini, R. (2007) 'Terrestrial vegetation as a carbon dioxide sink', in D. Reay, N. Hewitt, K. A. Smith and J. Grace (eds) *Greenhouse Gas Sinks*, CABI, Wallingford, pp11–30

International Fertilizer Association (2007) *Sustainable Management of the Nitrogen Cycle in Agriculture and Mitigation of Reactive Nitrogen Side Effects*, IFA, Paris

'CC (International Panel on Climate Change) (2006) *2006 IPCC Guidelines for National Greenhouse Gas Inventories*, Vol 4, H. S. Eggelston, L. Buenida, K. Miwa, T. Ngara and K. Tanabe (eds), IGES, Hayama, Japan

IPCC (2007) *Climate Change 2007: The Physical Science Basis. Contribution of Working Group I to the Fourth Assessment Report*, S. Solomon, D. Qin, M. Manning, Z. Chen, M. Marquis, K. B. Averyt, M. Tignor and H. L. Miller (eds), Cambridge University Press, Cambridge

New Zealand Government (2009) 'New Zealand's greenhouse gas inventory, 1990–2007', www.mfe.govt.nz/publications/climate/greenhouse-gas-inventory-2009/html/index.html

Ravishankara, A. R., Daniel, J. S. and Portmann, R. W. (2009) 'Nitrous oxide (N_2O): The dominant ozone-depleting substance emitted in the 21st century', *Science*, vol 326, pp123–125

Siebert, S., Döll, P., Feick, S., Hoogeveen, J., Faurès, J.-M. and Frencken, K. (2007) 'The global map of irrigation areas', www.fao.org/nr/water/aquastat/main/index.stm

Smil, V. (2001) *Enriching the Earth: Fritz Haber, Carl Bosch, and the Transformation of Food Production*, MIT Press, Cambridge, MA

Tilman, D., Cassman, G. K., Matson, P. A., Naylor, R. and Polasky, S. (2002) 'Agricultural sustainability and intensive production practices', *Nature*, vol 418, pp671–677

UNEP (United Nations Environment Programme) and Woods Hole Research Center (2007) *Reactive Nitrogen in the Environment: Too Much or Too Little of a Good Thing*, United Nations Environment Programme, Paris

World Bank (2010) '*World Development Indicators*', http://datafinder.worldbank.org/about-world-development-indicators?cid=GPD_WDI

Contributors

Elizabeth Baggs is a senior lecturer at the University of Aberdeen, UK. Her research interests are in rhizosphere biogeochemistry, plant–microbe–soil interactions, and linking greenhouse gas production in soils to the underpinning microbiology. She has developed stable isotope approaches for quantifying N_2O production from different microbial processes and for examining interactions between soil N and C cycles.

Hermann Bange is a chemical oceanographer in the Marine Biogeochemistry Research Division of the IFM-GEOMAR, Leibniz Institute for Marine Sciences, Kiel, Germany. His research interests include the oceanic emissions and pathways of trace gases such as nitrous oxide and methane. He is also interested in the oceanic nitrogen cycle and the distributions of short-lived intermediates such as hydroxylamine and hydrazine.

Lex Bouwman is a senior researcher at the Netherlands Environmental Assessment Agency (PBL). His main research interests are global modelling of land use and agricultural systems, and environmental problems associated with nutrient cycling in agriculture. During the last few years the focus has been on land–sea interactions through nutrient enrichment.

Klaus Butterbach-Bahl is a head of department at the Institute of Meteorology and Climate Research (IMK-IFU), Karlsruhe Institute of Technology in Garmisch-Partenkirchen, and a professor at the University of Freiburg, Germany. His main research interests are related to measurements and modelling of biosphere–atmosphere exchange processes of environmentally important trace gases at site and regional scales, with a specific focus on nitrogen compounds.

Franz Conen is a research fellow at the Department of Environmental Sciences at the University of Basel, Switzerland. He 'was inspired by Keith Smith to become a scientist and to study greenhouse gas emissions from soils, which he contentedly continues to do after 15 years'.

Paul Crutzen is a professor at the Max-Planck-Institute for Chemistry, Mainz, Germany. He was awarded the 1995 Nobel Prize in Chemistry, along with Professors Mario Molina and F. Sherwood Roland, for his work in atmospheric chemistry, particularly concerning the formation and decomposition of ozone. His main research interest is in atmospheric chemistry and its role in biogeochemical cycles and climate.

Cecile de Klein is a team leader and senior scientist with AgResearch, New Zealand. She has been working in the N_2O research area for almost 20 years and was a lead author of the N_2O chapter of the Intergovernmental Panel on Climate Change's 2006 *Revised Guidelines for Greenhouse Gas Inventories*. Her main research areas focus on the development of on-farm tools and technologies for reducing N_2O emissions from pastoral systems.

Richard Eckard is an associate professor with the University of Melbourne and Principal Scientist of Climate Change Research with the Victorian Department of Primary Industries, Australia. He is a science adviser to the Australian government on their mitigation policy and research investments into mitigation and adaptation research. He serves on the science committee of the International Greenhouse and Animal Agriculture Conference, having published papers on both enteric methane and nitrous oxide mitigation and modelling.

Alina Freing studied mathematics before she came to the IFM-GEOMAR, Leibniz Institute for Marine Sciences, Kiel, Germany, to start a PhD on the production and emissions of oceanic nitrous oxide. She successfully completed her PhD in December 2009 and is now working as a postdoctoral researcher at IFM-GEOMAR.

Annette Kock is a PhD student at the IFM-GEOMAR, Leibniz Institute for Marine Sciences, Kiel, Germany. Her work involves the investigation of production and transport pathways of nitrous oxide in the tropical oceans.

Carolin Löscher studied biological oceanography at IFM-GEOMAR, Leibniz Institute for Marine Sciences, Kiel, Germany. She started her PhD project on the sensitivity of the oceanic biological nitrogen cycle to changes in dissolved oxygen in 2008. Her main research focus is on the biological formation of nitrous oxide by microorganisms in tropical ocean areas.

Arvin Mosier retired as a research chemist with USDA/ARS in Fort Collins, Colorado, in 2004. During his last 20 years with Agricultural Research Service, he conducted research in the area of soil nitrogen transformations and its relationship to gaseous losses of nitrogen compounds from the soil to the atmosphere. He currently serves in an advisory capacity on N-related projects in the US and internationally.

Albrecht Neftel is a senior research scientist at the Swiss Agroscope Reckenholz-Tänikon Research Station in Zürich. A physicist by training, with a background in ice core analysis, his main interests are biosphere–atmosphere exchange process of agroecosystems, with a focus on reactive nitrogen species.

Laurent Philippot is Director of Research at the French Institute for Agricultural Research (INRA), in Dijon. His main research interest is to understand the ecology of denitrifiers and their role in ecosystem functioning. Recent work focuses on the importance of the size and diversity of the denitrifier community in regulating the denitrification process and production of the greenhouse gas N_2O in soils.

Keith Smith, the editor of this book, is an honorary professorial fellow in the School of Geosciences of the University of Edinburgh, UK, though he now lives in south-west England. His main research work in recent years has been on the exchange of greenhouse gases between soils and plants and the atmosphere, and mitigation of emissions. He was recognized by the Intergovernmental Panel on Climate Change as having contributed to the IPCC's shared award of the 2007 Nobel Peace Prize.

Elke Stehfest is a researcher in the IMAGE team at the Department of Climate and Global Sustainability of the Netherlands Environmental Assessment Agency (PBL). She is working on applications and further development of the land and climate part of the IMAGE model. Her current research focuses on modelling of crop production and land use, land-related mitigation options, biofuels, and projections and policy options for global land use.

Tony van der Weerden is a senior scientist with AgResearch, New Zealand. While he began his research career on ammonia emissions from grassland systems, his current interest lies with nitrous oxide emissions from pastoral systems with a focus on development of mitigation strategies.

Chris van Kessel is a professor and Chair of the Department of Plant Sciences at the University of California, Davis, USA. His main research over the past 25 years has been on soil nitrogen and carbon cycling in agro-ecosystems under ambient and elevated carbon dioxide concentrations and its impact on greenhouse gases and crop productivity.

Reinhard Well is a soil scientist at the Institute of Agricultural Climate Research in Braunschweig, Germany. His main research work has been on carbon and nitrogen cycling in soils and aquifers with special emphasis on denitrification and N_2O turnover. Currently, his responsibilities focus on the role of reactive nitrogen in the greenhouse gas budgets of agricultural ecosystems.

Peter Wiesen is Professor of Physical Chemistry at the University of Wuppertal, Germany. His main research work in recent years has been on the emissions from vehicular traffic, the oxidation capacity of the atmosphere and the development of ultra-sensitive analytical instrumentation for the detection of atmospheric reactive nitrogen species.

Wilfried Winiwarter is a senior research scholar at the International Institute for Applied Systems Analysis (IIASA) in Laxenburg, Austria, and also a senior scientist with the Austrian Institute of Technology, Vienna. A chemical engineer by training, his main interests include the release of trace constituents into the atmosphere and their subsequent fate, and their treatment in integrated assessment models, with a particular focus on reactive nitrogen compounds.

Acronyms and Abbreviations

AGWP	absolute global warming potential
AMO	ammonia mono-oxygenase
AOA	ammonium-oxidizing archaea
AOB	ammonia-oxidizing bacteria
AOU	apparent oxygen utilization
ATP	adenosine triphosphate
ATS	ammonium thiosulphate
BAT	best available technique
BMP	best management practice
BNF	biological nitrogen fixation
C-BNF	cultivation-induced biological nitrogen fixation
CEF	conceptual emission factor
CO_2-eq	carbon dioxide equivalent
CP	crude protein
CT	condensed tannins
DAYCENT	daily version of Century
DCD	dicyandiamide
DIN	dissolved inorganic nitrogen
DM	dry matter
DMPP	3,4-dimethyl pyrazole phosphate
DNDC	DeNitrification/DeComposition
DNRA	dissimilatory nitrate reduction to ammonium
EF	emission factor
ERT	expert review team
ETS	emission trading scheme
FAO	Food and Agriculture Organization of the United Nations
FPCM	fat and protein corrected milk production
FTIR	Fourier transform infrared

GDP	gross domestic product
GHGI	greenhouse gas intensity
GWP	global warming potential
IFA	International Fertilizer Industry Association
IFDC	International Fertilizer Development Center
IMAGE	Integrated Model for the Assessment of Global Environment
IPCC	Intergovernmental Panel on Climate Change
Mtoe	million tons of oil equivalents
NADH	nicotinamide adenine dinucleotide (reduced form)
NADPH	nicotinamide adenine dinucleotide phosphate (reduced form)
NCAS	National Carbon Accounting System
NMHC	non-methane hydrocarbons
NMOG	non-methane organic gases
NOB	nitrite-oxidizing bacteria
NP	nitrapyrin
NSCR	non-selective catalytic reduction
OMZ	oxygen minimum zone
PM	particulate matter
PNL	progressive nutrient limitation
ppb	parts per billion
ppm	parts per million
ppt	parts per trillion
REML	residual empirical maximum likelihood
RP	reaction progress
SCR	selective catalytic reduction
SE	standard error
SIP	stable isotope probing
SOM	soil organic matter
SP	site preference
TDLAS	tunable-diode laser absorption spectroscopy
TDR	time domain reflectometry
UNFCCC	United Nations Framework Convention on Climate Change
US EPA	United States Environmental Protection Agency
VSMOW	Vienna standard mean ocean water
WFPS	water-filled pore space
yr BP	years before the present

Index